Engineering Materials

The “Engineering Materials” series provides topical information on innovative, structural and functional materials and composites with applications in optical, electronical, mechanical, civil, aeronautical, medical, bio and nano engineering. The individual volumes are complete, comprehensive monographs covering the structure, properties, manufacturing process and applications of these materials. This multidisciplinary series is devoted to professionals, students and all those interested in the latest developments in the Materials Science field.

More information about this series at http://www.springer.com/series/4288

Md Rezaur Rahman

Wood Polymer Nanocomposites

Chemical Modifications, Properties and Sustainable Applications

Springer

Md Rezaur Rahman
Faculty of Engineering
Universiti Malaysia Sarawak
Kota Samarahan, Sarawak
Malaysia

ISSN 1612-1317 ISSN 1868-1212 (electronic)
Engineering Materials
ISBN 978-3-319-65734-9 ISBN 978-3-319-65735-6 (eBook)
DOI 10.1007/978-3-319-65735-6

Library of Congress Control Number: 2017949506

Printed on acid-free paper

This Springer imprint is published by Springer Nature
The registered company is Springer International Publishing AG
The registered company address is: Gewerbestrasse 11, 6330 Cham, Switzerland

Preface

Recently, nanoclay-dispersed polymeric materials exhibit better utilization in diverse applications. The nanoclay as inorganic fillers consequences from the exfoliation or the dispersion at nanoscale into polymeric matrices, which enhance the improvement of nanocomposites properties by adding slight amounts of clay due to the high specific area. Current studies reported about Wood Polymer Nanocomposites-Chemical Modifications, Properties, and Sustainable Applications, which demonstrated better mechanical and thermal properties as compared to nanoclay-reinforced polymer composites. The proposed book is focused on chemically dispersed nanoclay-impregnated wood polymer nanocomposites properties and applications. It will also include the introduction and reinforcing potential various clay and monomers dispersed wood nanocomposites. The readers will find complete information about preparation and characterizations of various clay and monomers dispersed wood nanocomposites, combined styrene/mma/nanoclay crosslinker effect, oxidation of wood species by phenyl hydrazine, *N,N*-dimethylacetamid impregnation, urea formaldehyde impregnation, epoxy/nanoclay impregnation, nanoclay/phenol formaldehyde resin impregnation, clay-dispersed styrene-co-glycidyl methacrylate impregnation, styrene-co-ethylene glycol dimethacrylate impregnation, styrene-co-3-(trimethoxysilyl)propyl methacrylate with clay impregnation, acrylonitrile/butyl methacrylate/halloysite nanoclay impregnation, furfuryl alcohol-co-glycidyl methacrylate/nanoclay impregnation, furfuryl alcohol-co-ethyl methacrylate-impregnated wood polymer nanocomposites, and sustainable application of various monomer/clay-dispersed wood polymer nanocomposites. I am thankful to all co-authors who contributed book chapters and provided their valuable ideas and knowledge to this edited book. I attempt to gather all the information of co-authors from same fields in chemically modified nanoclay-dispersed nanocomposites and finally produce this project that will hopefully become a success. I impressively appreciate co-authors' support to formulating ours idea in reality. I thank Springer International Publishing AG, Gewerbestrasse 11, 6330 Cham, Switzerland, team for their substantial cooperation at every stage of the book production.

Kota Samarahan, Malaysia Md Rezaur Rahman

Preface

Contents

Introduction to Reinforcing Potential of Various Clay and Monomers Dispersed Wood Nanocomposites'

M.R. Rahman and J.C.H. Lai

Abstract In this chapter, various clay and monomer were introduce to produce high strength wood polymer nanocomposites and discussed how they play important role on nonocomposites to enhance their properties. Low-quality wood can be modified through suitable chemical treatments to improve the physical, mechanical, and thermal properties to meet specific end-use requirements. The in situ polymerization is one of the most popular technique to modified and produce wood polymer nanocomposites. Some common chemical treatments have proven to be effective in improving wood hardness, dimensional stability, stiffness, fire resistance, UV resistance, biological resistance, and aesthetic appeal. Due the further improvement, dual monomer mixture along with reinforcing filler was significant improve the physical, mechanical, thermal, hardness, dimensional stability, stiffness, fire resistance, UV resistance, and biological resistance. The various monomers were used, namely styrene, methyl methacrylate, phenol formaldehyde, phenolic resins, urea formaldehyde, and melamine formaldehyde. However, monomer impregnation and chemical modification were also investigated using new chemical formulations in the present study.

Keywords Mechanical properties · Stability · Chemical treatment · Wood polymer nanocomposites

1 Introduction

Wood is a natural resource and one of the most attractive materials because of its multidimensional assembly and its extensive exhibition on globe. Wood is made up of cellulose, hemicellulose, and lignin to make their multifaceted structures which are biologically originated (Bowyer et al. 2007). The chemically modified or monomer incorporation of three cell wall polymer can be changed the

M.R. Rahman (✉) · J.C.H. Lai
Faculty of Engineering, Universiti Malaysia Sarawak,
94300 Kota Samarahan, Sarawak, Malaysia
e-mail: rmrezaur@unimas.my

M.R. Rahman, *Wood Polymer Nanocomposites*, Engineering Materials,
DOI 10.1007/978-3-319-65735-6_1

physico-mechanical, thermal, and chemical properties of wood (Rowell 2005). The performance of the modified wood would be changed due to the modification or monomer incorporation in the wood cell wall. The main objective of chemical modification or monomer impregnation of wood to revolute its properties and to increase its performance.

Wood has a diversity of completion uses such as construction, furniture, and tools due to its gorgeous features. Solid wood is the greatest desired one for the building and construction material due to its high-physical strength, low-processing cost, and attractive desirability. Researchers are motivated and produced to hunt for substitute resources, such as softwood and some low-density hardwoods, for value-added application due to a deficiency of high-quality hardwoods. In Malaysia, especially Borneo Island has abundantly available softwood and low-density hardwood. To accomplish this objective, appropriate technologies are desired to enhance the wood quality such as mechanical properties, thermal properties, durability, decay resistance, and hardness, and to satisfy specific end-use necessities.

The biodegradability and high-moisture uptake have negative impact on the wood materials if handled incorrectly. Various defects could be ascribed the chemical assembly and some functional features in wood (Rowell et al. 1982). Besides, the main drawback and limitation of the wood is deteriorated by environmental variation which impacts on its physical and mechanical properties (Galperin et al. 1995; Hill 2006). Chemically, wood contains plentiful reactive sites. The most common responsive sites are the hydroxyl groups, which are abundantly available in the three major chemical components of wood, i.e., cellulose, hemicellulose, and lignin. These OH groups absorb humidity from humid environments, which then fluctuate the physical properties of wood. Dynamic humidity conditions effect in alternate swelling and in physical degradation of wood. The moisture absorption easily changes the mechanical properties of wood due to its hydrogen bonding (Kumar 1994). On the other hand, wood is an amorphous material. The porosity of both cell voids and micro pores in the cell walls occurs due its structure. The main paths for the moisture movement in wood are cell cavities. (Schneider 1994). The chemical modification and monomer impregnation were enhanced mechanical, thermal as well as biodegradation properties and to resist the changes in moisture content. The reactive chemicals and polymer could be the blockage of these reactive sites or plugging the pores of wood cell wall.

Many scientists have made boundless efforts to strengthen solid wood with single monomer, dual monomer, and polymers in the past decades. Both thermoplastic and thermosetting polymer systems have been introducing to strengthen the solid wood (Ayer et al. 2003; Schneider 1994). It is obvious that the production of wood polymer composites (WPC) is a favorable approach to improved wood properties. Impregnation method or technique is the most collective and cheap technique to prepare wood polymer composites with a polymeric monomer or prepolymer. This technique easily can fill the empty lumens in the wood by polymers or monomer. This leads to a mixture of two materials of the wood and polymer rather than a true composite.

Wood polymer composites have the capability to transmit energy efficiently from the polymer matrix to the inflexible cellulose constituent. Polymer and the wood cell wall interaction and chemical bonding that will improve polymers performing an active role in the composite. The strong interaction of wood cell wall and polymer resulted in mechanical strength, thermal stability, and resistance to decay, and degradation of wood. It has been observed that wood impregnated with thermosetting resins such as water-soluble phenolic resin, urea formaldehyde, and melamine formaldehyde prepolymers, increases its compressive strength and moisture related behaviors. The thermosetting resins impregnation polymerizes bulks of the cell wall and reacts with hydroxyl groups of wood constituents to prevent the shrinkage of the wood upon drying. But in most cases, the resulting wood polymer composites are quite stiff of the original wood due to the thermosetting polymer impregnation (Deka and Saikia 2000).

Organophilic layered silicates introduce in situ nanoreinforcement technique to manufacturing the nanocomposite. The various thermoplastic and thermoset nanocomposites at small silicate content have been enhanced the physical properties, flexural modulus and strength, static Young's modulus, and thermal stability (Ray and Okamoto 2003). Polymer–clay nanocomposites were first reported in the literature as early as 1961(Blumstein 1961). Polymer/layered silicate nanocomposites were first used in industrial materials by researchers from Toyota based on the thermoplastic polyamide 6 (Usuki et al. 1993a, b). Since then, a great deal of research has been carried out in the field of polymer nanocomposites with various thermoplastic and thermoset polymers over the past decade (Ray and Okamoto 2003).

In this study, conventional and nanotechnology have been introduced on the selective tropical wood species. According to the literature finding, there was no work on tropical wood species with selective chemicals. Taking into account of all these observations and to overcome the above-mentioned problems, the nanotechnology technique and several kinds of chemicals modification have been considered in this study.

2 Problem Statement

Wood has many necessary features for a diversity of end uses such as furniture, building materials, and tools. A lack of high-quality hardwoods has driven researchers and manufactures to search for substitute resources, such as softwood and some low-density hardwood, for significant application. In addition, the dwindling supplies and rising costs of the heavy hardwoods have made attention in the consumption of lower quality woods such as tropical light hardwoods, whose usage can be prolonged by transforming into wood polymer composites. Low-quality wood can be treated through appropriate chemical treatments to improve their properties such as physical, mechanical, and thermal properties to meet specific end-use requirements. In addition, low quality of wood properties will be better than hardwoods after the chemical modification of them.

The wood hardness, durability, dimensional stability, and mechanical properties will be improved by chemical treatments. However, it has been established that most of the chemicals or monomer does not react or bond with wood even though only filled the empty lumens in the wood, which leads to a mixture of two monomers rather than a single monomer. Therefore, the resulting products were still subject to physical and mechanical properties changes with environmental variations. Poor chemical and physical interfacial interactions between the wood surface and chemical are two of the most important mechanisms of bond failure.

3 Literature Review

3.1 Enhancement of Wood Quality

The presence of abundant hydroxyl groups and various cavities in wood components have greatly been attributed to the changes of dimensional and biological degradation (Ghosh et al. 2008; Papadopoulos 2006; Singha and Thakur 2009; Tunc et al. 2010). These hydroxyl groups are considered to be reactive sites and the cell cavities are main paths for moisture movement in wood. The quality of wood will be changed if these reactive sites could be substituted by some suitable monomer or chemicals and the pores in the wood were also blocked by the polymer. The dimensional stability and physical and biodegradation properties have been improved by blockage of these reactive sites with some reactive chemicals or monomer. The pores have been plugging with a polymer will make the wood more resistant. Wood has been modified by different methods such as chemical, thermal, or enzymatic modification.

The idea of chemical modification of wood has been developed primarily to improve the dimensional stability of wood when subjected to change of moisture. Monomers or prepolymers for in situ polymerization or chemical treatments range have been introduced for the chemical composition of wood by chemical reactions which made product for sustainable application. The necessary requirements of chemical modifications are that the reacting chemicals must be penetrated into the wood cell wall and react with the available hydroxyl groups of the cell wall polymer with neutral or mild alkaline conditions at 120 °C temperatures (Ghosh et al. 2008; Papadopoulos 2006; Singha and Thakur 2009; Tunc et al. 2010). The thermoplastics and thermosets polymer are used to improve the quality product.

3.2 Modern Policies to Growth Wood Quality

Wood contains numerous hydroxyl groups and various cavities which create some unfavorable features such as high-moisture uptake, biodegradation, and dimensional instability that were permitted to modification of wood through various chemical treatments, single or dual monomer impregnation. (Hill 2011; Rowell 2005; Singha and Thakur 2009; Tunc et al. 2010). These hydroxyl groups are considered to be reactive sites and the cell cavities are main paths for moisture movement in wood. These reactive sites will be substituted by suitable or monomers to block the pores and react with hydroxyl and it changed their properties to improve the quality of wood product. Due the chemicals modification or polymer impregnation, the reactive sites blocked by reactive chemicals or the pores have been plugging by the polymer which enhanced the moisture-resistant content and improve its dimensional stability, physical, and biodegradation properties.

A number of studies have been engaged at improving numerous wood attributes through chemical modification or monomer impregnation. The chemical modification or polymer impregnation concept of wood was developed to improve their properties to become stability of wood product when exposed to change of moisture. The in situ polymerization is the most powerful technique to produce wood polymer composites using different chemicals reactions or monomers dispersion. The suitable chemicals or monomers are that the reacting chemicals which have been penetrated into the cell wall and react with the hydroxyl groups in alkaline conditions at 120 °C temperatures (Hill 2006; Papadopoulos 2006).

There are two types of polymers that used in improvement the quality of wood product. Thermoset polymers are the most widely used matrix in composite materials. The various types of thermoset polymer matrices are used for making reinforced composites such as Bis-Maleimids (BMI), Epoxy (Epoxide), Phenolic (PF), Polyester (UP), Polyimide, Polyurethane (PUR), Silicone, Polyetheretherketone (PEEK), and Cyanate ester (Salah and Raman 2014). A thermoplastic is a plastic polymer material, which can be flexible or moldable above a specific temperature and concentrate upon cooling. Thermoplastics become soft when heat is applied and have a smooth, hard finish when cooled. Thermoplastic polymer changed properties when heated and cooled. The different types of thermoplastic polymer matrix used to produce reinforced composites are Polyethylene (PE) (low and high density), Polypropylene (PP) (low and high density), Polystyrene (PS), Polybutylene (PB), Poly (vinyl fluoride) (PVF), Poly (vinyl chloride) (PVC), Poly (vinyl acetate) (PVAC), Poly (vinyl butyral) (PVB), and Nylon (Valente et al. 2011). In the case of thermoplastics, above a certain temperature, all polymers become softer and are able to melt. When the temperature increased the plastics frequently melt in the range of 100–250 °C. The polymers have liquid like order (they are shapeless and in the melting state). The density of polymer will enhance and specific volume decreases when reducing the temperature (Koutsos 2009).

In the past decay, most of the people are used expensive wood but durable hard wood species such as cheshnut or tropical hardwoods were the simplest solution to create strong wood products. As the availability of naturally durable species has declined, the industry has turned to softwoods plantations because of the first-growing species. Softwood species need to perform treatments modification or polymer impregnation in order to achieve satisfactory longevity under services conditions. Several approaches are introduced to increase wood properties for soft non-durable species. There are four important techniques to develop the wood product, namely chemical modifications, impregnation modifications, thermal modifications, and surface modifications.

Chemical modifications can be defined as the reaction of a chemical component with hydroxyls group from the cell wall of wood. They involve the formation of covalent bonds with OH groups from cellulose, hemicellulose, or lignin. The chemical composition of wood will be changed which confer to enhance the properties of wood species. Impregnation modifications define as solid specimens have been impregnated with various types of monomer or polymer to give biological resistance or in filling cell voids by polymers (bulk effect) to give dimensional stability. Thermal modification is one of the modifications that the heat is applied to the wood specimens that results in degradation associated with chemical composition in the wood material. If carefully controlled, the property changes on wood due to thermal modification can be interesting for certain applications. Surface modifications are applied to change the wood surface involving chemical modifications, biological modifications using enzymes, or physical processes such as plasma modifications. The main focusing areas of interest are modifications have been improved bonding between wood surfaces with monomer directly or between wood surfaces and coatings or to improve the weathering performance of wood. These wood treatments are proposed to increase at least one of the properties of wood. However, the chemical modifications or impregnations have an impact on the other wood properties as well, sometimes it will be positive while sometimes negative depend on the nature of the chemicals or monomers. For instance, the metal salts with reinforcing filler impregnation have improved light resistance against fungi decay, on the other hand, it brings the positive impact on the dimensional stability. Equally, thermal treatment has a negative impact on the mechanical properties while they increase the dimensional stability of wood. The next paragraphs present a short description of the past technologies to improve the characteristics of non-durable wood and detailed the new technologies that are applied in industry or are under current investigation in research laboratories.

3.3 Chemical Modification of Wood

The alcoholic function group can create by the hydroxyl groups of the wood macromolecules. The most of the wood modification has been conducted by acylation reactions with hydroxyl groups of the wood. Due to the acylation reactions, the functionalization of the OH groups and the natural hydrophilicity of wood are reduced. In addition, the dimensional stability and the biological resistance have been improved which reported in most cases (Militz et al. 1997). This modification increase of properties is due to covalent bonding and is therefore durable and not subject to leaching by water. Depending on the aim of this study, various properties of the chemically modified wood might be reported but they are not often comprehensively determined. They are laboratory biological tests to determine the decay and termite insect resistances, dimensional stability, mechanical properties, water repellency, and more rarely fire resistance and photodegradation.

The following paragraphs present the most frequency reported chemical reactions which enhanced the wood properties. The basic technical processes for wood quality improvement can be classified into the following, namely (Ayer et al. 2003; Fry 1976; Georgieva et al. 2008) the impregnation of wood with reactive chemicals, second is the compression of wood, and third is the chemical impregnation and compression of wood. Only few of them have the potential to be conducted in an industrial scale.

3.3.1 Reaction with Acid Chlorides

Wood has been acylated using acid halides. Hydrogen chloride is released as a by-product leading to degradation of the wood fibers if a base such as pyridine is not used. Therefore, the use of this technique is limited when treating lumber because of the necessity to remove the by-products present in wood after treatment.

$$\mathrm{Wood{-}OH + R{-}C({=}O){-}Cl \rightarrow Wood{-}O{-}R{-}C({=}O){-}R + HCl}$$

It has been reported that reaction of wood with palmitoyl chloride permits to obtain a treated wood presenting an anti-swelling efficiency (ASE) of 48% for a weight percentage gain (WPG) of 21% (Prakash et al. 2006).

3.3.2 Esterification of Wood by Carboxylic Acids

The esterification of wood by carboxylic acids in presence of a catalyst is very effective. An in situ carboxylic acid chemical reagent is often used to convert cellulose esterification. The cellulose esterification is more reactive entity compare to anhydride or an acid chloride. The formation of anhydrides is effectively achieved with trifluoroacetic anhydride or with *N*,*N*-dicyclohexylcarbodiimide.

$$\mathrm{Wood{-}OH + (CH_3)_2{-}C{=}CH{-}COOH \rightarrow Wood{-}O{-}C({=}O){-}CH{-}C(CH_3)_2}$$

An in situ method was used to increase the oven-dry specimen's volume without a change in color or a decrease in either crystallinity or moisture content which cause by unsaturated carboxylic acids esterification (Nakagami and Yokota 1975; Rowell 2005). Reaction with α-methylcrotonic acid gave a higher degree of substation which enough to make the reacted wood soluble in acetone and chloroform to the extent of 30% (Nakagami et al. 1976).

3.3.3 Wood Modification by Ketene

Ketenes are very toxic and reactive chemicals which penetrating in the wood cell and its tendency to polymerize. The particular case of ketene has been used to acetylate wood. This reaction does not create any by-product provided that the wood contains no moisture.

$$\mathrm{Wood{-}OH + CH_2{=}C{-}O \rightarrow Wood{-}O{-}C({=}O){-}CH_3}$$

Ketene modification of wood has been established to increase dimensional stability but it is not as actual at improving this property as acetic anhydride. Oven-dry aspen and southern pine wood flakes react with ethenone at 50–60 °C leading to WPG of the order of 20% after 10–15 h of reaction. Furthermore, wood acetylation with ketene is not highly resistant to decay at a WPG of 17% (Rowell et al. 1986).

3.3.4 Wood Modification by Aldehydes

A hemiacetal is produced by the addition of an aldehyde to a hydroxyl group. The hemiacetal was further reacted with another OH group of the cell wall polymers producing an acetal crosslinking bond. This bonding is vulnerable to do the hydrolysis.

$$\mathrm{Wood{-}OH + H{-}C({=}O){-}H \rightarrow Wood{-}O{-}C(OH){-}H_2 + Wood{-}OH}$$
$$\mathrm{Wood{-}O{-}C(OH){-}H_2 + Wood{-}OH \rightarrow Wood{-}O{-}CH_2{-}O{-}Wood}$$

Formaldehyde is a very popular aldehydes compound which frequently used and it's reported in literature (Stevens et al. 1979). Due to the formaldehyde modification, the mechanical properties of such modified wood are poor and severe embrittlement because of the rigidity induced by the crosslinking and the acidic conditions of the modification. Formaldehyde treatment has been carryout under vapor phase in the presence of SO_2 as a catalyst with a minimal loss of strength compared to other treatments in liquid phases. The equilibrium moisture content (EMC) of wood product has been significant reduce even low levels of formaldehyde introduce in the wood specimens. Yasuda et al. (1995) found that a WPG level of only 3.5% resulted in a 50% reduction of the EMC compared to unmodified wood. Decay resistance of formaldehyde treated with wood showed better results at only 2% of weight gain. It was reported that strong resistance showed against by white rot fungi while poor resistance show against by brown rot fungi (Minato et al. 1992). Glyoxal, glutaraldehyde, and other aldehydes have also been investigated as reagents for wood modification. It was concluded that none of them formed stable crosslinks in the wood cell wall.

3.3.5 Wood Modification by Isocyanates

Isocyanates and thio-isocyanates rapidly react with wood hydroxyl groups to produce wood–urethane bonds (Deka and Saikia 2000).

$$\text{Wood}{-}\text{OH} + \text{R}{-}\text{N}{=}\text{C}{=}\text{O} \rightarrow \text{Wood}{-}\text{O}{-}\text{C}({=}\text{O}){-}\text{NH}{-}\text{R}$$

4,4′-diphenylmethane diisocyanate (MDI) has been frequently used wood modification. Isocyanates swell wood and react with it at 100 to 120 °C without catalyst or with a mild alkaline catalyst such as trimethylamine (Gao and Gu 2007). The resulting urethane bond is very stable to acid and base hydrolysis. There are no by-products produced from the chemical reaction of isocyanate with dry wood. However, isocyanates react rapidly with water to yield a di-substituted urea. For this reason, it is significant that moisture is thoroughly omitted during reaction. Wood modified with butyl isocyanate shows a threshold for decay protection at around 15% of WPG with all the tested fungi. There is no major change in performance related to the chain length of isocyanate (Cardias and Hale 1999). Unlike mono-isocyanates, a reaction of wood with di- and poly-isocyanates can result in homopolymerization and subsequently in bulking effect.

3.3.6 Wood Modification by Epoxides

The epoxides modification occurred in the reaction between wood cell wall and epoxide to formation of an ether linkage with OH group. Therefore, graft-polymerization reactions are possible (Cetin and Hill 1999; Kumar 1994).

$$\text{Wood}{-}\text{OH} + \text{R}{-}\text{CH}({-}\text{O}{-})\text{CH}_2 \rightarrow \text{Wood}{-}\text{O}{-}\text{CH}_2{-}\text{CH}(\text{OH}){-}\text{R}$$

Several epoxides have been used in the past decades for wood modification purposes. They include ethylene oxide, propylene oxide, and butylene oxide (Norimoto et al. 1992). Usually, the reaction is catalyzed under mild basic conditions. In most experiments, trimethylamine is used as a catalyst. Wood modified with propylene oxide was ineffective toward *G. trabeum* decay resistance whereas butylene oxide modification proved to be effective at 23% WPG (Ibach et al. 2001).

3.3.7 Wood Modification by Cyanoethylation

Wood cell wall and OH group react with Acrylonitrile in the presence of alkaline catalyst. NaOH modified treated wood get higher WPG up to 30% which has been reported giving ASE in the region of 60% (Stamm and Baechler 1960).

$$\text{Wood}-\text{OH} + \text{CH}_2{=}\text{CH}-\text{CN} \rightarrow \text{Wood}-\text{O}-\text{CH}_2-\text{CH}_2-\text{CN}$$

Biological resistance due to bulking, rather than toxicity, of such treated wood have also been reported. Cyanoethylated wood in ground contact at 15% of WPG has an average life of almost 8 years, compared with 4 years for untreated samples (Stamm and Baechler 1960).

3.3.8 Wood Modification with Alkyl Halide

Wood has been etherificated by the alkyl halides in the presence of a strong base

$$\text{Wood}-\text{OH} + \text{NaOH} \rightarrow \text{Wood}-\text{Ona} + \text{R}-\text{X} \rightarrow \text{Wood}-\text{O}-\text{R}$$

Reactions of wood with crotyl chloride, methyl iodide, and butyl chloride have led to an improvement in dimensional stability. Decay resistance has also been investigated for wood treated with fatty diakyldimethylammonium chlorides and bromides. Modified wood exhibited better resistance against brown rot fungi (Hill 2006).

3.3.9 Wood Modification by β-Propiolactone

The wood OH group react with β-propiolactone in presence of acids or bases catalyst to yield two different products as shown in Fig. 1.

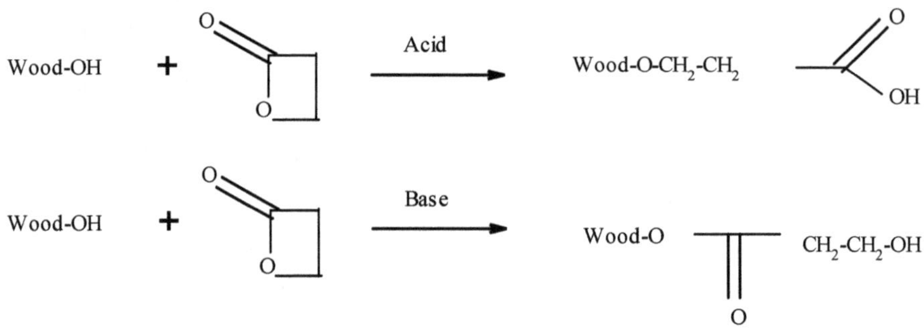

Fig. 1 Schematic chemical reaction with β-propiolactone

The modified wood in acidic environments with 30% WPG resulted in better decay resistance and ASE of 60%. However, a strong degradation of wood has been protected by the β-propiolactone (Goldstein et al. 1959).

3.3.10 Wood Modification by Cyclic Anhydride

The hydroxyl groups of wood react with cyclic anhydrides do not yield a by-product. The anhydride covalently bonded on wood by an ester function yields a free carboxylic group at its end as shown in Fig. 2. The free carboxylic group undergoes with another OH group to crosslink the cell wall polymers of wood (Matsuda 1987).

Cyclic anhydrides have been used as chemical reagents for wood modification for example phtalic anhydride, maleic anhydride, glutaric anhydride, succinic anhydride, and alkenyl succinic anhydride (ASA) (Chauhan et al. 2001; Morard et al. 2007). Scots pine modified samples with petrochemical octenyl succinic anhydrides (ASAs) dissolved in pyridine. The succinic anhydrides modification did not present enough resistance against fungi decay but increased dimensional stability (Chauhan et al. 2001).

3.3.11 Wood Modification by Acrylic Anhydride

The acetic anhydrides react with the wood cell wall OH groups to produce an ester and acetic acid as a by-product. The following reaction is conducted at 70 °C without catalyst (Bongers and Beckers 2003) which shown in Fig. 3.

It is known that the rate of reaction is promoted by wood-swelling agents such as pyridine that can be used only at laboratory scale. The improved dimensional

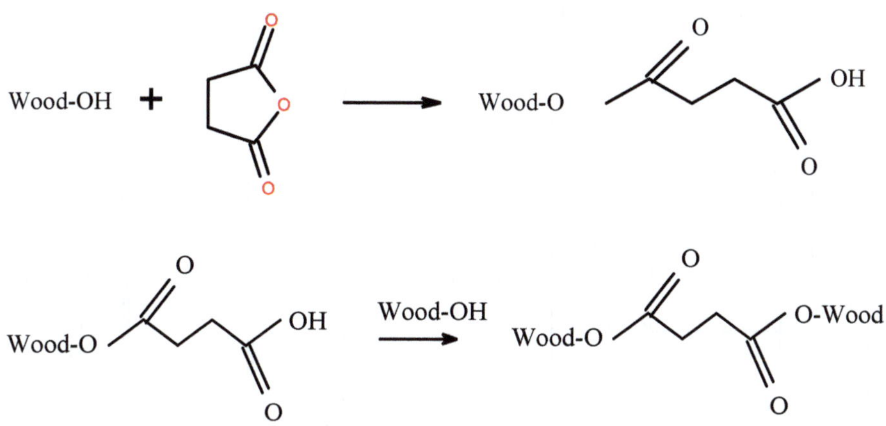

Fig. 2 Schematic chemical reaction with cyclic anhydride

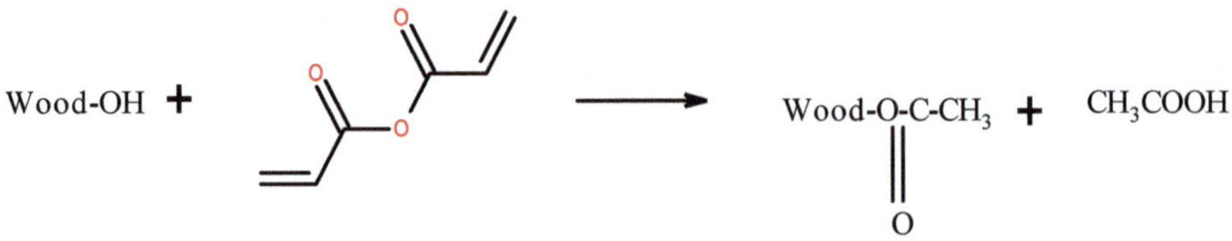

Fig. 3 Schematic chemical reaction with acrylic anhydride

stability of wood as a result of anhydride modification has been found to be a function of WPG only, regardless of the anhydride used for modification.

3.3.12 Oxidation of Wood

Wood has been oxidized using oxidizing agent with acidic catalyst as shown in Fig. 4. Oxidation reactions applied to cellulose in wood for chemical modifications (Hon and Shiraishi 2001). Oxidation reactions occur on cellulose selectively at particular position. The reaction of sodium metaperiodate with cellulose in wood fiber in the presence of sulfuric acid catalyst at 120 °C and 85 kPa pressure yielded the oxidized product. Sodium metaperiodate reacts with hydroxyl groups of cellulose and produces 2,3-dialdehyde cellulose which improved the mechanical and biological properties of wood (Rahman et al. 2010).

3.4 *Wood Modifications by Impregnation Technique*

Wood impregnation by various type of chemicals or polymer has been used to improve wood hardness, durability, dimensional stability, and mechanical properties (Ayer et al. 2003; Schneider 1994). The wood impregnation process

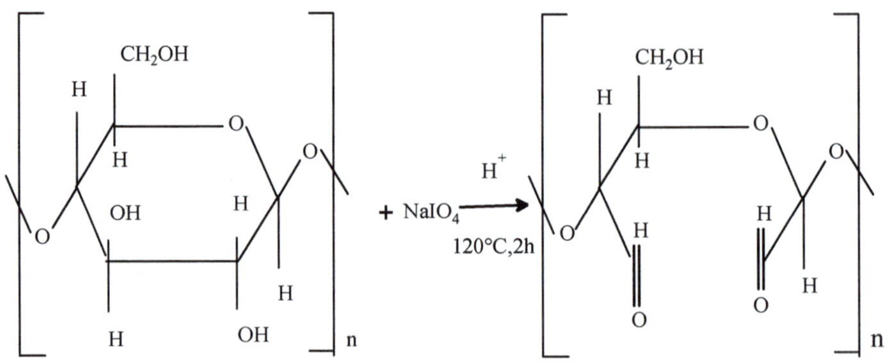

Fig. 4 Schematic chemical of oxidation

significantly depends on diffusion of the gas or liquid into the wood and the permeability of the wood cell wall. The specimens must be free from air, moisture, and water before chemical impregnation because it make the space for the treating chemicals to fill the pore. Treating chemical or monomer or polymer is required to be low viscosity, which normally corresponds to low or medium molecular weight. The treating chemicals are forced into the wood by vacuum and pressure. Vinyl monomers such as styrene and methyl methacrylate (MMA) are the examples of low viscosity monomer often used in wood polymer composites preparation through the impregnation technique. Some mechanical properties such as hardness of wood can be greatly enhanced by filling the empty spaces in the cell tissue with various type of chemicals or monomer or polymer. The filling process is done by impregnating the wood specimens in a closed chamber with a suitable chemical or monomer or polymer that can be hardened inside the wood. To penetrate the chemical or monomer or polymer evenly into the cell tissue of the wood is a stimulating problem due to the fact that the permeability of most species is poor and varies significantly in various sections of the wood (e.g., heartwood vs. sapwood). Another problem is how to polymerize the impregnated substances inside the wood without causing splits or burning of the treated wood or reducing its machinability. The wood structure and density, the properties of the treating chemicals, viscosity, and concentration of the treating solution, temperature, and pressure are all significant issues that affect the outcome of wood impregnation. Many chemicals have been used for wood impregnation, as briefly labeled in the following.

3.4.1 Wood Impregnated by Monomer and Prepolymer

The monomer and prepolymer are subsequently polymerized with the cell lumens of wood and filled their void and block the hydroxyl group in wood cell wall which resulting wood polymer material normally retains most of the required characteristics of wood and shows various improved properties. Most wood species can be readily impregnated with vinyl monomers or prepolymers by using equipment and techniques similar to those used in the conventional wood preservation in industry (Chao and Lee 2003; Meyer 1981). There is a big difference of viscosity among monomer and prepolymers.

3.4.2 Enhancement of Wood Properties by Polymers Impregnation

The thermoplastic and thermosetting polymers are used to produce wood-polymer composites. The well-known thermoplastic polymer is a poly vinyl chloride or similar monomers or modified vinyl monomers and polar monomers. Vinyl polymers have a large range of properties from soft rubber to hard. The brittleness of vinyl polymers depending upon the groups attached to the carbon–carbon backbone. Some examples of vinyl monomers are as follows: vinyl chloride, vinyl acetate, acrylonitrile, ethylene oxide, acrylates (especially methyl methacrylate), *t*-

butyl styrene, styrene, and chlorostyrene (Meyer 1981; Georgieva et al. 2008). The vinyl monomers are non-polar and poor interaction with wood hydroxyl groups.

In general, vinyl monomers fill the capillaries, vessels, and other void spaces in the wood structure (Georgieva et al. 2008). Polar monomers can swell cell wall to provide high-chemical resistance and dimensional stability in water. Thermosetting polymer is a type of polymers which include phenol formaldehyde, urea formaldehyde, melamine formaldehyde, polyurethanes, epoxides, silicones, and unsaturated polyesters (Ghosh et al. 2008). Phenol-formaldehyde resin is plasticized the cell wall of wood and yields dimensionally stability of the wood polymer composites (Tunc et al. 2010).

3.4.3 Wood Impregnated by Vinyl Monomers

The vinyl monomer impregnated wood polymer composites significantly improve their moisture resistance, thermal properties, mechanical properties, hardness, and other qualities of wood. The vinyl monomers are polymerized in presence of catalysts or radiation techniques (Siau et al. 1965). It is required to have the vinyl polymer successfully stabilize the wood by bulking the cell wall, but with the exception of acrylonitrile, vinyl monomers are generally poor swelling agents for wood (Ellwood et al. 1972; Loos and Robinson 1968; Siau 1969). Loos et al. investigated the dimensional stabilization of wood with vinyl monomers. Their work showed that without additional swelling agent vinyl monomers penetrate in the cell walls only to a limited extend. They are not predictable to impart any substantial dimensional stabilization to wood. The impregnation of wood with acrylic or vinyl type monomers showed less-dimensional stability in the presence of moisture. This was due to the detention of the monomer in the cell lumen instead of the cell wall. MMA and styrene impregnated wood polymer composites showed better dimensional stability compared to other vinyl monomer impregnated wood polymer composites (Brebner et al. 1988; Chao and Lee 2003; Khan et al. 1992a; Langwig et al. 1968; Raff et al. 1965).

3.4.4 Wood Impregnated by Methyl Methacrylate (MMA)

Methyl methacrylate (MMA) is the most commonly used vinyl monomers to produce wood polymer composites (WPCs). It is one of the least expensive and most readily available monomers that are used alone or in combination with other monomers to produce the crosslink the polymer system. MMA impregnated wood polymer composites showed fewer swells compared to other vinyl monomers (Loos and Robinson 1968). MMA monomer filled cell lumens but unchanged cell walls which showed in WPC. This type of WPC has time-dependent dimensional stability improvement on the other hand, and it might not be sustainable for long term. Over the years, MMA has been extensively used monomer for production of WPC (Duran and Meyer 1972; Langwig et al. 1968, 1969; Noah and Foudjet 1988; Siau and Meyer 1966; Siau et al. 1978; Yalinkilic et al. 1998, 1999a, b). The use of a crosslinker rapidly increased the initial viscosity of the system and an auto-acceleration of the polymerization was observed for the crosslinker, trimethyl propane trimethacrylate (TMPTMA), reduced the time to the exothermic peak and dramatically increased the exothermic peak temperature. The effect of the crosslinking agent on the mechanical properties of wood polymer composites was uncertain (Islam et al. 2011). Siau and Meyer (1966) compared the properties of MMA impregnated wood polymer composites cured by heat and radiation, respectively. There is no significant change in yellow birch-polymethyl methacrylate combinations which were found between heat and radiation when tested for compression parallel to the grain direction, shear strength, permeability, diffusion coefficient, and anti-shrink efficiency. The material polymerized by radiation was about 25% better than that of the material polymerized by heat; this is due to the superior evaporation of monomer from near the surface during heat-induced polymerization which increases the surface hardness. The dimensional distortion and volumetric changes were observed if the wood impregnated by MMA and polymerized by heat (Siau 1969). The distortion of basswood was much more severe than that of maple wood. Siau (1969) also investigated the swelling behavior of basswood impregnated with MMA and styrene. Anti-shrink efficiency of up to 40% indicated the entry of monomer into the cell wall before polymerization. The rate of swelling by monomers improved with temperature and moisture content of the wood. The amount of swelling at steadiness usually reduced with temperature. Other studies report a large difference in swelling properties between wood species. Beall et al. (1966) investigated the polymerization of basswood impregnated with MMA containing a peroxide initiator by direct heat and radio-frequency radiation. Three types and two concentrations of organic peroxides were used for the initiation of the polymerization of MMA. Heating the wood-monomer composite by radio-frequency energy which temperature between 90 and 98 °C caused a rate of free radical generation satisfactory to overcome the inhibitor concentration and to stimulate the achievement of polymerization in about 20% of the time required for direct heating. The radio-frequency heating polymer retention was approximately 80% compared to direct heating. The radio-frequency treatment techniques were used to handle the accounted of monomer loss. The heat catalyst method applies for polymerization and impregnations of MMA monomer to heartwood and sapwood before and after which compare their properties (Young and Meyer 1968). Vazo free radical catalyst was used to polymerize the MMA in wood cell wall. Eight species of wood including yellow birch and red maple were investigated. After treatment, the sapwood exhibited greater increase in compressive strength perpendicular to the grain direction, tangential hardness and density than did the heartwood. Sapwood also showed a greater reduction in permeability and it had higher polymer retention. Beall et al. (1973) compared the hardness of WPC of red oak, aspen, and hard maple that were impregnated with MMA. They found that the hardness of untreated wood was mainly related to density. Schneider (1995) proposed a method to prepare cell wall and cell lumen wood polymer composites by using two monomers, one of which can swell the cell wall. Furfuryl alcohol-based

(FA) and MMA-based mixture preparations were used. The outcomes recommended that treating solution departure occurred during impregnation and the resulting inhomogeneous chemical distribution yielded the inclines in properties. A combination of wood cell wall modification with acetic anhydride and lumen filled with MMA treatments were studied for their usefulness in decreasing the rate of moisture absorption and the degradation effects of accelerated weathering (Feist et al. 1991). The treatment of acetylation followed by MMA impregnation was the most effective in reducing the rate and extent of swelling and reducing erosion produced by accelerated weathering (85%). Both of acetylation and MMA treatments or a combination of the two monomers which subsequently reduced the loss of surface lignin with improvement in durability. Compared to untreated wood, treatment of wood with MMA polymers decreased the rate of moisture uptake and improved mechanical properties, including modulus of elasticity and rupture. The stress at proportional limit, maximum crushing strength, and hardness index improve their properties (Langwig et al. 1968; Rowell et al. 1982). However, treatment with MMA polymers results in high-polymer weight gain and is more expensive.

3.4.5 Wood Impregnated by Styrene

Styrene is one of the powerful monomers that is usually used for manufacturing of WPCs. Styrene impregnation WPCs has been extensively studied and the styrene wood product has developed water repellence, compression, and bending strength of WPCs (Baki et al. 1993; Siau et al. 1965). Styrene and wood cell wall polymerization were occurred by heat or radiation in presence of catalyst. The investigation by Siau et al. (1965) was primarily focused on bulking wood with styrene, followed by polymerization with gamma irradiation. Impregnation was done by solvent exchange and high-vacuum methods. The gamma irradiation produced graft copolymers in the wood that were more effective than homopolymer in improving dimensional stability and tangential tensile strength.

Brebner et al. (1985) explored impregnating eastern white pine heartwood with a combination of styrene monomer and a crosslinking agent. The polymerization of styrene was then accomplished using a heat-catalyst technique. Styrene uptake was dependent on the growth rate of different wood samples, but the strength did not suffer in evaluation with more uniformly loaded low-density woods. At air-dry moisture content and equal density, the pine-polystyrene composite product was found to be mechanically inferior to sugar maple. Devi et al. (2003) examined chemical modification of rubberwood which impregnated by the styrene in combination with a crosslinker glycidyl methacrylate (GMA). The polymerization was took place by a catalyst and heating. The dimensional stability and anti-shrink efficiency of treated wood were improved by the combined monomer impregnation. Water absorption was decreased due to the combined monomer impregnation. Mechanical properties, including modulus of rupture (MOR) and modulus of elasticity (MOE), were also improved. An anti-shrink efficiency value of 53% was obtained for styrene-GMA treated wood samples compared to 23% for styrene treated wood samples after 24 h soaking in water. Hardness was similar for these samples but 33% greater compared to untreated samples. Chao and Lee (2003) examined enhancement in hardness of southern pine by regular diffusion and vacuum impregnation with styrene which was polymerized in situ by a catalyst and heating. The density, water repellence and hardness of the southern pine increased after styrene impregnation. However, the vacuum method facilitated absorption of a greater amount of styrene into the wood cell wall in a greatly shorter time. The modified wood showed a greater increment in density, less water absorption and twofold hardness increased by 1 min of vacuum.

3.4.6 Wood Impregnated by Vinyl Chloride

The wood has been impregnated and polymerized by the vinyl chloride under a wide variety of conditions (Juneja and Hodgins 1970). The substrate will distortion and discoloration because of the polymerization reaction is highly exothermic. The hydrogen chloride is released from degradation of the polyvinyl chloride (PVC) and leaving behind conjugated double bonds along the chain which is responsible for the polymer discoloration. Vinyl chloride was considered unsuitable as a monomer for producing acceptable WPC material mainly for the reasons: (i) its polymer is a combination of powder and smooth polyvinyl chloride. (ii) Wood mechanical properties are only slightly improved. (iii) Reaction heats are so high that discoloration and deformation occur. (iv) Polyvinyl chloride is insoluble in its monomer and during the reaction, it precipitates as a fine powder which contributed nothing to the enhancement of mechanical properties. (v) PVC has a negative environment image.

3.4.7 Wood Impregnated by Hydroxymethylacrylate and Ethyl-α-Hydroxymethylacrylate

According to Mathias and Wright (1989), WPCs were developed by hydroxymethylacrylates impregnation and in situ polymerization. The monomer can penetrate the cell walls of the wood fibers, thus forming an entirely filled wood polymer composites. Wright and Mathias (1993) also established a lightweight material based on balsa wood-polymer composites by ethyl-α-(hydroxymethylacrylate) acrylate (EHMA) and styrene impregnation. The dimensional stability and mechanical properties were improved by the impregnation of EHMA-styrene copolymer. Improvements in specific modulus and specific toughness (absolute properties divided by specific gravity) were also achieved using an EHMA-styrene monomer mixture with polybutadiene diacrylate as a cross-linker and toughening agent. The best results were obtained at low weight gain (10–40%).

These enhancements in modulus and toughness were attributed to effective diffusion of cell walls by the monomers.

3.4.8 Wood Impregnated by Phenolic Resin

Wood was impregnated by water-dispersible phenolic resins. The wood cell walls and swell can be diffused by the water-dispersible phenolic resin. The water can then be removed by drying the wood at low temperature and the resin is cured by heating. Bryant (1966) examined the effects of phenolic resin impregnation on mechanical properties of wood. The dimensional stabilization of wood depends on the relative molecular size of phenolic resin and concentration. The compressive strength, moisture-related shrinking, and swelling were improved by the phenolic resins impregnation. Phenolic resins penetrate and bulk the cell wall structure, preventing shrinkage of the wood upon drying.

3.4.9 Wood Impregnated by Polyurethane

According to the Hartman (1969), wood was modified by aqueous polyurethane emulsion because of its accessibility as low molecular weight components in aqueous media. Four different wood species were impregnated with different molecular polyurethanes. The aqueous polyurethane systems have a high solid content and low viscosity which are cured readily at elevated temperature. Polyurethanes are used repeatedly without affecting polymerization. Lower molecular weight polyurethane showed better diffusion and dispersion into the wood cell wall (Hartman 1969; Khan et al. 1992a; Khan and Idriss Ali 1992b). The mechanical strength of wood composites also improved by the urea modification. They suggested that the use of urea in the wood polymer composite formation open up a wide scope of preparation technique of WPC when high polymer loading along with strong tensile strength is considered.

3.4.10 Wood Impregnated by Melamine Formaldehyde

Polymers of melamine (1,3,5-triamino-2,4,6-triazine) and formaldehyde form an vital class of amino resins, which have been commercially used for over 60 years. Melamine formaldehyde (MF) itself is one of the hardest and stiffest isotropic polymeric materials used in attractive laminates, moulding compounds, adhesives, coatings, and other products. It has many advantageous such as high hardness and stiffness and low flammability. MF resins have potential to improve properties of solid wood. Impregnation of solid wood with water-dispersible MF resin has led to a significant development in surface hardness and modulus of elasticity (MOE) (Deka and Saikia 2000; Inoue et al. 1993; Miroy et al. 1995). Aqueous melamine-formaldehyde (MF) resins can penetrate into the wood cell wall (Gindl et al. 2002; Gindl and Gupta 2002; Rapp et al. 1999) and the amorphous region of cellulose fibrils. The strength and stiffness of spruce in transverse compression were investigated by the melamine-formaldehyde impregnation (Gindl et al. 2003). Due to the melamine-formaldehyde polymerization, the tangential compression strength of modified samples was 83% higher than that of the unmodified one. On the other hand, radial compression strength had increased by 290%. Less than 2% of the cell cavities (lumina) were found to be filled with resin. Thus the improvement of strength and stiffness obtained are attributed to modification of the cell wall but not to filling of tracheid lumina. Yielding behavior under excessive compression load changed from plastic bulking in the reference samples to brittle fracture of cell walls in the modified samples. According to the Gindl and Gupta (2002) the cell wall hardness and Young's modulus of melamine-modified spruce wood by nanoindentation. Their study demonstrated that modification of wood with melamine formaldehyde resin changes cell wall mechanical properties. Gindl et al. (2003) also observed the selected factors influencing the uptake of MF resin into the cell wall of softwood. They used UV-microspectroscopy, to determine that water-dispersible MF diffused well into the cell wall. For the cell wall impregnation, few factors are favorable such as moisture content, high-water content of the resin used for impregnation, and low-extractive content. For dry cell walls, solvent exchange drying improved resin uptake to a similar extent, as was the case when cell walls were soaked in water.

3.5 Wood Impregnated by Inorganic Substance

It is remarkable that wood-inorganic composites have been developed by inorganic substances which are filled with the wood's cell lumens and/or cell walls (Furuno et al. 1991, 1992a, b, 1993; Ogiso and Saka 1993; Saka et al. 1992; Saka and Yakake 1993; Yamaguchi 1994a; Zollfrank and Wegener 2002). The fire and decay resistance of the wood and other properties is improved by impregnation of inorganic substance. Silicified wood is a type of fossil produced by the replacement of some or all of the wood substances with SiO_2 mineral (silica) during burial over extremely long periods, and it is almost fossilized. A combination of acetylated and propionylated wood-silicate composites was investigated by Li et al. (2001). The wood was impregnated with an aqueous sodium silicate solution. The manufacturing composites showed good dimensional stability. The silicate gels protective in composites capable them with flame-resistance, while the oxygen index of the composites increased with weight percent gain of silicate gels. Yalinkilic et al. (1998) established a new technique in which boron treatment was combined with vinyl polymerization to improve leaching resistance of boron from wood, as well as dimensional stability, biological, and thermal resistance of wood. Wood specimens (sapwood of Cryptomeria japonica) were impregnated with boric acid (BA) as 1% aqueous solution prior to vinyl monomer treatment. Styrene, MMA and their mixture (50:50, v/v) were impregnated in the presence of catalyst and crosslinker agent which conducted by heat radiation at 90 °C for 4 h. Treated specimens were then subjected to decay and termite tests, as well as oxygen index determination. Anti-swelling efficiency and water absorption levels were also measured. Vinyl monomers succeeded in reducing water absorption of wood to minimum level and deferring boron leaching significantly. The treated wood proved to be resistant against two decay fungi such as white and brown root fungi. It is well known that vinyl monomers contribute to dimensional stability and strength properties of wood, and they are expected to provide a delayed leaching of boron from wood. Boron can improve the biological and fire resistance of vinyl polymerized wood in a mixture treatment system whereas vinyl monomers alone are not toxic to fungi after polymerization. Wan (2004) has investigated the addition of an inorganic component to the monomer/prepolymer impregnation of eastern Canadian wood species. The work has shown that this addition can increase the anti-fungal, anti-mold properties.

3.5.1 Wood Impregnated by Combination of Different Monomers System

Gaylord (1973) established a technique for impregnating wood in situ with styrene and maleic anhydride monomers that formed a 1:1 charge transfer complex. At low temperatures, heating produced uncatalyzed polymerization. US Forest Products Laboratory researchers investigated bonding of fluorophenyl isocyanates to wood cell polymers and reaction of 8-hydroxyquinoline, pentachlorophenol, or β-napthnol with ethylene-maleic anhydride copolymer to cell wall polymers (Chen and Rowell 1986). Their goal was to lower chemical load by using bonded biocides as controlled-release compounds. However, filling cell lumens with bioactive polymers led to problems with solubility and penetration because generally, polymers have low solubility and high viscosity. Therefore, they diffuse with difficulty into wood. It is preferable, then, to synthesize a monomer with a bioactive group to fill the wood with monomer, and then in situ polymerize or copolymerize with a carrier monomer using a catalyst. This will effect in higher loading of the bioactive polymer.

A styrene-acrylonitrile-wood composite was developed by Juneja and Hodgins (1970). It was found that the styrene-acrylonitrile-wood composite had improved properties that are comparable with the best materials produced by radiation polymerization. Ellis and O'Dell (1999) established wood composites which made by a combination of acrylic monomers, isocyanate, and maleic anhydride. Pine, maple, and oak solid wood were treated with different combinations of hexanediol diacrylate, hydroxyethyl methacrylate, hexamethylene diisocyanate, and maleic anhydride. It was found that the treatments reduced the rate of water vapor and liquid water absorption. Although the resultant dimensional stability was not permanent, the rate of swelling of WPC specimens was less than that of unmodified wood specimens. In general, WPC prepared with hydroxyethyl methacrylate were tougher than specimens made without hydroxyethyl methacrylate. They also resist water and moisture more efficiently. Using swelling solvents to dilute vinyl monomers and allow the monomer to enter cell walls in an approach that has been used to form vinyl WPC with moisture-resistant cell walls (Furuno and Goto 1973). Wood swelling caused by the approaches uses polar vinyl monomer molecules or additives mixed with the monomer (Loos and Robinson 1968; Rowell et al. 1982; Schneider and Brebner 1985). Such combination treatments (treating cell lumens and walls) show reunite for producing WPC with the benefits of vinyl-wood composite but improve permanent dimensional stability.

3.5.2 Methyl Methacrylate (MMA) Based Wood Polymer Nanocomposites (WPNCs)

The combination of methyl methacrylate (MMA) and hexamethylene diisocyante (HMDIC) monomers mixture in the ratio of 1:1 were introduction to manufacturing the nanocomposite based on selected wood species. All wood species cell wall reacted with HMDIC and crosslinked with MMA which enhanced the hydrophobic (restrained water) nature of wood. The properties of wood species were showed that the monomer mixture loading achievable or not which dependent on lower loading for high-density wood species. The FTIR spectroscopic analysis confirmed interaction among the wood cell and polymer. Morphological properties of WPNCs were evaluated by SEM and XRD analysis. In addition, the modified WPNCs had lower moisture absorption and higher water repellent efficiency compared to raw wood. The storage modulus for WPNCs was enhanced by dynamic mechanical analysis compare to raw wood. Mechanical strength such as modulus of elasticity (MOE) and modulus of rupture (MOR) of fabricated WPNCs was established to be considerably improved. Thermal properties of manufactured WPNCs in terms of TGA and DSC analysis were also calculated, and a development in thermal stability was established for fabricated WPNCs. Decay resistance of WPNCs was measured through the loss of weight percentage (%). Decay resistance result showed that a significant enhancement was found in the impregnated wood compared to the raw wood.

Structural wood has very aesthetically pleasing character. It has been always and continues to be a very significant and multipurpose material with many uses. However, this type of wood species has some drawbacks such as high-moisture uptake, biodegradation, and physical, mechanical, and thermal properties changes due to the environmental issues (Hill 2006). These troublesome inherent properties of wood can be minimized by appropriate chemical treatment which produces the WPNCs (Kamden et al. 2002). The main issue is accountable the negative properties which showed by the presence of hydrophilic hydroxyl groups (–OH) in the wood components. Wood attracts moisture through hydrogen bonding, making it physically unstable. The physico-mechanical and thermal properties of wood can be developed by using an impregnation technique with appropriate chemicals that can react with cell wall constituents (Feist et al. 1991). The enhancement in properties of wood may also be improved preparing WPNCs with different monomers (Boding and Goodman 1973; Ellis and Dell 1999; Kyziol 1999; Mathias et al. 1991). Manufactured WPNCs generally displays actual dimensional stability and outstanding mechanical properties (Hamdan et al. 2009, 2010). WPNCs can also improve many properties of solid wood such as surface hardness, toughness, abrasion resistance, moisture exclusion, weather resistance, and others. In general, the enhancement in property can be recognized to the polymer content, which is dependent on the type of wood, the polymer, and the processing technology applied. Wood properties such as density, moisture content, direction of the grain, and the chemical compositions of cell wall constituent are the main accountable features for the development WPNCs and its properties. In the literature, it can be seen that the heavy hardwoods species gain lower amount of monomer than those from the medium and light wood species (Yap et al. 1990). The lower monomer loading of hardwoods can be attributed by their some essential properties such as high density and specific gravity including internal vessel diameter, the number of vessel percent per unit area, and its high extractive. Therefore, it has been well-known that the properties of WPNCs depend on the wood species and the physical and mechanical properties of WPNCs enhance with the monomer loading (Islam et al. 2010).

Currently, significant interest has been demonstrated in wood impregnation with a diversity of monomers such as styrene, epoxy resins, urethane, phenol formaldehyde, methyl methacrylate, vinyl, or acrylic and their combination to change the specific properties of WPNC (Holyle and Woeste 1989; Rowell 2005). However, it has been recognized that most of the monomers do not form bonds with hydroxyl groups of the wood constituent. They only bulk the void spaces within the wood structure (Elvy et al. 1995). It can, therefore, be removed that if bonding was to take place between the impregnated monomers and the hydroxyl groups on the wood constituent, and the physical and mechanical properties of WPNCs may be advance upgraded. HMDIC is a difunctional reagent which has two reactive functional groups and also has been extensively used as a wood adhesive, cross-linker, and copolymer (Frihart 2005). HMDIC modification of wood depends on modifying the principal wood components like cellulose, hemicellulose, and lignin by reacting wood hydroxyl groups with a diisocyanate group to form wood-urethane derivatives (Frazier 2003). Some researchers consider that the isocyanates also react with available –OH groups according to the following proposed chemical reaction.

$$\text{Wood}-\text{OH}+\text{R}-\text{N}=\text{C}=\text{O} \rightarrow \text{Wood}-\text{O}-\text{C}(=\text{O})-\text{NH}-\text{R}$$

This reaction can also produce the new configurations in the WPNCs that effect on the morphology, crystallization, mechanical, thermal, biological, and other properties of wood (Collier et al. 1996; Quilin et al. 1993). Many studies have been carried out on physical, mechanical, and morphological properties of wood and WPNC (Matsunaga et al. 1996; Pandey 2005; Pandey et al. 2009; Rowell 2006).

However, very limited work has been devoted to Malaysian tropical light hardwood species and their chemical modification with the combination of two monomers.

3.5.3 Alkali Pretreated Wood Polymer Nanocomposites (WPNCs)

WPNCs were prepared by the combination of MMA and styrene (ST) vinyl monomer mixture (50:50 v/v). Initially, wood species were chemically pretreated with 5% sodium hydroxide (NaOH) solution for the decrease of hydrophilic hydroxyl groups and impurities from the cellulose fiber in wood and to increase adhesion and compatibility of wood fiber to polymer matrix before the specimens being impregnated with an MMA/ST monomer mixture. Monomer mixture (MMA/ST) was impregnated into raw wood and NaOH pretreated wood specimens to manufacture WPNC and pretreated wood polymer nanocomposites (PWPNC). Comparison has been through among the properties of raw wood, WPNC, and PWPNC. The result showed that PWPNC produced better mechanical and morphological properties compared to raw wood and WPNC.

Due to its high physical strength, low processing cost, and appealingly pleasing character, solid wood is the most preferred for the building and construction material. However, wood has some drawback properties as mentioned in Sect. 3.4.12 (Hill 2006; Kamden et al. 2002). Recently, wood has been treated with a variety of chemicals such as styrene, epoxy resins, urethane, phenol formaldehyde, MMA, vinyl, or acrylic monomers to improve its physical, mechanical, and biological properties (Hamdan et al. 2010; Islam et al. 2010; Yalinkilic et al. 1991). The physical and mechanical properties of wood can also be significantly improved by the impregnation of vinyl monomer mixture (Baysal et al. 2007). Thermosetting and thermoplastic monomers have been extensively used and accomplished positive improvements in wood properties, but both showed boundaries (Kumar 1994). Thermosetting-related polymer such as phenolic resins, urea formaldehyde and melamine formaldehyde, and others shows improvement in compressive strength properties and moisture-related shrinking and swelling behavior. However, WPNC with these types of polymers may be more brittle, and display only marginal development in morphological properties. Acrylate or methacrylates do not improve dimensional stability because of their thermoplastic type monomers behavior. It has been recognized that the monomer and its mixture does not form bonds with hydroxyl groups of the cellulose fibers. They merely bulk the void spaces within the wood pore and cell wall (Rowell et al. 1994; Yamaguchi 1994a, b; Yasuda and Minato 1995). Since most of the vinyl monomers are non-polar, there is slight interaction between these monomers and the hydroxyl groups of the cellulose fiber.

There are two most important chemical mechanisms work to failure the bond which is the poor chemical and physical interfacial interactions between the wood surface and polymer (Rowell 1995). Therefore, the polymer component of the WPNC purely bulks the wood cell wall construction by filling the capillaries, vessels, and other void spaces within the wood. Therefore, one can deduce that if there was the development of chemical bonds between the impregnated monomers and the hydroxyl groups on the cellulose fibers. The physical and mechanical properties of WPNC could be further improved due to the polymerization. It has been noted that better adhesion and interaction between hydrophilic wood fibers and polymer can be better quality by pretreatment using diversities of chemicals and reagents such as alkoxysilane coupling agents, diazonium salt, sodium perchlorate, dinitrophenyl hydrazine (DNPH), sodium periodate, and others (Favaro et al. 2010; Haque et al. 2009, 2010; Qutubuddin et al. 2002; Rowell 1983). However, sodium hydroxides (NaOH) are extensively used to modify the interface between dissimilar materials such as raw fibers and thermoplastic or thermosetting polymer (Mwaikambo and Ansell 2000; Mohanty et al. 2000). Alkaline sodium hydroxide removes natural fats and waxes from the cellulose fiber surface thus revealing chemically reactive functional groups like hydroxyl groups. The removal of the surface impurities from the cellulose fibers also advances the surface unevenness of the fibers or particles thus opening more hydroxyl groups and other reactive functional groups on the surface (George et al. 2001). Sodium hydroxide also reacts with reachable –OH groups permitting to the subsequent projected chemical reaction (George et al. 2001; Mohanty et al. 2000; Sreekala and Thomas 2003).

$$\text{Cellulose}{-}\text{OH} + \text{NaOH} \rightarrow \text{Cellulose}{-}\text{O}^{-}\text{Na}^{+} + \text{H}_2\text{O} + \text{impurities}$$

The effects of alkali pretreatment on mechanical and morphological properties of selected tropical WPNCs in Malaysia have been investigated (Islam et al. 2012). Major drawback of using light wood species is their high-moisture uptake, and physical and mechanical properties that change with environment issues. Hydroxyl groups are fundamental groups of cellulose that are accountable for the water uptake, a negative characteristic to this purpose, but these groups are responsible for the water uptake, a negative characteristic to this purpose, but these groups are answerable for general features of the wood. In order to overcome this problem and to improve the adhesion and compatibility of polymer to the cellulose of wood, the wood samples were chemically pretreated with a 5% alkaline NaOH solution and then impregnated with an MMA/ST monomer mixture to manufacture pretreated wood polymer nanocomposites (PWPNC) (Thakur et al. 2014).

3.5.4 Benzene Diazonium Salt Modified Wood Polymer Nanocomposites (WPNCs)

Based on the FT-IR results, benzene diazonium salt reacted with cellulose in wood. This reaction produced 2,6-diazocellulose by a coupling reaction. Through the coupling reaction, great improvement was observed on the dimensional properties especially on volumentric swelling coefficient and anti-swelling efficiency. Mechanical properties of WPNC such as dynamic Young's modulus (E_d), modulus of elasticity (MOE), and compressive Young's modulus were significantly enhanced while modulus of rupture (MOR) was greatly reduced. Fourier Transform

Infrared Spectroscopy (FT-IR) and Scanning Electron Microscopy (SEM) which used to study the physical and morphological properties showed that the properties were greatly enhanced for WPNCs. In addition, the thermal stability of WPNCs were investigated and improved through Thermogravimetric Analysis (TGA) and Differential Scanning Calorimetry (DSC). Most of the WPNCs had improved thermal stability compared to raw wood due to the formation of 2,6-diazo cellulose compound. Furthermore, this WPNCs greatly improved their hardness compared to raw wood. All the decay resistance property of WPNCs were significantly higher compared to raw wood and the results were assessed through percentage (%) of weight loss.

From the nature, wood is classified as the renewable resource and one of the most attractive materials due to the complex structure. Besides, wood can be well applied in many fields. Most of the solid wood will be processed and applied in building and construction material as it has better physical and mechanical properties. In addition, this type of wood also performed well in decay resistance compared to raw wood. Although solid wood seems to bring abundant benefits, the properties of wood can be readily changed through environmental factors, namely light, water, temperature, and biological organisms. Due to this, solid wood could not be applied widely outdoors and indoors application (Brelid et al. 2000; Chao and Lee 2003; Deka et al. 2002; Yalinkilic et al. 1999a, b). The limitation of solid wood was significantly due to the presence of hydroxyl groups (–OH) in the cellulose, hemicelluloses, and lignin within the wood. These hydroxyl groups changed the physical, chemical, and thermal properties of wood by attracting water molecules from the surrounding environment. The water molecules attracted would form hydrogen bonding, which led to swelling. The swelling process could be minimized through suitable chemical treatment, which would greatly improve the properties of wood (Halabe et al. 1995; Hartley and Schneider 1993; Rowell 2005; Schneider 1994).

Due to the disadvantages observed from the raw wood, wood undergoes treatment with the introduction of various chemicals. Based on the research of, the cell wall does not directly react with modified chemicals in a so-called substitution reaction. The research proved that most of the polymers including the in situ polymers would only fill the empty lumens in the wood by mixing two materials rather than a real interaction. One of the chemicals, namely benzene diazonium salt was widely applied in fabrication for WPNCs to reduce hydrophilicity of raw fibers in the synthesis of WPNCs (Haque et al. 2009; Rahman et al. 2009). Besides, benzene diazonium salt undergoes coupling reaction with –OH groups of diazo cellulose fiber. According to the previous research and literature, studies on raw fiber such as jute, coir, and abaca were performed through coupling reaction. However, there was no work carried out on wood modification using diazonium salt. In addition, there was very little work has been developed on Malaysian tropical wood species as well as the chemical modification (Yap et al. 1990).

In the present work, five types of Malaysian tropical light hardwood species were used to undergo modification. This study was carried out as the shortage of hardwoods had been growing rapidly as well as the rising costs. These limitations created interest within the researchers to utilize the lower grade woods especially the abundantly available tropical light hardwoods. These lightwoods are expected to be modified into good quality hardwoods. However, the major problems of light hardwoods such as high-moisture uptake, thermal instability, and physical and mechanical property change with environmental variations should be solved through chemical treatment of benzene diazonium salt.

3.5.5 Nanotechnology for Wood Polymer Nanocomposites

In the recent years, wood modification on nanotechnology with silicate and nanoclays through in situ reinforcement has been thoroughly investigated. The application of nanotechnology modification on wood managed to improve the properties of wood effectively (Cai et al. 2007a, b, 2008). Few common types of silicates nanoclay such as montmorillonite, hectorite, and saponite have a layered structure. This layered structure created high stiffness and strength on the wood composites formed. According to Ray and Okamoto (2003), the addition of low silicate loading on thermoplastic and thermoset nanocomposites significantly improved the mechanical properties (tensile modulus and strength, flexural modulus, and strength), thermal stability, flame retardant, and barrier resistance. Therefore, nanotechnology technique on wood managed to improve its advantages properties such as physical, mechanical, thermal, and barrier resistance.

Based on Caseri (2007), nanotechnology was developed since the ninth century. An iridescent glaze in pottery can be generated by the preparation of nanoparticles prepared from silver, copper, and iron salts. In the middle of sixteenth century, metal nanoparticles have also been used to dye glass windows which named as gold ruby glasses. Gold nanoparticles were successfully embedded in gum Arabic 1833. Besides, the nanocomposites prepared from silver and gum used the same principle in year 1899. At the end of the nineteenth century, natural polymers were treated with solutions of gold or a silver salt by penetrating the metal ions into the fibrils. These metal ions were subsequently reduced to metal nanoparticles within the polymer matrix due to the exposure of light. The resulting composite materials exhibited pronounced dichroism. The colors observed in the dichroic materials were found to be dependent on two matters, namely the incorporated chemical element and particle size in nanoscale.

In year 1950, carbon black nanoparticles extended their weathering life of polyethylene until 20 years through the introduction of nanoparticles in composites. The application of nanoparticles in carbon black nanoparticles took place as they were able to absorb UV and visible light efficiently. Hence, the retardation of light-induced polymer degradation processes was happened. Besides, carbon black composites showed distinct electrical conductivity especially when the polymers are insulators. However, the particle size influenced the properties of nanocomposites which differed from those of analogous composites with larger particles. This was greatly due to the interactions of numerous atoms that directly influenced the physical characteristics of the nanocomposites. The two main factors that influenced the properties of nanoparticles were the size of the primary particles as well as the distance to neighboring particle changes. The effects of using the very small particle sizes included the exhibitions of some unique and desirable properties. The properties were conductivity, color, ductility, fluorescence, transparency, reactivity, surface energy, and stiffness. All these special properties created many novel applications that could be very useful in our daily life.

Polymer composites were well known to be applied as catalysts, gas sensors, materials with improved flame retardance, optical filters, electronics, biomedical research, and others while for nanotechnology, it has been found to be well applied in coating materials especially on wood coating materials due to the color, transparency, protection from light, and scratch resistance that created a successful wood coating. With this, the light scattering by particles with dimension below the wavelength of visible light is reduced. Wood coating materials usually appear translucent. On the other hand, larger particles provided opacity in composites of considerable thickness and particle content. The refractive index differences between matrix and particles could vary due to the particle diameter, and this particle diameter was important in determining the transparency of the polymer composites. For example, very small particle composites managed to provide high transparent even at high-refractive index differences. Due to the advantage of the properties of nanocomposite materials, many new opportunities for wood coating and wood products improvements could be analyzed. One of the well-known products based on alumina, silica, and zinc oxide nanoparticles was wear-scratch resistance UV coatings for wood. This wood-based UV coating provided significant improvements through alumina and silica nanoparticles that provided scratch resistance while zinc oxide nanoparticles provided UV coatings when compared to normal wood (Du et al. 2008). This was due to the strongly connected polymer-particle bond that had been created when appropriate amount of modified surfaces especially the reactive organic molecules that allowed more binding of the particles within the polymer matrices. According to Wang (2007), the bonding strength as well as strong adhesive properties could be greatly enhanced through the dispersion of a small amount of inorganic nanoclay into phenolic resin in oriented strand board manufacturing.

In depth, polymer nanocomposites can be classified as polymer–clay nanocomposites. This type of nanocomposites was first reported by Blumstein (1961) in the literature on the demonstration on the polymerization of vinyl monomers intercalated into montmorillonite clay. Besides, the group of researchers from Toyota also worked on the nanocomposites field with specialization on the potential of nanocomposites (Kojima et al. 1993a, b; Usuki et al. 1993a, b; Border et al. 2009; Gao and Gu 2007) with the combinations of academic and industrial researchers. Polymer nanocomposites are defined as a new class of composites with at least one nanoscale dimension dispersed phase compared to conventional polymer nanocomposites, namely glass fiber and carbon fiber reinforced polymer nanocomposites. For example, the incorporation of nanoscale clay usually functioned as the reinforcing phase in the polymer matrix to form organic–inorganic nanocomposites. This type of nanocomposites is also known as polymer–clay nanocomposites. The addition of little amount of nanofiller such as nanoclay into the polymer matrix, physical properties, mechanical properties such as tensile modulus and strength as well as flexural modulus and strength, swelling resistance, ionic conductivity and also flame resistance could be significantly enhanced (Byun et al. 2001; Gilman 1999; Kornmann et al. 2001). Besides, thermal expansion coefficient as well as gas permeability of the polymer–clay nanocomposites were decreasing (Lan et al. 1995; Su and Wilkie 2003). All the properties were enhanced due to the large interfacial area per unit of weight of the dispersed phase (750 m^2/g) and the high-aspect ratio. The clay particles introduced are bundles of nanoclay sheets with approximately 1 nm in thickness and 100–1000 nm in breadth. For example, 8-μ m clay particle has more than 3000 nanoclay sheets of 1 nm in thickness and 20–200 nm in diameter.

The shortage of the introduction of hydrophilic clay sheets into hydrophobic polymer matrices was usually incompatible. This led to the narrowing of the spacing between the clay sheets that reduced the chance for the direct diffusion of polymer chains in the clay galleries. Due to this, aggregation of clay particles frequently happened as the aggregated clay sheets became the stress-concentrated sites in the polymer matrix. Therefore, adsorption of organic molecules through chemical treatment of clay to weaken the interlayer cohesive energy and to create better interaction between the nanoclay and the polymer matrix. WPNCs were prepared with the introduction of hydrophilic and hydrophobic nanofillers that were grounded with a ball mill through impregnation of solid aspen wood with water-soluble melamine–urea-formaldehyde (MUF) resin (Cai et al. 2007a). Impregnation of MUF resin and nanoclay showed significant improvements in physical and mechanical properties especially on density, surface hardness, and MOE. Ball mill treatment was found to favor dispersion of the nanofillers into wood to improve hardness of the wood from 1.09 to 3.25 MPa. The average MOE of treated wood was almost twice as much as that of raw wood while the MOR did not differ significantly between treated and raw woods.

Based on the research of Cai et al. (2007b), wood impregnated with melamine–urea-formaldehyde resin and wood impregnated with different nanofiller/MUF resin formulations were investigated especially on the water absorption and dimensional stability. It was clearly showed that the water repellence and dimensional stability showed significant improvements for the nanofiller/MUF resin-treated wood. After 24 h, the raw wood absorbed approximately 63% of moisture when soaking in water while water uptake after 1 week immersion in water was about 125%. The moisture percentage of the MUF resin-impregnated wood absorbed was about 8.3% and 38.5% after 24 h and 1 week immersion in water, respectively. The water absorption of organophilic nanoclay/MUF resin-impregnated wood showed about 5% water uptake in 24 h and 22% after 1 week. The nanofiller/MUF-treated wood showed anti-swelling efficiency (ASE) improved from 63.3 to 125.6%. Both water resistance and dimensional stability of the resulting WPNCs were showed better properties with the introduction of MUF and nanofillers into the wood. Based on the research, X-ray fluorescence photo showed that some nanoparticles had migrated into the wood cell wall (Cai et al. 2007a, b, 2008). The significantly improved MUF resin and nanofiller/MUF resin showed no significant influence on the color of the wood to be used in flooring. Based on the literature from Forest Product Laboratory (2002), flame retardants for wood are commonly used with phosphates, ammonium compounds, zinc chloride, sodium tetraborate, boric acid, and polymeric materials are commonly used in Europe and the USA. Aluminum oxide is used to improved the fire performance in the research of Ximenes and Evans (2006). Lately, the incorporation of silicon dioxide or titanium oxide improved fire performance (Miyafuji and Saka 1997, 2001; Simkovic et al. 2005) as well as other properties, such as UV and microbial protection, water repellence, color stability, and mechanical resistance (Mahltig et al. 2008). The combination of various advanced chemicals led to better fire resistance compared to the conventional treatments (Saka and Ueno 1997; Simkovic et al. 2005).

4 Summary

In conclusion, the hydroscopic wood has some adverse drawbacks especially high-moisture uptake, low biodegradation and dimensional variations. These drawbacks occurred due to the presence of abundant numbers of hydroxyl groups within the wood. To reduce and minimize the drawbacks, wood undergoes different methods of medication such as by impregnation with reactive chemicals, combination of two or three monomers, or by a nanotechnologies technique. Through the modification, physical and mechanical properties were expected to be greatly enhanced. From the previous researches, few studies had proven that chemical modification on raw wood could improve its hardness, mechanical properties, and dimensional stability. Both thermoplastic and thermosetting systems have been applied in this study. The most common technique developed is in situ polymerization of monomers. The monomers used included styrene and methyl methacrylate, or prepolymers such as phenol formaldehyde. Besides, vinyl monomer is another monomer that can be applied in the modification. However, commercial vinyl monomers do not bond strongly to the wood component as they simply bulk the void spaces within the wood structure. As most of the vinyl monomers are non-polar, bond failure could only happen when there are little chemical and physical interfacial interactions between the wood surface and chemical. These polymers only fill the empty lumens of the wood. This situation can be, namely as mixture of two polymers instead of a composite. Compared to the polymer existed in the voids like the cell lumen, polymer existing in the cell wall of a WPNC performed better in dimensional stability of WPC. Wood impregnated with thermosetting resins such as water-soluble phenolic resins, urea formaldehyde and melamine-formaldehyde prepolymers improves its compressive resins. The improved resins can penetrate and bulk the cell wall to react with the hydroxyl groups of the wood components. This can prevent the wood from shrinking and drying. However, the resulting WPNCs are quite brittle. Not only that, the toughness properties of raw wood are deteriorated.

Conventional chemical treatments have proven to be effective in improving wood hardness, dimensional stability, and stiffness. Besides, fire resistance, UV resistance, biological resistance, and aesthetic appeal can be greatly improved. However, nanotechnology offers new opportunities for further improving wood product. This was due to unique and desirable properties of chemical materials in the form of particles in nanoscale. The advantages of nanotechnology are to create polymer nanocomposites for example, wood coatings. Polymer nanocomposites consist of a continuous polymer matrix. The polymer matrix contains inorganic particles of a size below approximately 100 nm at least in one dimension. A continuous polymer matrix combined with inorganic nanoparticles inside wood cells and lumens effectively would be very difficult. This was because the nature of the solid wood contained numerous of hydroxyl groups. Through the application of nanotechnology, the areas that can be involved included wood coatings and wood adhesives. Therefore, it is summarized that research should be carried out to explore the potential of nanotechnology in wood coatings and adhesives. Besides, the potential nanotechnology was apply to develope the tropical wood species for value-added applications.

Among all the methods described above, nanotechnological technique was chosen as it is a multipurpose technique that aided to improve the wood durability simultaneously. The present work is to investigate the feasibility to prepare WPNCs using tropical light hard wood. Besides, some conventional technique, namely monomer impregnation and chemical modification were investigated using new chemical formulations in this study.

References

Ayer SW, Fell D, Wan H (2003) Hardening of solid wood: Market opportunities and review of existing technologies. Forintek Canada Corp, Canada

Baki H, Yalcin Ors M, Hakki A (1993) Improvement of wood properties by impregnation with macromonomeric initiators. J Appl Polym Sci 47:1097–1103

Baysal E, Yalinkilic MK, Altinok M, Huseyin Peker AS, Colak M (2007) Some physical, biological, mechanical and fire properties of wood polymer composite (WPC) pretreated with boric and borax mixture. Const Build Mater 21(9):1879–1885

Beall FC, Meyer JA, Skaar C (1966) Direct and RF heat curing of wood-plastic composites. Forest Prod J 16(9):99–106

Beall FC, Witt AE, Bosco LR (1973) Hardness and hardness modulus of wood-polymer composites. Forest Prod J 23(1):55–60

Blumstein A (1961) Etude de polymerization on couche adsorbee I. Bull Soc Chim Fr 899–905

Boding J, Goodman IR (1973) Prediction of elastic parameters for wood. Wood Sci 4:378–385

Bodirlau R, Teaca CA, Spiridon I (2009) Preparation and characterization of composites comprising modified hardwood and wood polymers/poly(vinyl chloride). BioRes 4(4): 1285–1304

Bongers HPM, Beckers EPJ (2003) Mechanical properties of acetylated solid wood treated on pilot plant scale. In: Van Acker J, Hill CAS (ed). Proceedings of the first European conference on wood modification, Chent, Belgium, 3–4 April 2003
Border P, Pollet E, Averous L (2009) Nano-biocomposites: biodegradable polyester/nanoclay system. Prog Polym Sci 34:125–155
Bowyer JL, Shmulsky R, Haygreen JG (2007) Hardwood structure: in forest product and wood science an introduction. Blackwell publishing Ltd, United States
Brebner KI, Schneider MH, Jones RT (1988) The influence of moisture content on the flexural strength of styrene-polymerized wood. Forest Prod J 38(4):55–58
Brebner KI, Schneider MH, St-Pierre LE (1985) Flexural strength of polymer-impregnated eastern white pine. Forest Prod J 35(2):22–27
Brelid PL, Simonson R, Bregman O, Nilsson T (2000) Resistance of acetylated wood to biological degradation. Holz Roh Werkst 58:331–337
Bryant BS (1966) The chemical modification of wood-from the point of view of wood science and economics. Forest Prod J 16(2):20–27
Byun HY, Choi MH, Chung IJ (2001) Synthesis and characterization of resol type phenolic resin/layered silicate nanocomposite. Chem Mater 13:4221–4226
Cai X, Riedl B, Zhang SY, Wan H (2007a) Formation and properties of nanocomposites made up from solid aspen wood, melamine-urea—formaldehyde and clay. Holz 61:148–154
Cai X, Riedl B, Zhang SY, Wan H (2007b) Effects of nanofillers on water resistance and dimensional stability of solid wood modified by melamine-urea-formaldehyde resin. Wood Fiber Sci 39(2):307–318
Cai X, Riedl B, Zhang SY, Wan H (2008) The impact of interphase between wood, melamine-urea-formaldehyde and layered silicate on the performance of wood polymer nanocomposites. Compos Part A 39:727–737
Cardias WF, Hale MD (1999) The resistance of wood chemically modified with isocyanates. Part I. Brown rot, white rot and acid chlorite delignification. Holz 53:230–236
Caser W (ed) (2007) Hybrid materials: synthesis, characterization, and applications. Wiley-VCH Verlag, Gernany
Cetin NS, Hill CAS (1999) An investigation of the reaction od epoxides with wood. J Wood Chem Technol 19:247–264
Chao WY, Lee SWC (2003) Properties of southern pine wood impregnated with styrene. Holz 57(3):333–336
Chauhan SS, Aggarwal P, Karmarkar A, Pandey KK (2001) Moisture adsorption behavior of esterified rubber wood. Holz als Roh- und Werkstoff 59:250–253
Chen GC, Rowell R (1986) Approaches to the improvement of biological resistance of wood through controlled-release technology. In: Chaudry I (ed) Proceedings of the 13th international symposium on controlled-release of bioactive materials, Norfolk, 3–6 Aug 1986
Collier JR, Fahrurrozi M, Collier BJ (1996) Cellulosic reinforcement in reactive composite system. J Appl Polym Sci 61:1423–1430
Deka M, Saikia CN (2000) Chemical modification of wood with thermosetting resin: Effect on dimensional stability and strength property. Biores Technol 73:179–181
Deka M, Saikia CN, Baruah KK (2002) Studies on thermal degradation and termit resistant properties of chemically modified wood. Biores Technol 84:151–157
Devi RR, Maji TK, Banerjee AN (2003) Studies on dimensional stability and thermal properties of rubber wood chemically modified with styrene and glycidyl methacrylate. J Appl Polym Sci 93:1938–1945
Du WL, Xu YL, Xu ZR, Fan CL (2008) Preparation, characterization and antibacterial properties *E. coli* K88 of chitosan nanoparticle loaded copper ions. Nanotechnol 19:1–5
Duran JA, Meyer JA (1972) Exothermic heat released during catalytic polymerization of basswood-methyl methacrylate composites. Wood Sci Technol 6:59–66
Ellis WD, O'Dell JL (1999) Wood-polymer composites made with acrylic monomers, isocyanate and maleic anhydride. J Appl Polym Sci 73:2493–2505

Ellwood EL, Gilmore RC, Stamm AJ (1972) Dimensional stabilization of wood with vinyl monomers. Wood Sci 4(3):137–141
Elvy SB, Dennis GR, Loo-teck NG (1995) Effect of coupling agent on the physical properties of wood-polymer composites. J Mater Process Technol 48:365–372
Favaro SL, Lopes MS, Neto AGVC, Santana RRR, Dovanovic E (2010) Chemical, morphological and mechanical analysis of rice husk/post—customer polyethylene composites. Comp Part A 41(1):154–160
Feist WC, Rowell RM, Ellis DW (1991) Moisture sorption and accelerated weathering of acetylated and methacrylated aspen. Wood Fiber Sci 23(1):128–136
Forest Products Laboratory (2002) Appearance application for BC coastal hemlock and Douglas-fir: a market fit analysis. PFInnovations, Canada
Frazier CE (2003) Isocyanate wood binders. In: Pizzi A, Mittal KL (eds) Handbooks of adhesive technology, 2nd edn. Marcel Dekker, New York, pp 63–135
Frihart CR (2005) Wood adhesion and adhesives. In: Rowell RM (ed) Handbook of Wood chemistry and wood composite. CRC Press, United Kingdom, pp 215–278
Fry HJ (1976). US 3950577 A, 13 Apr 1976
Furuno T, Goto T (1973) The penetration of MMA monomer into the woody cell wall. Mokuzai Gakkashi 19(6):271–274
Furuno T, Shimada K, Uehara T, Jodai S (1992a) Combinations of wood and silicate II. Wood-mineral composites using water glass and reactants of barium chloride, boric acid and borax, and their properties. Mokuzai Gakkashi 38(5):448–457
Furuno T, Uehara T, Jodai S (1991) Combinations of wood and silicates I. Impregnation by water glass and applications of aluminum sulfate and calcium chloride as reactants. Mokuzai Gakkashi 37(5):462–472
Furuno T, Uehara T, Jodai S (1992b) The role of wall polymer in the decay durabilities of wood-polymer composites. Mokuzai Gakkashi 38(3):285–293
Furuno T, Uehara T, Jodai S (1993) Combinations of wood and silicate III. Some properties of wood-mineral composites using the water glass-boron compound system. Mokuzai Gakkashi 39(5):561–570
Galperin AS, Kuleshov GG, Tarashkevich VI, Smtov GM (1995) Manufacturing and properties of modified wood dimensional stability. Wood Sci Technol 18:225–240
Gao Z, Gu J (2007) CN 1944009 A, 11 Apr 2007
Gaylord NG (1973) US 3765934 A, 16 Oct 1973
George J, Sreekala MS, Thomas S (2001) A review on interface modification and characterization of natural fiber reinforced plastic composites. Polym Eng Sci 41(9):1471–1556
Georgieva M, Harata M, Miloshev G (2008) The nuclear actin-related protein Act3p/Arp4 influences yeast cell shape and bulk chromatin organization. J Cell Biochem 104(1):59–67
Ghosh A, Shuman S, Lima CD (2008) The structure of FCP1, an essential RNA polymerase II CTD phosphatase. Mol Cell 32(4):478–490
Gilman JW (1999) Flammability and thermal stability studies of polymer layered silicate (clay) nanocomposites. Appl Clay Sci 15:31–49
Gindl W, Dessipri E, Wimmer R (2002) Using UV-microscopy to study diffusion of melamine-urea-formaldehyde resin in the cell walls of spruce wood. Holz 56:103–107
Gindl W, Gupta HS (2002) Cell-wall hardness and Young's modulus of melamine-modified spruce wood by nano-indentation. Compos Part A 33:1141–1145
Gindl W, Zargar-Yaghubi F, Wimmer R (2003) Impregnation of softwood cell walls with melamine-formaldehyde resin. Biores Technol 87:325–330
Goldstein IS, Dreher WA, JEroski EB, Nielson JF, Oberley WJ, Weaver JW (1959) Wood processing; inhibiting against swelling and decay. Ind Eng Chem 51(10):1313–1317
Halabe UB, Bidigalu GM, Ganga Rao HVS, Ross RJ (1995) Nondestructive evaluation of green wood using stress wave and transverse vibration techniques. Mater Eval 55(9):1013–1018
Hamdan S, Sedik Y, Jusoh I, Hasan M, Talib ZA (2009) Dynamic Young's modulus and glass transition temperature of selected tropical wood species. Mater Sci Technol 25(6):805–808

Hamdan S, Talib ZA, Rahman MR, Ahmed AS, Islam MS (2010) Dynamic Young's modulus measurement of treated and post-treated tropical wood polymer composites (WPC). BioRes 5(1):324–342

Haque MM, Hasan M, Islam MS, Ali ME (2009) Physico-mechanical properties of chemically treated palm and coir fiber reinforced polypropylene composites. Biores Technol 100: 4903–4906

Haque MM, Islam MS, Islam MDS, Islam MN, Huque MDM, Hasan M (2010) Physico-mechanical properties of chemically treated palm fiber reinforced polypropylene composites. J Reinf Plast Comp 29(11):1734–1742

Hartley ID, Schneider MH (1993) Water vapour diffusion and absorption characteristics of sugar maple (*Acer saccharum*, Marsh.), wood polymer composites. Wood Sci Technol 27:421–427

Hartman S (1969) Modified wood with aqueous polyurethane systems. Forest Prod J 19(5):39–42

Hill CAS (2006) Modifying the properties of wood. In Hill CAS (ed) Wood modification: chemical, thermal and other processes. Wiley, United Kingdom

Hill CAS (2011) Wood modification: An update. BioRes 6(2):918–919

Holyle RJ, Woeste FE (1989) Wood technology in the design of structure. IOWA State University Press, United States of America

Hon DNS, Shiraishi N (2001) Cellulosic chemistry, 2nd edn. Marcel Dekker Inc., New York, pp 443–501

Ibach RE, Rowell RM, Lange SE, Schumann RL (2001) Effect of wet-dry cycling on the decay properties of aspen fiber high density polypropylene composites. Paper presented at International conference on wood fiber- plastic composites, Madison, Wisconsin, 15–16 May 2001

Inoue M, Ogata S, Kawai S, Rowell RM, Norimoto M (1993) Fixation of compressed wood using melamine-formaldehyde resin. Wood Fiber Sci 25(4):404–410

Int Scholar Res Notices 2014, pp 1–13

Islam MS, Hamdan S, Jusoh I, Rahman MR, Ahmed AS (2011) The effect of crosslinker on mechanical and morphological properties of tropical wood material composites. Mater Des 32(4):2221–2227

Islam MS, Hamdan S, Jusoh I, Rahman MR, Ahmed AS (2012) The effect of alkali pretreatment on mechanical and morphological properties of tropical wood polymer composites. Mater Des 33:419–424

Islam MS, Hamdan S, Rahman MR, Jusoh I, Ibrahim NF (2010) Dynamic Young's modulus and dimensional stability of Batai tropical wood impregnated with polyvinyl alcohol. J Sci Res 2(2):227–236

Juneja SC, Hodgins JW (1970) The properties of thermo-catalytically prepared wood polymer composites. Forest Prod J 20(12):24–28

Kamden DP, Pizzi A, Jermannaud A (2002) Durability of heat-treated wood. Holz als Roh-Werkstoff 60:1–6

Khan MA, Idriss Ali KM (1992) Effect of urea on the mechanical strength of wood-plastic composites. Rad J Phys Chem 40(1):69–70

Khan MA, Idriss Ali KM, Ahmad MU (1992) Radiation-induced wood plastic composites under combinations of monomers. J Appl Polym Sci 45:2113–2119

Kojima Y, Usuki A, Kawasumi M, Okada A, Kurauchi T, Kamigaito O (1993a) Sorption of water in nylon 6-clay hybrid. J Appl Polym Sci 49:1259–1264

Kojima Y, Usuki A, Kawasumi M, Okada A, Kurauchi T, Kamigaito O (1993b) Mechanical properties of nylon 6-clay hybrid. J Mater Res 8:1185–1189

Kornmann X, Lindberg H, Berglund LA (2001) Synthesis of epoxy-clay nanocomposites: Influence of the nature of the curing agent on structure. Polym 42:4493–4499

Koutsos V (2009) Polymeric materials: An introduction. In: Forde M (ed) ICE manual of construction materials. Thomas Telford, London, pp 571–577

Kumar S (1994) Chemical modification of wood. Wood Fiber Sci 26(2):270–280

Kyziol L (1999) Modified wood (ligonomer)—a promising material for ship building. Pol Marit Res 2:6–10

Lan T, Kaviratna PD, Pinnavaia TJ (1995) Mechanism of clay tactoid exfoliation in epoxy-clay nanocomposites. Chem Mater 7:2144–2150

Langwig JE, Meyer JA, Davidson RW (1968) Influence of polymer impregnation on mechanical properties of basswood. Forest Prod J 18(7):33–36

Langwig JE, Meyer JA, Davidson RW (1969) New monomers used in making wood-plastic. Forest Prod J 19(11):57–61

Li JZ, Furuno T, Katoh S (2001) Preparation and properties of acetylated and propionylated wood-silicate composites. Holz 55(1):93–96

Loos WE, Robinson GL (1968) Rates of swelling of wood in vinyl monomers. Forest Prod J 18(9):109–112

Mahltig B, Swaboda Roessler A, Bottcher H (2008) Functionalizing wood by nanosolapplication. J Mater Chem 18:3180–3192

Mathias LJ, Wright JR (1989) New wood-polymer composites: Impregnation and in situ polymerization of hydroxymethylacrylates. Polym Reprints 30(1):233–234

Mathias LL, Lee S, Wright JR, Warren SC (1991) Improvement of wood properties by impregnation with multifunctional monomers. J Appl Polym Sci 42:55–67

Matsuda H (1987) Preparation and utilization of esterified woods bearing carboxyl groups. Wood Sci Technol 21(1):75–88

Matsunaga M, Sugiyama M, Minato K, Norimoto M (1996) Physical and mechanical properties required for violin bow materials. Holz 50(6):511–517

Meyer JA (1981) Wood-polymer materials: State of the art. Wood Sci 14(2):49–54

Militz H, Beckers EPJ, Homan WJ (1997) Modification of solid wood: research and practical potential. Paper presented at 28th annual meeting, Vancouver, Canada, 25–30 May

Minato K, Yusuf S, Imamura Y, Takahashi M (1992) Resistance of formaldehyde-treated wood to biological attacks. Mokuzai Gakkaishi 38:1050–1056

Miroy F, Eymard P, Pizzi A (1995) Wood hardening by methoxymethyl melamine. HolzRoh Werkst 53:276–278

Miyafuji H, Saka S (1997) Fire-resisting properties in several TiO_2 wood—inorganic composites and their topochemistry. Wood Sci Technol 31:449–455

Miyafuji H, Saka S (2001) Na_2O–SiO_2 wood-inorganic composites prepared by the sol-gel process and their fire-resistant properties. J Wood Sci 47:389–483

Mohanty AK, Khan MA, Hinrichsen G (2000) Surface modification of jute and its influence on performance of biodegradable jute-fabric/biopol composites. Comp Sci Technol 6:1115–1124

Morard M, Vaca-Garcia C, Stevens M, Van Acker J, Pignolet O, Borredon E (2007) Durability improvement of wood by treatment with methyl alkenoate succinic anhydrides (M-ASA) of vegetable origin. Int Biodeter Biodegrad 59:103–110

Mwaikambo LY, Ansell MP (2000) The effect of chemical treatment on the properties of hemp, sisal, jute and kapok for composite reinforcement. Angew Macromol Chem 272:108–116

Nakagami T, Ohta M, Yokota T (1976) Esterification of wood with unsaturated carboxylic acids. III. Dissolution of wood by TFAA esterification method. Bull Kyoto University Forests 48:198–205

Nakagami T, Yokota T (1975) Esterification of wood with unsaturated carboxylic acids. II. Reaction conditions of esterification and properties of the prepared esters of wood. Bull Kyoto University Forests 47:178–183

Noah JN, Foudjet A (1988) Wood-polymer composites from some tropical hardwoods. Wood Sci Technol 22:115–119

Norimoto M, Gril J, Rowell RM (1992) Rheological properties of chemically modified wood: Relationship between dimensional and creep stability. Wood Fib Sci 24:25–35

Ogiso K, Saka S (1993) Wood-inorganic composites prepared by sol-gel process II. Effects of ultrasonic treatments on preparation of wood—inorganic composites. Mokuzai Gakkashi 39(3):301–307

Pandey KK (2005) Study of effect of photo-irridation on the surface chemistry of wood. Polym Degrad Stabil 90:9–20

Pandey KK, Jayashree Nagaveni HC (2009) Study of dimensional stability, decay resistance and light stability of phenylisothiocyanate modified rubberwood. BioRes 4(1):257–267
Papadopoulos AN (2006) Decay resistance of cement-bonded oriented strand board. BioRes 1(1):62–66
Prakash GK, Pandey KK, Ram RKD, Mahadevan KM (2006) Dimensional stability and photostability of octanoylated wood. Holz 60:539–542
Quilin DT, Caulfield DF, Koutsky JA (1993) Crystallinity in the polypropylene/cellulose system. I. Nucleation and crystallinity morphology. J Appl Polym Sci 50:1187–1194
Qutubuddin S, Fu X, Tajuddin Y (2002) Synthesis of polystyrene-clay nanocomposites via emulsion polymerization using a reactive surfactant. Polym Bull 48(2):143–149
Raff RAV, Herrick IW, Adams MF (1965) Polymerization of styrene and styrene-divinylbenzene in wood. Forest Prod J 15(7):260–262
Rahman MR, Hamdan S, Ahmed AS, Islam MS (2010) Mechanical and biological performance of sodium metaperiodate-impregnated plasticized wood (PW). BioRes 5(2):1022–1035
Rahman MR, Haque MM, Islam MN, Hasan M (2009) Mechanical properties of polypropylene composites reinforced with chemically treated abaca. Compos Part A 40:511–517
Rapp AO, Bestgen H, Adam W, Peek RD (1999) Electron energy loss spectroscopy (EELS) for qualification of cell-wall penetration of a melamine resin. Holz 53(2):111–117
Ray SS, Okamoto M (2003) Polymer/layered silicate nano-composites: A review from preparation to processing. Prog Polym Sci 28:1539–1641
Rowell RM (1983) Chemical modification of wood. Forest Prod Abst 6(12):363–382
Rowell RM (1995) Chemical modification of wood for improved adhesion in composites. USDA Forest Products Laboratory Madison, Wisconsin, pp 56–60
Rowell RM (2005) Chemical modification of wood for improved adhesion in composites. USDA forest Products Laboratory Madison, Wisconsin, pp 56–60
Rowell RM (2006) Chemical modification of wood: a short review. Wood Mater Sci Eng 1:29–33
Rowell RM, Moisuk R, Meyer JA (1982) Wood-polymer composites: cell wall grafting with alkylene oxides and lumen treatments with methyl methacrylate. Wood Sci 15:90–96
Rowell RM, Simonson R, Hess S, Plackett DV, Cronshaw D, Dunningham E (1994) Acetyl distribution in acetylated whole wood and reactivity of isolated wood cell wall components to acetic anhydride. Wood Fiber Sci 26(1):11–18
Rowell RM, Tillman AM, Simonson R (1986) Vapour phase acetylation of southern pine, douglas-fir and aspen wood flakes. J Wood Chem Technol 6:293–309
Saka S, Sasaki M, Tanahashi M (1992) Wood-inorganic composites prepared by sol-gel processing I. Wood-inorganic composites with porous structure. Mokuzai Gakkashi 38 (11):1043–1049
Saka S, Ueno T (1997) Several SiO_2 wood-inorganic composites and their fire- resisting properties. Wood Sci Technol 31:457–466
Saka S, Yakake Y (1993) Wood-inorganic composites prepared by sol-gel process III. Chemically-modified wood-inorganic composites. Mokuzai Gakkashi 39(3):308–314
Salah UH, Raman PS (2014) Effect of hygrothermal aging on the mechanical properties of fluorinated ad nonfluorinated clay-epoxy nanocomposites. Int Scholarly Res Notices 2014: 1–13
Schneider MH (1994) Wood polymer composites. Wood Fiber Sci 26(1):142–151
Schneider MH (1995) New cell wall and cell lumen wood polymer composites. Wood Sci Technol 29:121–127
Schneider MH, Brebner KI (1985) Wood-polymer combinations: The chemical modification of wood by alkoxysilane coupling agents. Wood Sci Technol 19:67–73
Siau JF (1969) The swelling of basswood by vinyl monomers. Wood Sci 1(4):250–253
Siau JF, Meyer JA (1966) Comparison of the properties of heat and radiation cured wood-polymer combinations. Forest Prod J 15(10):426–434
Siau JF, Meyer JA, Skaar C (1965) Wood-polymer combinations using radiation techniques. Forest Prod J 15(10):426–434

Siau JF, Smith WB, Meyer JA (1978) Wood-polymer composites from southern hardwoods. Wood Sci 10(3):158–164
Simkovic I, Martvonova H, Manikova D, Grexa O (2005) Flame retardance of insolubilized silica inside of wood material. J Appl Polym Sci 97:1948–1952
Singha AS, Thakur VK (2009) Study of mechanical properties of urea formaldehyde thermosets reinforced by pine needle powder. BioRes 4(1):292–308
Skreekala MS, Thomas S (2003) Effect of fiber surface modification on water—sorption characteristics of oil palm fibres. Compos Sci Technol 63:861–869
Stamm AJ, Baechler RH (1960) Decay resistance and dimensional stability of five modified woods. Forest Prod J 10:22–26
Stevens M, Schalck J, Van Raemdonck J (1979) Chemical modification of wood by vapor phase treatment with formaldehyde and sulphur dioxide. Int J Wood Preservation 1(2):57–68
Su S, Wilkie CA (2003) Exfoliated poly(methyl methacrylate) and polystyrene nanocomposites occur when the clay contains a vinyl monomer. J Polym Sci Part A 41:1124–1135
Thakur VK, Islam MS, Hasan M, Ahmad MB (2014) Chemical Modification and properties of cellulose-based polymer composites. Scrivener Publishing LLC, Beverly
Tunc MS, Lawoko M, Heiningen A (2010) Understanding the limitations of removal of hemicelluloses during autohydrolysis of a mixture of southern hardwoods. BioRes 5(1): 356–371
Usuki A, Kawasumi M, Kojima Y, Fujushima A, Okada A, Kamigaito O (1993a) Swelling behaviour of montmorillonite cation exchanged for v-amino acids by ε-caprolactam. J Mater Res 8:1174–1178
Usuki A, Kojima Y, Kawasumi M, Okada A, Fujushima A, Kurauchi T, Kamigaito O (1993b) Synthesis of nylon 6-clay hybrid. J Mater Res 8:1179–1184
Valente M, Sarasini F, Marra F, Tirillo J, Pulci G (2011) Hybrid recycled glass fiber/wood flour thermoplastic composites: manufacturing and mechanical characterization. Compos Part A 42:649–657
Wan H (2004) Report: wood hardening technologies. FPInnovations, Canada
Wang SG (ed) (2007) Personal communication with Martin Feng. FPInnovations, Canada
Wright JR, Mathias LJ (1993) New lightweight materials: Balsa wood-polymer composites based on ethyl a-(hydroxylmethyl) acrylate. J Appl Polym Sci 48:2241–2247
Ximenes FA, Evand PD (2006) Protection of wood using oxy-aluminum compounds. Forest Prod J 56:116–112
Yalinkilic MK, Imamura Y, Takahashi M, Demirci Z, Yalinkilic AC (1999) Biological, mechanical and thermal properties of compressed-wood polymer composites (CWPC) pretreated with boric acid. Wood Fiber Sci 31(2):151–163
Yalinkilic MK, Takahashi M, Imamura Y, Gezer ED, Demirci Z, Iihan R (1991) Boron addition to non or low formaldehyde cross-linking reagents to enhance biological existence and dimensional stability for wood. Holz als Roh-und werkstoff 57(1):151–163
Yalinkilic MK, Tsunoda K, Takahashi M, Gezer ED, Dwianto W, Nemoto H (1998) Enhancement of biological and physical properties of woof by boric acid-vinyl monomer combination treatment. Holz 52(6):667–672
Yamaguchi H (1994a) Properties of silicic acid compounds as chemical agents for impregnation and fixation of wood. Mokuzai Gakkashi 40(8):830–837
Yamaguchi H (1994b) Preparation and physical properties of wood fixed with silicic acid compounds. Mokuzai Gakkashi 40(8):838–845
Yap MGS, Chia LHL, Teoh SH (1990) Wood polymer composites from some tropical hardwood. J Wood Chem Technol 10(1):1–19
Yasuda R, Minato K (1995) Chemical modification of wood by non-formaldehyde crosslinking reagents. Wood Sci Technol 29:243–251
Young RA, Meyer JA (1968) Heartwood and sapwood impregnations with vinyl monomers. Forest Prod J 18(4):66–68
Zollfrank C, Wegener G (2002) FTIR microscopy and ultrastructural investigation of silylated solid wood. Holz 56(1):39–42

Preparation and Characterizations of Various Clay- and Monomers-Dispersed Wood Nanocomposites

M.R. Rahman and S. Hamdan

Abstract In this chapter, different types of wood polymer nanocomposites (WPNCs) with various clay and monomer were prepared through curing methods, wood-hardening process, and chemical impregnation as well as compression of wood. The samples were ensured to dry at 105 °C up to constant weight before treatment. The dimensions and weights were measured. The samples were undergoing impregnation process in an impregnation vacuum chamber. WPNCs produced were characterized by Fourier Transform Infrared Spectroscopy (FT-IR), compression test, Thermogravimetric Analysis (TGA), and Scanning Electron Microscopy (SEM).

Keywords Mechanical properties · Stability · Morphology · Wood polymer nanocomposites

1 Overview

In the past research, many efforts have been applied on the wood reinforcement. Based on the previous literature, it has been clearly showed that the raw wood can be greatly modified and improved its properties through chemical modification. One of the most common and inexpensive techniques developed is in situ polymerization technique on wood polymer nanocomposites (WPNCs). Besides, nanotechnological modification has also been proven to be very effective in improving the various properties of raw wood. From the previous research, there are some research works of polymer/nanoclay system onto wood. Therefore, with the combination of nanotechnological modification and impregnation process, more chemical formulations could be formed and were chosen for this study.

M.R. Rahman (✉) · S. Hamdan
Faculty of Engineering, Universiti Malaysia Sarawak,
94300 Kota Samarahan, Sarawak, Malaysia
e-mail: rmrezaur@unimas.my

M.R. Rahman, *Wood Polymer Nanocomposites*, Engineering Materials,
DOI 10.1007/978-3-319-65735-6_2

In this study, there are five types of wood species selected as raw materials as these materials are abundantly available in the local forest in Malaysia. All the selected raw materials are large in quantities but none of them are fully utilized in many applications due to their low qualities in properties. Furthermore, from the literature finding, there was few or no work had been carried out on tropical wood species with various chemicals.

In this study, the experimental work is divided into three sections. The first and second sections included the preparation of wood specimen and the chemical fabrications of wood polymer nanocomposites (WPNCs). The third section of the experimental work is the analytical testing which conducted to investigate the properties to ensure the WPNCs produced would be greatly enhanced. Both the raw and treated wood samples are synthesized in all experiments. To ensure the accuracy of the experimental work, a total of ten wood samples for each species are used in every treatment. This chapter also includes the materials as well as methods to fabricate WPNCs. Besides, all the characterizations involved throughout the whole book will be included in this chapter.

2 Methods Related for Wood Polymer Nanocomposites (WPNC)

2.1 Curing Methods for Wood Polymer Nanocomposites (WPNC) Preparation

With the various curing processes, it can well convert a monomer or a prepolymer that was used to impregnate in wood. All these techniques can be explained in detail.

The first technique is radiation curing. It is a very commercial production of wood polymer materials that began in the mid-1960s using the radiation process. There are few examples of this type of radiation which included atomic particles, neutrons, photons, gamma rays, X-rays, and electrons (Betty 1976; Dobreva et al. 2006). Gamma radiation treatment appears to be the cheapest and most practical due to the uniform distribution of the polymer within the sections of wood when gamma rays penetrate the wood. However, radiation curing had a main drawback which was the safety and environment concerns. In addition, the transportation of the various wood products to and from the irradiation site was prohibitive (Dobreva et al. 2006; Meyer et al. 1965). According to Meyer et al. 1965, catalyst-heat technique was another method to treat WPNCs.

When it reached the early of 1980s, the process of producing WPNCs using catalyst-heat process was much more widely used than radiation process throughout the USA. The specific catalyst-heat process which was the microwave heating curing was the most widely used. This process occurred when heat transfers from the surface into the materials especially for the wood. However, the heating rate of a wood material was limited. Based on the researches carried out by Galperin et al. (1995) and Patyakin et al. (2008), microwave radiation had been found to be efficient for curing treatment of an impregnated wood. Urea and formaldehyde resin solutions were used to impregnate the wood species at various temperature which change the behavior of wood products (Galperin et al. 1995). This research showed that the birch samples used 150 s to heat up 100^0 C for microwave treatment, while the same samples used 1920s to reach the same temperature for convective heating process.

2.2 Wood-Hardening Process

In the past research, many efforts had been carried out to improve wood properties physically or chemically. Therefore, there was a great establishment in knowledge and different techniques for wood quality improvement to ensure the modified wood products can be widely applied in different niches. The techniques that were commonly used for hardening wood included impregnation with reactive chemicals, compression, or combination of chemical impregnation and compression. The new WPNCs produced showed better dimensional stability, compressive strength, and hardness compared to the original wood (Bodirlau et al. 2009; Schneider 1994; Ayer et al. 2003; Devi et al. 2003).

The introduction of a monomer into the empty voids within the cell tissue will greatly improve the hardness of raw wood. The filling can be done by impregnating the liquid into the wood in a closed space. This can have caused wood hardening. One of the examples monomers used in the impregnation was the styrene or methyl methacrylate (MMA) that becomes solid by polymerization. However, this type of impregnation showed one major challenge which was the proper technique for well impregnation of the liquid monomer into the cell tissue of wood due to the poor permeability of most species of wood. Besides, wood would be burnt or split once the polymerization took place due to the impregnation process. To overcome this challenge, the liquid monomer should be well impregnated into the cell tissue of the wood through vacuum impregnation or pressure impregnation to ensure the poor permeability of wood was greatly enhanced.

In general, reactive monomers or prepolymers can be impregnated into the raw wood through suitable techniques such as heat impregnation with the proper polymer initiator. As mentioned previously, wood splitting or wood burning as well as the changes in the wood dimensions can be happened due to the exothermal polymerization reaction occurred with the cell tissue of the wood. These can be happened because the temperature within the wood rises to a very high level during the polymerization reactions. Therefore, suitable method such as the application of radiation can be carried out to initiate and control the polymerization reactions. For this case, raw wood can be treated by gamma radiation. The polymerization reaction impact on the anatomic structure, chemical structure, density, viscosity, composition of the wood cell wall whereas the polarity of the monomer that can affect the wood impregnation process.

2.3 Monomer and Polymer Treatments

The introduction of suitable liquid monomer or prepolymer into the cell lumens of wood to produce polymerized wood usually remains their original desirable characteristics with the addition of improved properties. Similar to the conventional wood-treated product, most of the wood species can be readily impregnated with various monomers or prepolymers through suitable equipment and techniques. For example, the impregnation of vinyl polymers into wood contributes better performance especially in preservation, mechanical properties, and the water repellency. This type of impregnation by using curing greatly improved the moisture resistance and hardness of wood (Bodirlau et al. 2009; Meyer et al. 1978). Besides, catalysts should be applied in the polymerization of vinyl monomer with wood (Bodirlau et al. 2009; Yan et al. 2009; Meyer et al. 1965; Inoue et al. 1993). In addition, radiation technique was used for vinyl monomer to impregnate the wood species. (Siau 1969; Yan et al. 2009). According to Langwig et al. (1968), wood can be polymerized by *t*-butyl styrene, epoxy monomers, and methyl methacrylate via in situ impregnation.

With the introduction of monomer into wood through impregnation, the properties of WPNCs such as hardness and static bending strength significantly increased. Through this, the wood can be effectively stabilized with the impregnation of vinyl polymer into the wood by bulking the cell wall (Timmons et al. 1971). However, vinyl monomers except for acrylonitrile are all poor swelling agents for wood (Bodirlau et al. 2009; Yan et al. 2009; Loos and Robinson 1968). Based on the research carried out by Ellwood et al. (1972), it was clearly mentioned that the dimensional stabilization of wood with vinyl monomers had been gradually increased. For wood impregnated with styrene, this WPNC has been largely investigated which significantly improved the water repellency, compression, and bending strength (Baki et al. 1993). Moreover, the research of Meyer et al. (1965) investigated that wood–styrene combinations through radiation techniques showed the well bulking of wood with styrene as well as the with gamma irradiation.

Besides mechanical properties, the dimensional stability of wood was improved. However, an initial swelling of wood by water or some other polar solvent was required. There are two different types of impregnation, namely solvent exchange and high-vacuum methods. Besides, gamma irradiation impregnation wood showed better properties such as improving dimensional stability and tangential tensile strength. In addition, modulus of rupture and modulus of elasticity were improved through the gamma radiation impregnation. Based on the previous research by Bodirlau et al. (2009), the styrene/glycidyl methacrylate (GMA) WPNCs showed higher anti-shrink efficiency value with value 53%, while styrene-treated wood samples showed 23% anti-shrink efficiency value for 24 h in water. However, both styrene/GMA WPNCs and styrene-treated wood samples showed similar hardness with an increment by 33% compared to raw wood samples.

To synthesize WPNCs, methyl methacrylate (MMA) is another type of vinyl monomers. MMA is commonly used because it is an inexpensive, clear-colored, and commodity chemical. MMA is one of the most widely used monomers in making clear and colored WPNCs (Duran and Meyer 1972; Langwing et al. 1968, 1969; Noah and Foudjet 1988; Siau and Meyer 1966; Siau et al. 1968, 1978; Yalinkilic et al. 1998; Yan et al. 2009). According to Loos and Robinson (1968) and Yan et al. (2009), the wood swelled with the addition of MMA that led to a small change in dimensional stability. However, the cell walls remained unchanged for MMA-treated wood. MMA-treated WPNCs were time-dependent dimensional stability improvement which defined that these WPNCs would degrade over the years until the end; these WPNCs showed almost the similar properties as raw wood.

The properties of WPNCs produced are usually characterized through compression parallel to grain direction, static bending strength, dynamic modulus of elasticity, and dielectric constant. The mentioned properties are investigated through different theoretical equations which derived from an assumed cellular model. These theoretical equations can be used as a reference for the experimental works for the WPNCs formed. For raw wood, it was found that the hardness of wood was closely related to the density of raw wood. On the other hand, the hardness of the treated wood was more related to the hardness modulus, density, and polymer loading. From the impregnation or treatment, the impregnated or treated wood with methacrylate polymers showed better properties including the decrement in the rate of moisture uptake, higher mechanical properties such as modulus of elasticity and rupture, higher fiber stress at proportional limit, improved maximum load, and crushing strength limits as well as greater hardness index compared to the raw wood (Ates et al. 2009; Langwig et al. 1968; Rowell et al. 1982).

From the research carried out by Juneja and Hodgins (1970), it studied on the impregnation as well as the polymerization of vinyl chloride with different conditions in the raw wood. As the polymerization reaction was highly exothermic, the WPNCs produced were distorted and discolored. The discoloration of WPNCs occurred due to the conjugated double bonds left along the chain when the hydrogen chloride was released during the degradation of the polyvinyl chloride (PVC). Therefore, vinyl chloride was not a suitable monomer for producing WPNCs due to three reasons. First, the mechanical properties of PVC-related WPNCs were only slightly improved. Second is that the discoloration and deformation happened as the heat of reaction was higher. The third reason is that the insoluble polyvinyl chloride formed fine powder precipitate during the reaction. The fine powder would not improve the mechanical properties of the PVC-related WPNCs formed.

The method to develop WPNCs is important. WPNCs can be synthesized through in situ polymerization of ethyl α-hydroxymethylacrylates (EHMA) (Mathias and Wright 1989). EHMA is a good monomer to be impregnated into WPNCs. The monomer as reported can penetrate the cell walls of the wood fibers which formed a complete WPNC. Bryant (1966) investigated that the mechanical properties of phenolic resin-impregnated wood were greatly increased. Besides, he also investigated the changes in relative molecular size as well as the concentration of phenolic resin on dimensional stabilization of WPNCs. Water-soluble phenolic resins are usually used to impregnate wood by bulking the cell wall structure and at the same time preventing shrinkage of wood upon drying. However, the application of phenolic resin-modified wood products is limited due to their dark brown color on the surface of the products.

To ensure the improvement in modulus of elasticity, modulus of rupture, and higher thickness swelling, the compression of boards should be significantly increased. Besides phenolic resin mentioned above, aqueous polyurethane emulsion can be introduced into raw wood as it consists of low molecular weight components in aqueous media (Hartman 1969). This aqueous polyurethane system could be reused without affecting the polymerization at room temperature. It was clearly proven that better diffusion and penetration could be obtained with the impregnation of lower molecular weight resin (Yan et al. 2009). Apart from polyurethane, amino resin formed by melamine (1, 3, 5-triamino-2, 4, 6-triazine) and formaldehyde has been commercially used for over 60 years. Another polymer named melamine formaldehyde (MF) is one of the hardest and stiffest isotropic polymeric materials with low flammability. This type of polymeric material is applied in decorative laminates, molding compounds, adhesives, coatings, and other products. These advantageous properties helped to improve properties of raw wood such as surface hardness and modulus of elasticity (MOE) (Ates et al. 2009; Deka and Saikia 2000; Georgieva et al. 2008; Miroy et al. 1995).

In addition, poly(ethylene glycols) (PEG) can be widely used in the field of the dimensional stabilization of wet wood (Georgieva et al. 2008; Hoffmann 1988, 1990; Stamm 1959, 1964a, b). PEG is a water-soluble synthetic wax. The addition of heated PEG into the raw wood increased the strength by replacing the hydroxyl group within the raw wood with PEG. Not only PEG, alkylene oxide is another specific polymer that can be introduced into the raw wood. Based on the previous research, the impregnation of alkylene oxide may improve the dimensional stability of wood due to the changes in hydrophilic nature of wood (Ates et al. 2009; Georgieva et al. 2008; Guevara and Moslemi 1984; Rowell 1975; Rowell et al. 1976). WPNCs prepared by Rowell et al. (1982) were formed through the combination treatment of cell wall grafting with alkylene oxides and lumen treatments with methyl methacrylate.

WPNCs that were prepared using a two-stage treatment resulted a higher stabilized WPNCs. The first-stage treatment is that raw wood chemically modified the cell walls of southern pine and sugar maple with propylene oxide. Propylene oxide in this case was functioned to reduce the flexural strength of the wood. For second-stage treatment, methyl methacrylate (MMA) was impregnated into the void volumes of raw wood. Another research on the combination of wood with alkoxysilane was carried out by Su and Wilkie (2003). Alkoxysilane was used as coupling agent that helped to increase the chemical impregnation reaction within the raw wood tissue.

2.4 Other Treatments

Due to the advanced technology, wood inorganic nanocomposites can be fabricated through the introduction of inorganic substances into the wood's cell walls (Furuno et al. 1991, 1992a, b; Ogiso and Saka 1993; Saka et al. 1992; Saka and Yakake 1993; Tunc et al. 2010; Yamaguchi 1994a, b; Yan et al. 2009; Zollfrank and Wegener 2002). Wood inorganic nanocomposites were synthesized to improve the fire and decay resistance properties. For example, a combination of acetylated and propionylated formed wood silicate nanocomposites (Li et al. 2001). Besides, wood silicate nanocomposites could also be impregnated with an aqueous sodium silicate solution that improved the dimensional stability. The filler such as silicate gels were incorporated in WPNCs which enchanced the flame-resistance as it helped to increase the oxygen index of the WPNCs.

2.5 Combination of Two or Three Monomers

WPNCs can be produced with one or more monomers. From the researcher carried out by the cell wall and cell lumen of WPNCs were investigated. Both furfuryl alcohol-based (FA) cell wall and methyl methacrylate-based (MMA) combination formulations were introduced into the raw wood. Besides, the wood cell wall could be modified with acetic anhydride, while lumen could be filled with methyl methacrylate (MMA). Their effectiveness in reducing the rate of moisture sorption and degradative effects of accelerated weathering was investigated by Feist et al. (1991). This showed the effectiveness in reducing the rate of swelling and reducing erosion up to 85% due to accelerated weathering through the combined treatment of acetylation followed by methacrylate impregnation. On the other hand, cell lumens could be fully filled with bioactive polymer. However, this polymer might lead to solubility and penetration problem as the polymers had low solubility but high viscosity. Therefore, bioactive group polymer should be used together with a monomer to fill the wood. Besides, the combination of bioactive polymer with monomer should be carried out through in situ polymerization or copolymerization with the aid of a catalyst that led to higher loading of the bioactive polymer into the wood.

2.6 Chemical Impregnation and Compression of Wood

Chemical impregnation is one of the common methods to harden the raw wood for value-added applications. In year 1940s, compressed (compreg) high-density wood for manufacturing airplane propellers was produced with the impregnation of phenol formaldehyde (PF). Recently, the impregnated wood polymer nanocomposites are widely used in the industrial applications such as bolts, rivets, gears, high and low tension circuit breakers, turbo generators, sports goods and chair seats etc.

2.7 Summary of Wood Quality Improvement Methods and Technologies

Although wood is a common material to be applied in industrial application, it consists of some drawbacks that include high moisture uptake, biodegradation, and dimensional variations. All these drawbacks happened due to the presence of numerous hydroxyl groups within the raw wood components and different cavities in the wood. Due to this drawback, raw wood is necessary to be modified. There are some techniques that can be well applied in modifying wood, namely impregnation with reactive chemicals, compression, and the combination method of chemical impregnation and compression. Different techniques of wood modification were carried out to improve the properties of wood. Recently, chemical modification on wood is increasing due to the great improvement in the hardness, mechanical properties, and dimensional stability of WPNCs formed. Two major types of polymer which are thermoplastic and thermosetting systems have been used in the impregnation of raw wood through in situ polymerization of monomers such as styrene and methyl methacrylate, or prepolymers such as phenol formaldehyde as these methods are the most common and inexpensive technique. In the early 1960, chemical modification was widely performed to obtain better properties of WPNCs. As mentioned previously, vinyl monomers do not strongly bond with the component and thus, the monomers usually filled only the empty lumens of the wood. With the filling of monomer into lumens, the dimensional stability of WPNC can be greatly improved. On the other hand, the introduction of polar monomers led to better chemical resistance as well as dimensional stability of WPNCs.

To ensure the vinyl monomer can be applied in chemical modification, vinyl monomer itself should be modified by introduction of polar groups such as ethyl α-hydroxymethylacrylate (HMA). HMA is recognized to be good monomer for cell wall swelling. However, the monomer modification is a time-consuming process. Thermosetting resin such as water-soluble phenolic resins, urea formaldehyde, and melamine-formaldehyde prepolymers was all suitable to be applied in improving its compressive strength properties and at the same time reduced shrinking and swelling behaviors. These resins penetrated and bulked the cell wall to react with the hydroxyl groups of the raw wood. This caused the brittleness and toughness properties of the raw wood to deteriorate.

3 Methods

Solid wood is one of the most important structural and renewable raw materials that are abundantly available in the world. Natural beauty, durability, and versatility make wood the preferred material for many uses. Nevertheless, wood has a few drawbacks which limit its use, including high moisture uptake, biodegradation, and physico-mechanical properties which change with environmental variation (Vetter et al. 2009; Yalinkilic et al. 1998). Recently, raw wood is modified intensively through in situ reinforcement with the addition of silicate and nanoclays (Cai et al. 2007a, b, 2008). Besides, suitable treatment improved the physical and morphological properties of WPNCs (Bergman et al. 2009; Islam et al. 2011; Yildiz et al. 2005). Thermal and mechanical properties of wood could improve when treated with chemical treatment (Larsson and Simonson 1994). To ensure the properties could be improved, different kinds of chemicals, namely styrene, methyl methacrylate, sodium metaperiodate, phenyl hydrazine, *N*, *N*-dimethylacetamid, phenol-formaldehyde resin, maleic acid, ethylene glycol dimethacrylate, 3-(trimethoxysilyl)propyl methacrylate, acrylonitrile, butyl methacrylate, furfuryl alcohol, ethyl methacrylate, glycidyl methacrylate, and urea formaldehyde, have been extensively used in impregnation of raw wood. This was due to the addition of these chemicals improved the wood properties. However, some chemicals had their own limitations (Islam et al. 2010; Rowell 2005).

In the recent years, there are many researches on the improvement of properties of wood which was modified by nanotechnological modification effectively (Cai et al. 2007a, b, 2008). According to Border et al. (2009), the incorporation of different layered silicates and nanoclay (montmorillonite, hectorite and saponite) incorporate into the composites which led the composites to a higher stiffness and strength. With nanotechnological modification, WPNCs with organophilic-layered silicate via in situ nanoreinforcement improved the wood properties. Moreover, lower silicate loading thermoplastic and thermoset nanocomposites considerable improvements in physical properties, mechanical properties which included tensile modulus and strength, flexural modulus and strength, thermal properties such as thermal stability, flame retardant as well as barrier resistance (Ray and Okamoto 2003). Due to the advantageous properties, the nanotechnological technique is perceived as potential in improving the properties of solid wood to obtain better products.

Tropical light hardwood is abundantly available in quantity. However, their quality of raw wood is poor to be utilized on specific purposes due to their poor physico-mechanical and thermal properties. Suitable chemical modification is required before they can be utilized. The discussed newly developed nanotechnological modification technique on selected tropical woods can be promising for this approach. However, there is less research that has been carried out on the nanotechnological modifications. In addition, less research has been carried out on the chemical modification of Malaysian tropical light hardwood species especially with nanoclay and chemicals combination technique.

One of the characterizations that are used to study the effect of temperature on thermomechanical properties of WPNCs is dynamic mechanical thermal analysis (DMTA) (Luckenbach, Rheometrics Inc. (Sealseastem 1994)). DMTA managed to determine the viscoelastic behavior of WPNCs, and at the same time, it provided important information on the structure, morphology, and properties of WPNCs (Mancado and Arroyo 2000). According to Salmen (1984), the viscoelastic behavior of wood greatly depended on the response of the cellular structure to mechanical force. The temperature-dependent dynamic parameters such as dynamic modulus, storage modulus, and loss modulus are classified as temperature-dependent dynamic parameters, and mechanical damping (tan δ) is used to investigate the interface and interaction between the layered silicate or nanoclay on the polymer matrix. There are many researchers proved that WPNCs showed better thermomechanical properties through DMTA analysis (Hamdan et al. 2010; Sugiama et al. 1996).

Besides, free-free flexural vibration method is another method specifically used to investigate the dynamic Young's modulus of wood. According to Halabe et al. (1995), free-free vibration method is one of the nondestructive testing (NDT) methods that are widely applied in timber industry which are used to measure the elastic properties and energy dissipation of raw wood and WPNCs (Halabe et al. 1995). This testing provided accurate information on the stiffness of the WPNCs. Besides, three types of vibration, which are bending (flexural), longitudinal (axial), and torsion, are well determined by the nature of vibration. From the three mentioned vibration methods, the flexural vibration method is the most widely used method as it is easier to excite and detect the vibrations under investigation.

In this research, different types of tropical wood species were selected and fabricated with the aid of phenol-formaldehyde resin (PF) and halloysite nanoclay. All the prepolymer mixture was impregnated into the raw tropical wood by vacuum-pressure method and in situ polymerization. Based on the previous research, it was clearly shown properties of WPNCs formed affected the percentage of weight gain and the density. Nanoclay was incorporated into the wood, and this was confirmed through the characterization such as FTIR and SEM, respectively. Besides, mechanical properties of WPNCs especially the modulus of elasticity (MOE) and compressive modulus could be improved significantly. Dynamic mechanical thermal analysis (DMTA) was used to investigate the thermomechanical properties of raw wood and WPNCs over the temperature −100 to 200 °C. DMTA was also used to evaluate the intrinsic properties of the components, morphology of the system, and the nature of interface between the phases. In addition, the storage modulus (E') of both raw wood and WPNCs was investigated to determine the improvement in both glassy region and rubbery plateau over the temperature range. Furthermore, WPNCs were improved especially on the surface interphase through the investigation on the damping peaks (tan δ) which lowered after the modification or treatment. Dynamic Young's modulus (E_d) of wood was calculated using free-free vibration testing which proved the significant improvement on the impregnated WPNCs. Moreover, decay resistance of both raw wood and WPNCs was investigated through fungi test. All the wood samples were exposed to two types of fungi, namely white rot (polyporus versicolor) and brown rot (postia placenta), for 12 weeks, and the weight loss was presented in percentage (%) form. From this characterization, WPNCs showed higher decay resistance compared to raw wood.

3.1 Flowchart of Project

This project started with the title received from the department. Once the title was received, more research on the information about wood was carried out, followed by the identification of problem statement and objectives of the project. After clearly finalized the problem statement and objectives, the literature review study especially on mechanical properties, the pros and cons of raw wood and fabricated WPNCs as well as the method of wood modifications was carried out. Continuously, the experiment was investigated based on the research and was carried out consecutively. After carrying out laboratory work, the results obtained would be analyzed and discussed. Lastly, conclusion would be made to summarize the experimental results together with analysis and explanation of the project. Figure 1 showed the flowchart of project as the reference of this project to ensure the project went smoothly.

3.2 Preparation of WPNCs

The process to fabricate WPNC is shown in Fig. 2. All the samples were oven dried to constant weight at 105 °C before treatment. All the dimensions and weights before and after treatment were measured, respectively. The samples were then placed in an impregnation vacuum chamber to synthesize WPNCs. All the samples were undergoing vacuum chamber with pressure of 10 kPa for 30 min to remove air from the pores of the samples formed. An accurate amount of monomers and initiator was introduced into the chamber until the samples were completely covered. The samples were kept immersed in the monomer solution for 6 h. Besides, all the samples were ensured to be placed at ambient temperature and atmospheric pressure to obtain further impregnation. Continuously, the samples were removed from the chamber and all the excess chemicals were wiped off from wood surfaces. Lastly, the samples were wrapped with aluminum foil and placed in an oven for 24 h at 105 °C for polymerization to take place (Deka and Saikia 2000; Devi et al. 2003). All the polymerized samples formed were weighted to determine weight percentage gain (WPG).

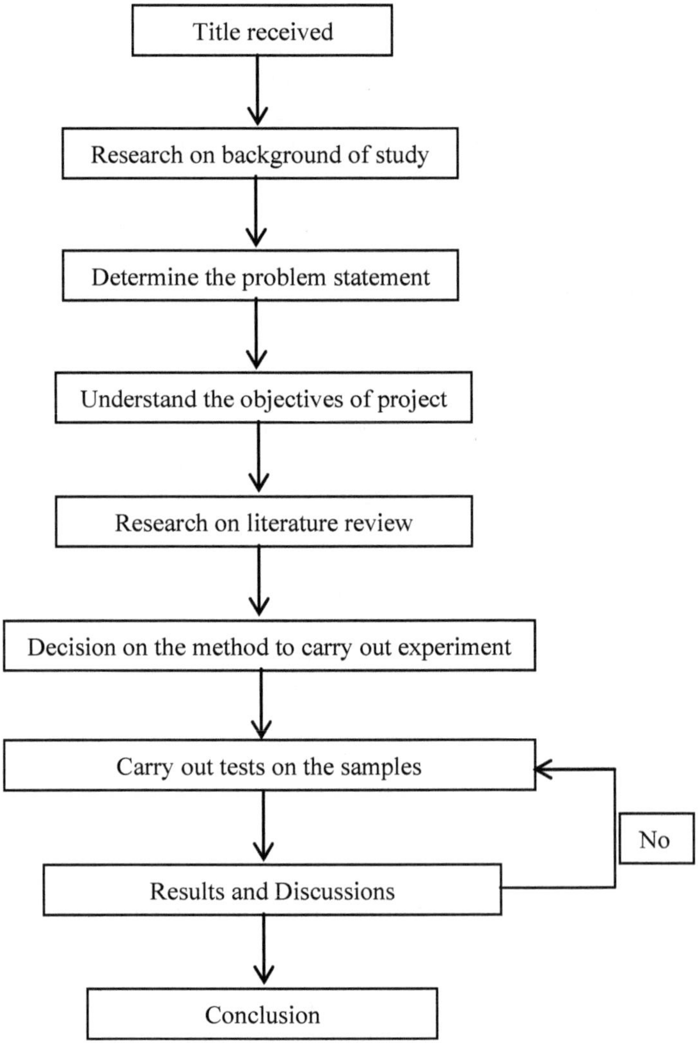

Fig. 1 Flowchart of project

3.3 Characterization of Wood Polymer Nanocomposites (WPNC)

3.3.1 Fourier Transform Infrared Spectroscopy (FT-IR)

One of the common analytical characterizations that have been applied in pharmaceutical, life sciences, and electronic industries is Fourier Transform Infrared Spectroscopy (FT-IR). This characterization included structural information of

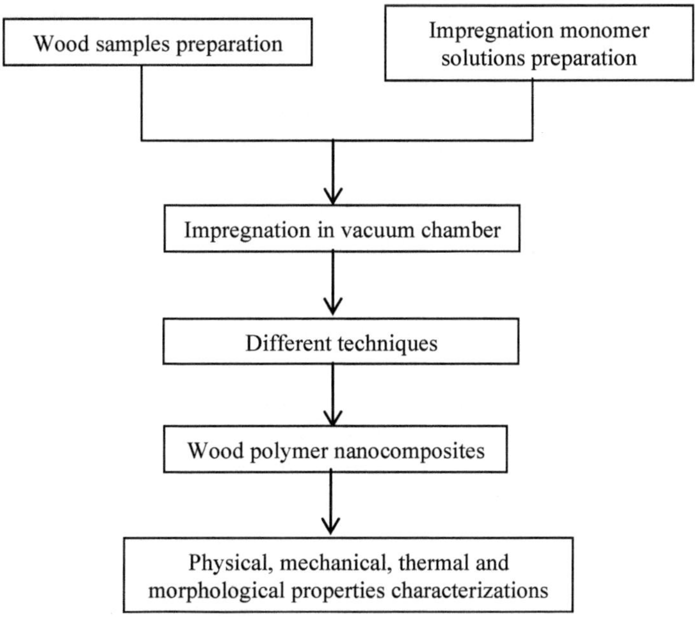

Fig. 2 Schematic diagram of the experimental procedure

Table 1 Functional groups with band position (FlashcardExchange.com, 2012)

Functional groups	Band position (cm^{-1})	Intensity
C–H (alkyl group)	2850–2960	Medium to strong
=C–H (alkene)	3020–3100	Medium
C=C	1650–1670	Medium
≡C−H	3300	Strong
C≡C	2100–2260	Medium
C–Cl	600–800	Strong
C–Br	500–600	Strong
C–I	500	Strong
O–H	3300–3540	Strong, broad
C–O	1050–1150	Strong
Aromatic ring	1600, 1500	Strong
N–H	3310-3500	Medium
C–N	1030, 1230	Medium
C=O	1670–1780	Strong
O–H (carboxylic acid)	2500–3100	Strong, very broad
C≡N	2210–2260	Medium

materials, analysis of materials as well as determination of contaminations IR Absorptions of Functional Groups (FlashcardExchange, United States of America 2006). The functional groups with its band position were shown in Table 1 (FlashcardExchange.com 2012). This equipment was used to determine the chemical change upon impregnation of the raw wood with impregnation of chemicals.

From Fig. 3, it showed FT-IR-branded Shimadzu IRAffinity-1 that was well applied with the guidance of operators. This FT-IR machine was kept in a clean and safe place to prevent any damage on it. Besides, chemical such as ethanol was used to remove stains and microorganisms. Both raw wood and WPNCs powders were taken out from the plastic tube using spatula and placed on top of the specimen in the FT-IR machine. The knob in the FT-IR machine was tightened. Continuously, the button “sample” appeared in the computer connected was then pressed. After 3 min, all the peaks in the graph form were shown. All the steps were repeated for all the raw wood and WPNCs samples. The infrared spectra of raw wood and WPNCs were recorded on a Shimadzu IRAffinity-1 at 20 scans with a resolution of 4 cm^{-1}. The intensity range used for this test is from 4000 to 600 cm^{-1}.

Fig. 3 FT-IR machine

3.3.2 Compression Test

Compression test can be carried out to determine the mechanical properties of both raw wood and WPNCs. Mechanical properties, namely modulus of rupture (MOR) and modulus of elasticity (MOE), could be tested through compression machine branded Shimadzu Universal Testing Machine. This machine functioned according to ASTM D-143 (1996) as shown in Fig. 4. It has the maximum loading capacity of 300kN with the crosshead speed of 10 mm/min. All the fabricated samples were stored at 85 °C inside a cool incubator for four days.

This test was carried out to ensure all the fabricated WPNCs withstand higher force compared to raw wood. For each experiment, 20 samples were prepared to undergo compression test. To obtain the result, the machine is well linked to the computer and all the compression test can be carried out using software named Trapezium2. The software Trapezium2 can be consistently applied as shown in Table 2 as well as in Figs. 5, 6, 7, 8, 9, 10, 11, 12, 13, 14, 15, 16, 17, 18, and 19.

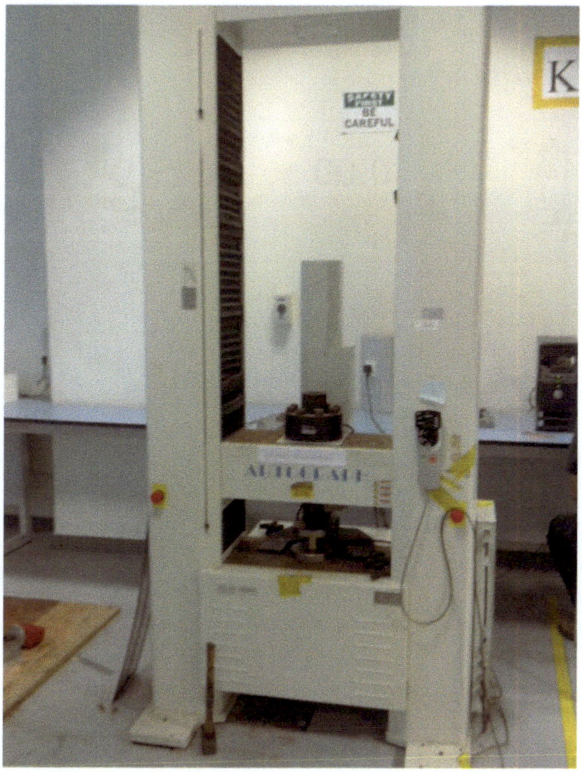

Fig. 4 Compression machine

Table 2 Thirteen steps used in compression testing software

Steps	Explanation
1	The overview of software
2	The icon "New" was clicked and followed by the clicking on "Method"
3	A command box was shown. The icon "No" should be clicked
4	Method Wizard will appear. The "Next" icon should be clicked
5	Icon "Comp." should be clicked, and "Next" icon was clicked after that
6	The "Next" icon should be clicked
7	V1's value was changed depending on the requirement, and "Next" icon was clicked
8	The value of wood sample's dimensions was changed and click "Next" icon
9	There are three figures that did not change by just clicking "Next"
10	Method Wizard completed by clicking the "Finish" icon
11	The position and force value should reset it at zero
12	A Navigation Bar was shown and click "Start" icon
13	Begin Test is pressed and navigation bar will show a "Stop" icon

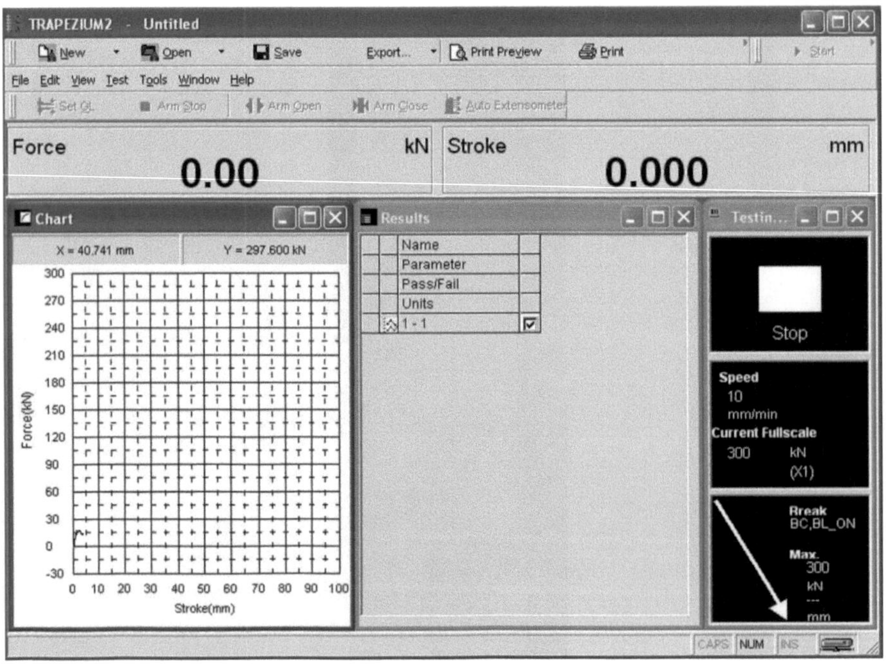

Fig. 5 Software overview

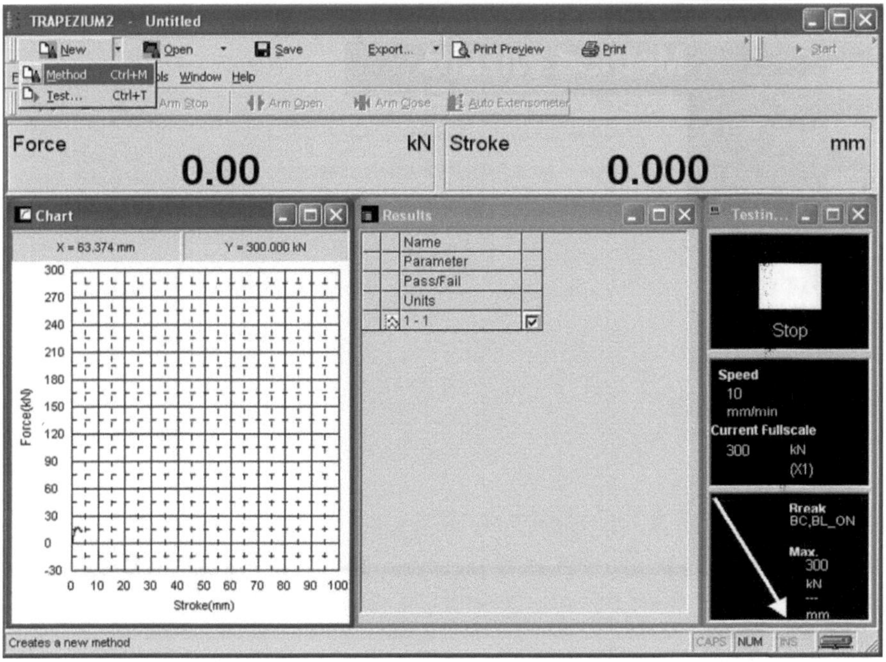

Fig. 6 Second step's overview

Fig. 7 Third step's overview

3.3.3 Thermogravimetric Analysis (TGA)

For thermal properties, Thermogravimetric Analysis (TGA) is carried out. This characterization is used to determine the degradation temperatures or absorbed moisture content of materials especially on polymer nanocomposites. TGA is a process which can determine the percent by mass ratio of a solute through heat and stoichiometry ratios. The process involved heating a mixture to a high enough temperature to decompose the samples into a gas, which dissociated into the air. It was plotting weight percentage against temperature. TGA was used to investigate the corrosion kinetics in high-temperature oxidation.

TGA started with the weighing of a finely ground sample. The samples were then exposed to a heated chamber in the presence of oxygen. The weight losses of

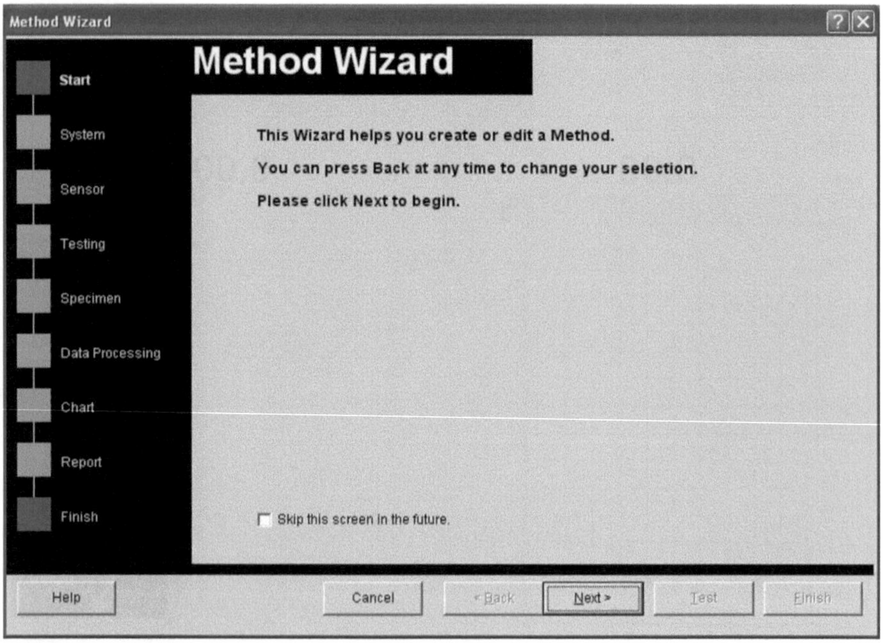

Fig. 8 Fourth step's overview

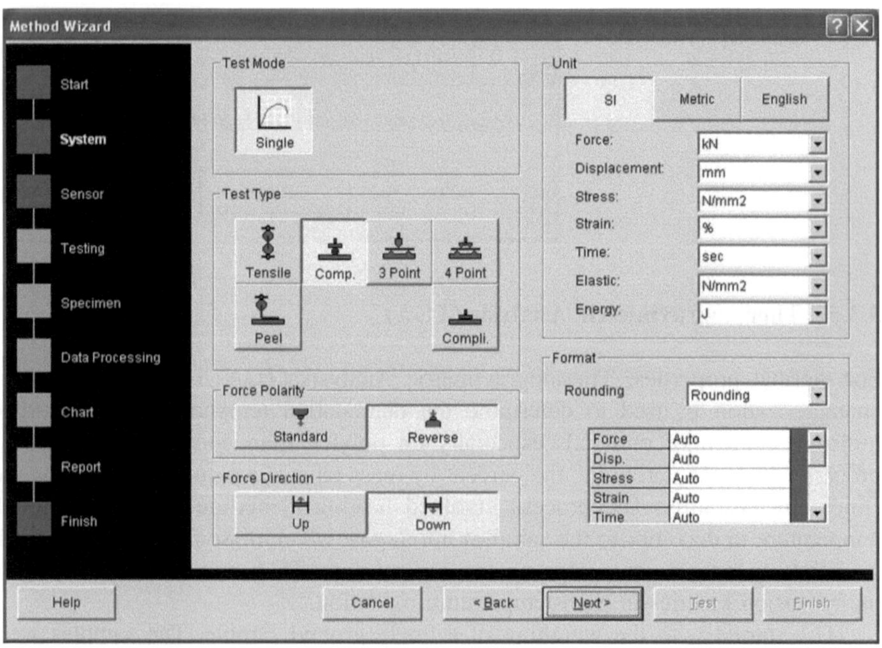

Fig. 9 Fifth step's overview

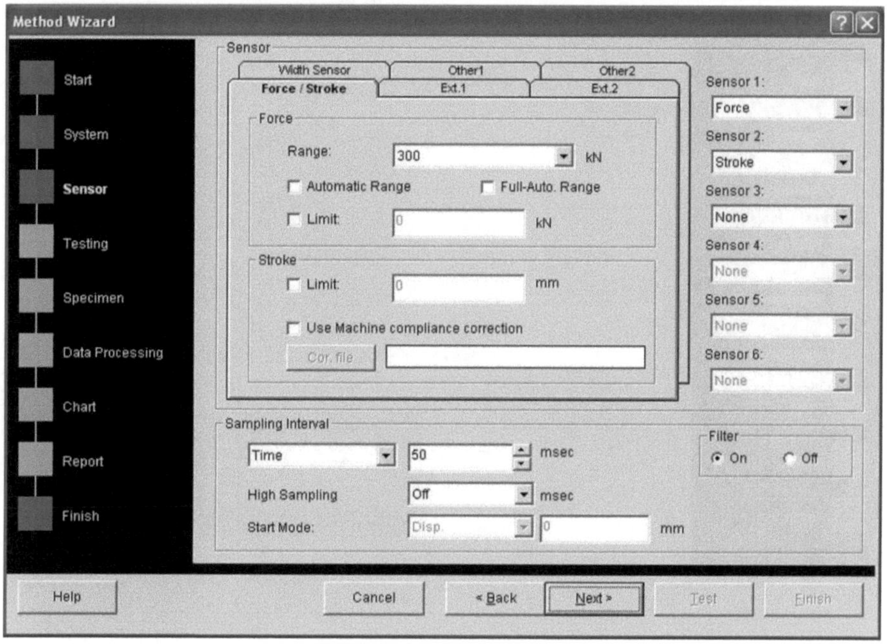

Fig. 10 Sixth step's overview

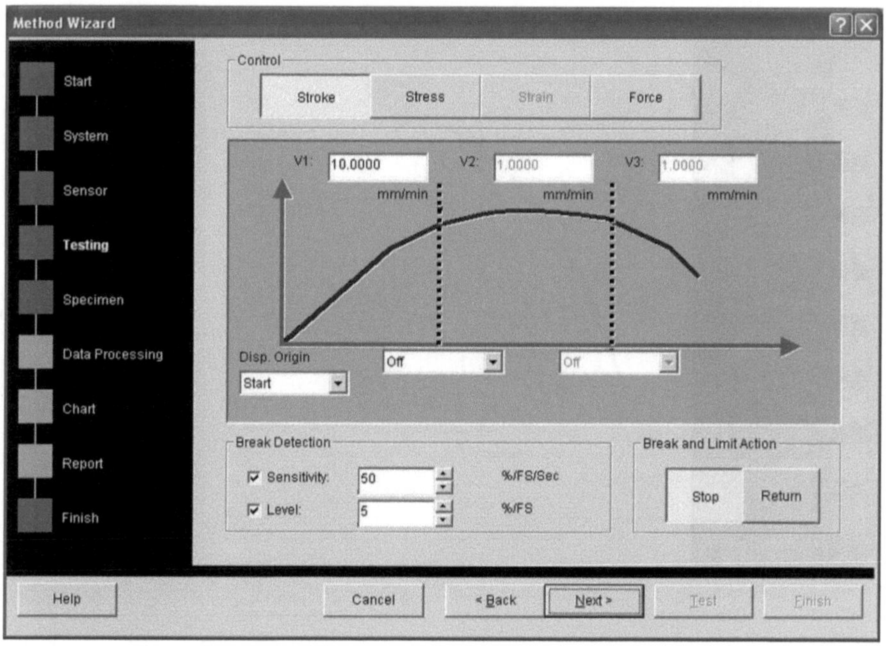

Fig. 11 Seventh step's overview

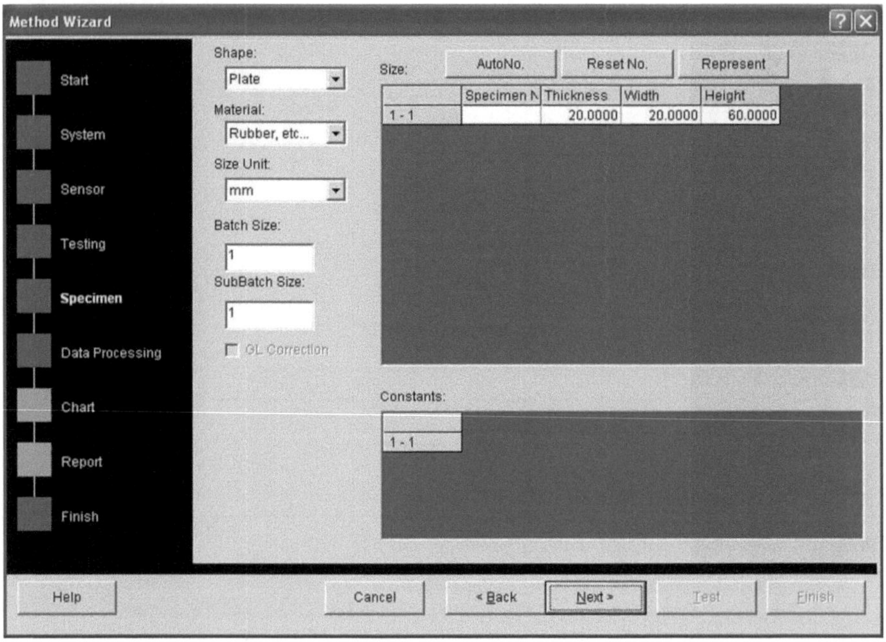

Fig. 12 Eighth step's overview

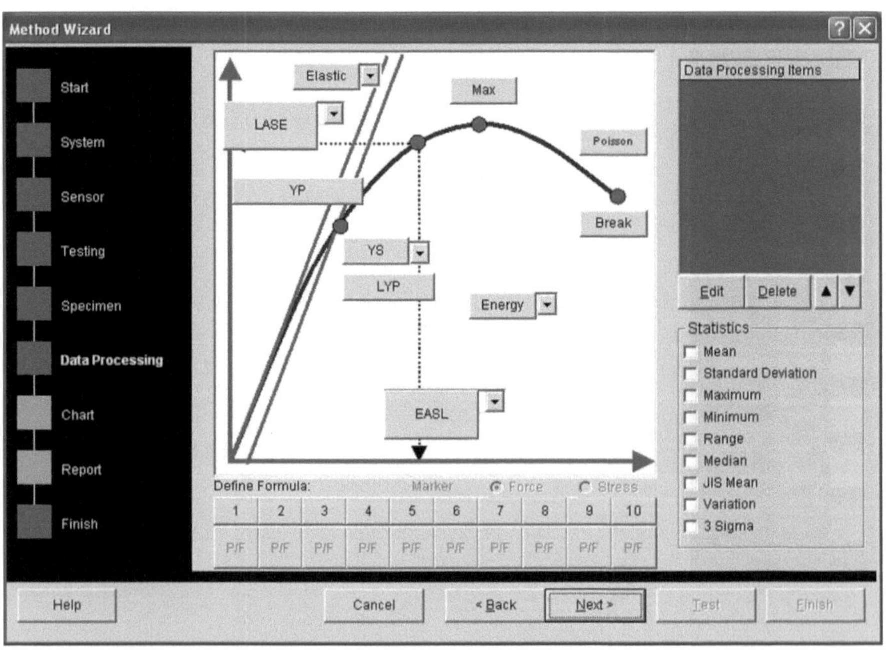

Fig. 13 Ninth step part 1's overview

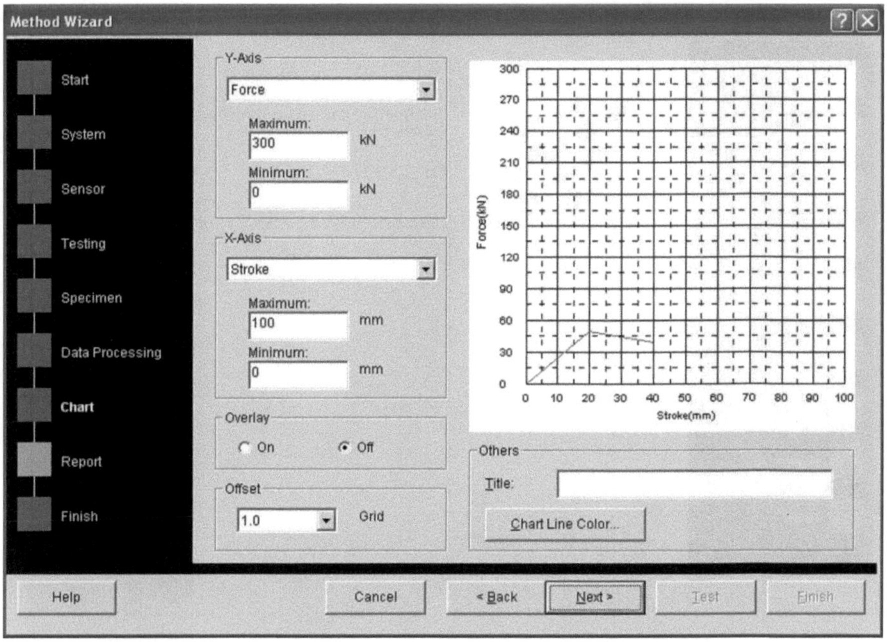

Fig. 14 Ninth step part 2's overview

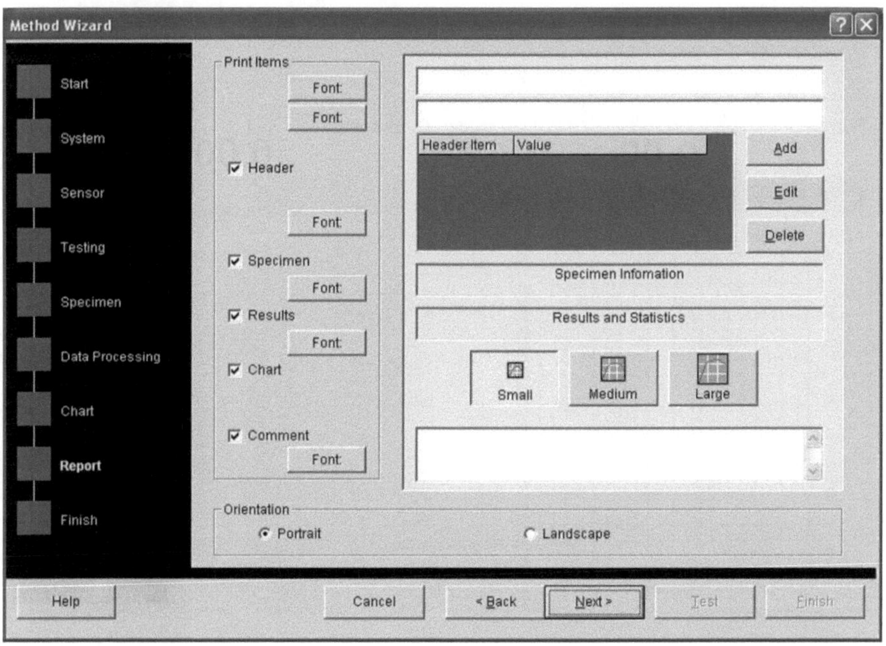

Fig. 15 Ninth step part 3's overview

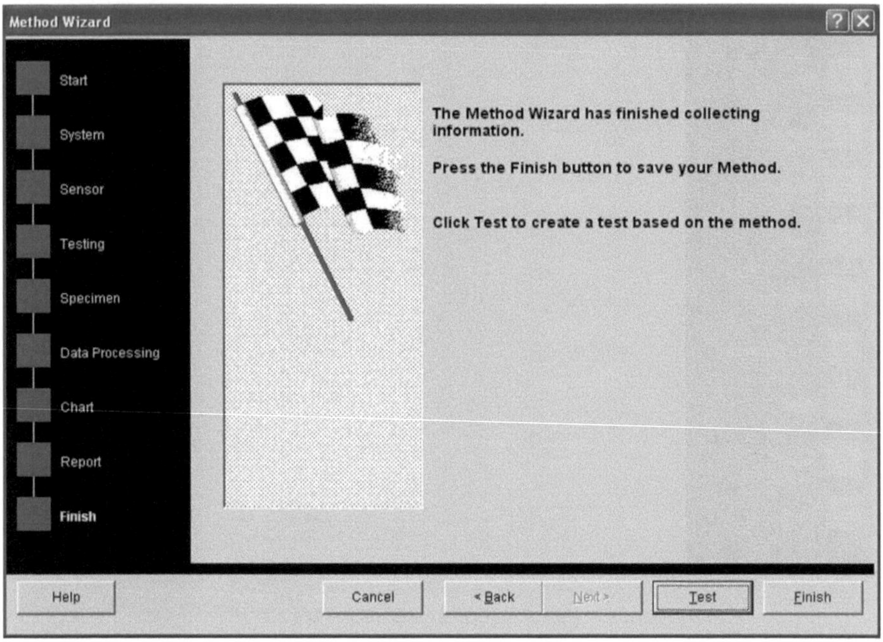

Fig. 16 Tenth step's overview

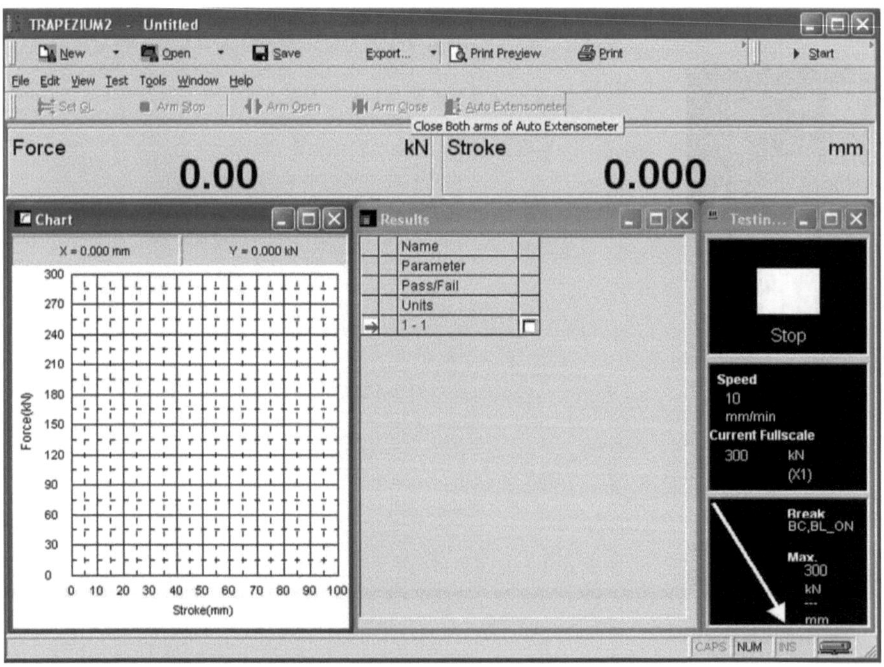

Fig. 17 Eleventh step's overview

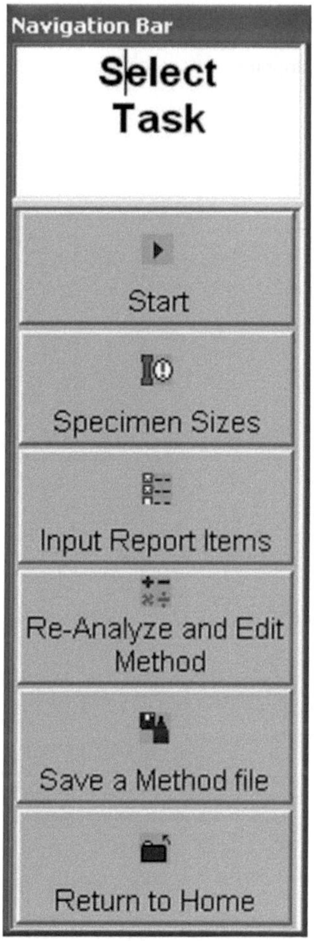

Fig. 18 Navigation bar for starting

the samples were measured before the samples were heated. The change in weight of the sample was recorded as a function of time with temperature ranging from 20 to 700 °C. The rate of weight loss as a function of time and temperature could be measured after setting up. The use of a derivative computer was needed to record the results. After the data were obtained, curve smoothing and other operations could be carried out to find the exact points of inflection. In this method, temperature increased which led to increased weight loss. This was to investigate the exact temperature where a peak occurred more accurately. The weight for wood powders was measured using electronic balance as shown in Fig. 20. DTG-60H branded Shimadzu was used to run this testing as shown in Fig. 21.

Fig. 19 Begin test command and navigation bar for stopping

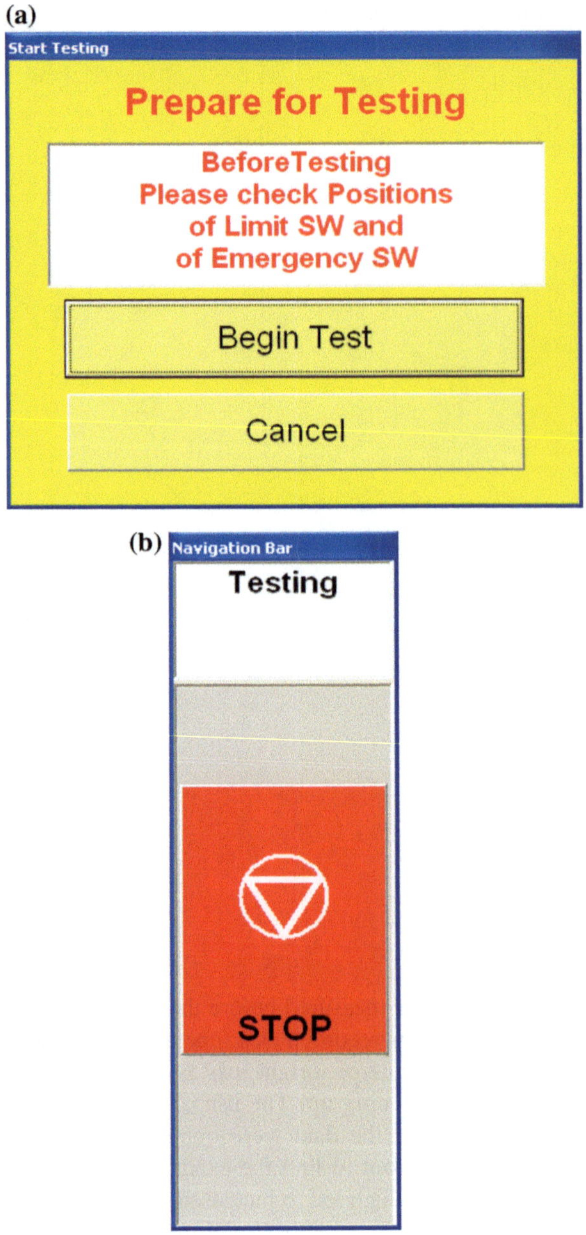

3.3.4 Scanning Electron Microscopy (SEM)

Besides FT-IR, compression test, and TGA, Scanning Electron Microscopy (SEM) was used to detect the pore size of the raw wood and WPNCs using

Fig. 20 Electronic balance

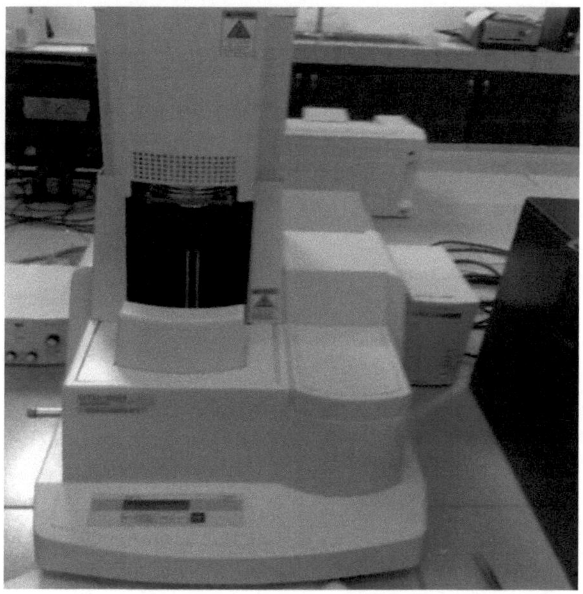

Fig. 21 DTG-60H

Scanning Electron Microscope under the warranty of Kosijaya Didactic Sdn. Bhd as shown in Fig. 22. The primary electron beam was produced by heating a metallic element under high vacuum at the top of the microscope. It was scanned across the surface of a specimen. The signal was generated when the electrons strike the

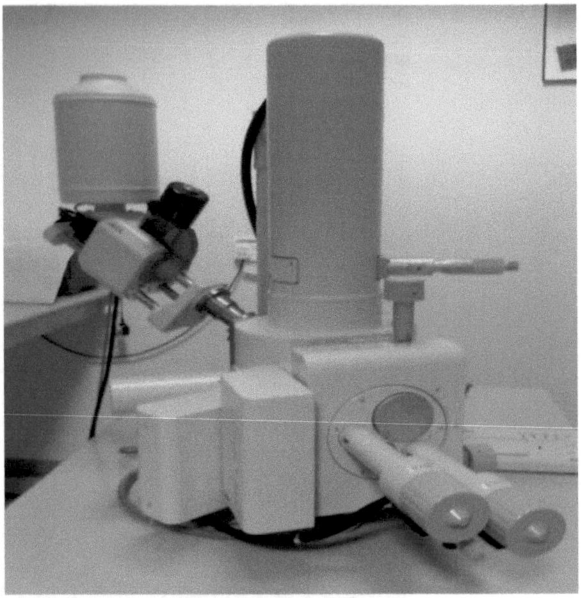

Fig. 22 Scanning electron microscope

Fig. 23 Coating machine

specimen. This was to detect the specific signals that produced an image of the surface. Three different types of signals that provided the greatest amount of information in SEM were the secondary electrons, backscattered electrons, and X-rays (Islam et al. 2011).

SEM was generally used to investigate the interfacial bonding between the cell wall polymer and reacted chemical. The specimens were first fixed with Karnovsky's fixative. It was then taken through a graded alcohol dehydration series. After the samples were dehydrated, all the samples undergo coating by a thin layer

Fig. 24 Samples on top of the magnetic substance

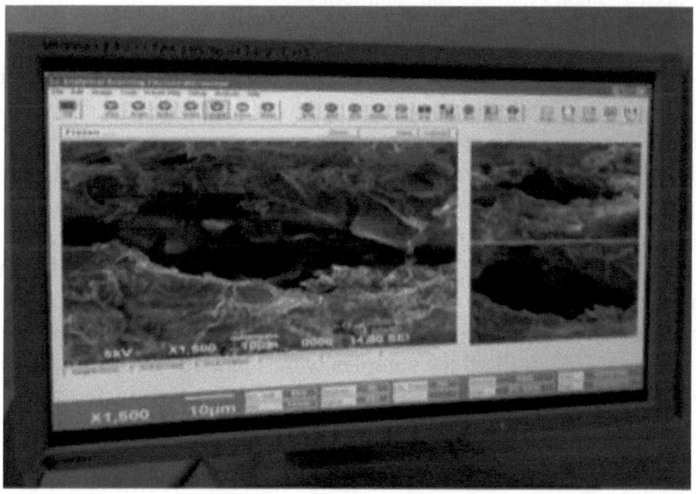

Fig. 25 Picture shown by SEM-linked computer

of gold before viewing on the SEM. The micrographs were taken at magnification between 500× and 1500×.

All the wood samples either in powder or solid form will be heated at temperature of 105 °C for an hour before SEM test. This step was to remove moisture content of wood. For solid wood samples, 3 cm × 3 cm × 1 cm of raw wood and WPNCs were prepared.

After heating, both raw wood and WPNCs either in powder form or solid form would be coated. Before coating, the carbon tape was placed on top of the aluminum plate and the sample powders were placed on top of the carbon tape as

Fig. 26 Collected samples (powder and solid form)

shown in Fig. 23 for 15 min. The aluminum plate was placed aside to prevent the tested samples mixed with non-tested samples. A pair of forceps was used to clip the aluminum plate into the sampling tube.

Scanning electron process could be started after coating process. There were five samples placed on top of a magnetic substance with double tape to ensure the aluminum plate samples did no move randomly as shown in Fig. 24. After the machine start-up, the electron laser beam started to scan the samples (powder or solid form) by manual for this experiment. The picture scanned would be shown in computer linked as shown in Fig. 25, and the images would be saved into a folder. This step was repeated until all the samples were scanned. After the experiment was done, the samples were collected in a clean plate as shown in Fig. 26.

4 Summary

This chapter covered the overview, methods related to WPNCs, methods such as flowchart of project, preparation of WPNCs as well as characterization included Fourier Transform Infrared Spectroscopy (FT-IR), compression machine, Thermogravimetric Analysis (TGA), and Scanning Electron Microscopy (SEM) in general.

References

Ates S, Akyildiz MH, Ozdemir H (2009) Effects of heat treatment on calabrian pine (Pinus brutia ten.) wood. BioRes 4(3):1032–1043

Ayer SW, Fell D, Wan H (2003) Hardening of solid wood: Market opportunities and review of existing technologies. Forintek Canada Corp, Canada

Baki H, Yalçin Örs M, Hakki A (1993) Improvement of wood properties by impregnation with macromonomeric initiators. J Appl Polym Sci 47:1097–1103

Bergman R, Ibach RE, LaPasha C, Denig J (2009) Evaluating physical properties changes for small diameter plantation-grown southern pine after *in situ* polymerization of an acrylic monomer. Forest Prod J 59(10):64–71

Betty WR (1976) US 3968318 A, 6 Jul 1976

Bodirlau R, Teaca CA, Spiridon I (2009) Preparation and characterization of composites comprising modified hardwood and wood polymers/poly(vinyl chloride). BioRes 4(4): 1285–1304

Border P, Pollet E, Averous L (2009) Nano-biocomposites: Biodegradable polyester/nanoclay system. Prog Polym Sci 34:125–155

Bryant BS (1966) The chemical modification of wood from the point of view of wood science and economics. Forest Product J 16(2):20–27

Cai X, Riedl B, Zhang SY, Wan H (2007a) Formation and properties of nanocomposites made up from solid aspen wood, melamine-urea- formaldehyde and clay. Holz 61:148–154

Cai X, Riedl B, Zhang SY, Wan H (2007b) Effects of nanofillers on water resistance and dimensional stability of solid wood modified by melamine-urea-formaldehyde resin. Wood Fiber Sci 39(2):307–318

Cai X, Riedl B, Zhang SY, Wan H (2008) The impact of interphase between wood, melamine-urea-formaldehyde and layered silicate on the performance of wood polymer nanocomposites. Compos Part A 39:727–737

Deka M, Saikia CN (2000) Chemical modification of wood with thermosetting resin: Effect on dimensional stability and strength property. Biores Technol 73:179–181

Devi RR, Maji TK, Banerjee AN (2003) Studies on dimensional stability and thermal properties of rubber wood chemically modified with styrene and glycidyl methacrylate. J Appl Polym Sci 93:1938–1945

Dobreva D, Nenkova S, St Vsileva (2006) Morphology and mechanical properties of polypropylene-wood flour composites. BioRes 1(2):209–219

Duran JA, Meyer JA (1972) Exothermic heat released during catalytic polymerization of basswood-methyl methacrylate composites. Wood Sci Technol 6:59–66

Ellwood EL, Gilmore RC, Stamm AJ (1972) Dimensional stabilization of wood with vinyl monomers. Wood Sci 4(3):137–141

Feist WC, Rowell RM, Ellis DW (1991) Moisture sorption and accelerated weathering of acetylated and methacrylated aspen. Wood and Fiber Sci 23(1):128–136

Furuno T, Shimada K, Uehara T, Jodai S (1992a) Combinations of wood and silicate II. Wood-mineral composites using water glass and reactants of barium chloride, boric acid, and borax, and their properties. Mokuzai Gakkaishi 38(5):448–457

Furuno T, Uehara T, Jodai S (1991) Combinations of wood and silicate I. Impregnation by water glass and applications of aluminum sulfate and calcium chloride as reactants. Mokuzai Gakkaishi 37(5):462–472

Furuno T, Uehara T, Jodai S (1992b) The role of wall polymer in the decay durabilities of wood-polymer composites. Makuzai Gakkaishi 38(3):285–293

Galperin AS, Kuleshov GG, Tarashkevich VI, Smtov GM (1995) Manufacturing and properties of modified wood dimensional stability. Wood Sci Technol 18:225–240

Georgieva M, Harata M, Miloshev G (2008) The nuclear actin-related protein Act3p/Arp4 influences yeast cell shape and bulk chromatin organization. J Cell Biochem 104(1):59–67

Guevara R, Moslemi AA (1984) The effect of alkylene oxides, furan resin and vinylpyrrolidinone on wood dimensional stability. Wood Sci Technol 18:225

Halabe UB, Bidigalu GM, Ganga Rao HVS, Ross RJ (1995) Nondestructive evaluation of green wood using stress wave and transverse vibration techniques. Mater Eval 55(9):1013–1018

Hamdan S, Talib ZA, Rahman MR, Ahmed AS, Islam MS (2010) Dynamic Young's modulus measurement of treated and post-treated tropical wood polymer composites (WPC). BioRes 5(1):324–342

Hartman S (1969) Modified wood with aqueous polyurethane systems. Forest Prod J 19(5):39–42

Hoffmann P (1988) On the stabilization of waterlogged oakwood with polyethylene glycol (PEG). Holz 42(5):289–294

Hoffmann P (1990) On the stabiliztion of waterlogged softwoods with polyethylene glycol(PEG). Four species from China and Korea. Holz 44(2):87–93

Inoue M, Ogata S, Nishikawa M, Otsuka Y, Kawai S, Norimoto M (1993) Dimensional stability, mechanical properties, and color changes of a low molecular weight melamine-formaldehyde resin impregnated wood. Mokuzai Gakkaishi 39:181–189

IR Absorptions of Functional Groups (FlashcardExchange, United States of America, 2006), http://www.flashcardexchange.com/cards/ir-absorptions-of-functional-groups-375010. Accessed 20 Nov 2012

Islam MS, Hamdan S, Rahman MR, Jusoh I, Ahmed AS, Idrus M (2011) Dynamic young's modulus, morphological, and thermal stability of 5 tropical light hardwoods modified by benzene diazonium salt treatment. BioRes 6(1):737–750

Islam MS, Hamdan S, Rahman MR, Jusoh I, Ibrahim NF (2010) Dynamic Young's modulus and dimensional stability of Batai tropical wood impregnated with polyvinyl alcohol. J Sci Res 2(2):227–236

Juneja SC, Hodgins JW (1970) The properties of thermo-catalytically prepared wood polymer composites. Forest Prod J 20(12):24–28

Langwig JE, Meyer JA, Davidson RW (1968) Influence of polymer impregnation on mechanical properties of basswood. Forest Prod J 18(7):33–36

Langwig JE, Meyer JA, Davidson RW (1969) New monomers used in making wood-plastics. Forest Prod J 19(11):57–61

Larsson P, Simonson R (1994) A study of strength, hardness and deformation of cetylated Scandinavian softwood. Holz Roh Werkst 52:83–86

Li JZ, Furuno T, Katoh S (2001) Preparation and properties of acetylated and propionylated wood-silicate composites. Holz 55(1):93–96

Loos WE, Robinson GL (1968) Rates of swelling of wood in vinyl monomers. Forest Prod J 18(9):109–112

Luckenbach TA, Rheometrics Inc. (Sealseastem 1994), http://www.sealseastem.com/PDF/DynamicMechThermalAnal.pdf. Accessed 12 April 2004

Manchado M, Arroyo M (2000) Thermal and dynamic mechanical properties of polypropylene and short organic fiber composites. Polym 41(21):7761–7767

Mathias LJ, Wright JR (1989) New wood-polymer composites: Impregnation and in situ polymerization of hydroxymethylacrylates. Am Chem Soc Polym Prepr Div Polym Chem 30:233–234

Meyer JA, Siau JF, Skaar C (1965) Dimensional stabilization of wood. Forest Product J 15(4): 162–166

Meyer JA, Siau JF, Smith WB (1978) Wood-polymer composites from southern hardwoods. Wood Sci 10(3):158–164

Miroy F, Eymard P, Pizzi A (1995) Wood hardening by methoxymethyl melamine. Holz Roh Werkst 53:276

Noah JN, Foudjet A (1988) Wood-polymer composites from some tropical hardwoods. Wood Sci Technol 22:115–119

Ogiso K, Saka S (1993) Wood-inorganic composites prepared by sol-gel process II. Effects of ultrasonic treatments on preparation of wood- inorganic composites. Mokuzai Gakkaishi 39(3): 301–307

Patyakin UU, Vasily I, Sugaipov AR, Birman SM, Bazarov YN, Pilshikov A, Spitsyn Howard DM (2008) Mechanical and chemical modification of wood materials- compressed wood and oxidized charcoal. BioRes 3(3):731–744

Ray SS, Okamoto M (2003) Polymer/layered silicate nano-composites: A review from preparation to processing. Prog Polym Sci 28:1539–1641

Rowell RM (1975) Chemical modification of wood, advantages and disadvantages. Am Wood Preservers Assoc 71:41–51

Rowell RM (2005) Chemical modification of wood for improved adhesion in composites. USDA forest Products Laboratory Madison, Wisconsin, pp 56–60

Rowell RM, Gutzmer DI, Sachs IB, Kinney RE (1976) Effects of alkylene oxide treatments on dimensional stability of wood. Wood Sci 9(1):51–54

Rowell RM, Moisuk R, Meyer JA (1982) Wood-polymer composites: Cell wall grafting with alkylene oxides and lumen treatments with methyl methacrylate. Wood Sci 15:90–96

Saka S, Sasaki M, Tanahashi M (1992) Wood-inorganic composites prepared by sol-gel processing I. Wood-inorganic composites with porous structure. Mokuzai Gakkaishi 38(11): 1043–1049

Saka S, Yakake Y (1993) Wood-inorganic composites prepared by sol-gel process III. Chemically-modified wood-inorganic composites. Mokuzai Gakkaishi 39(3):308–314

Salmen NL (1984) Viscoelastic properties of in situ lignin under water saturated conditions. J Mater Sci 19:3090–3096

Schneider MH (1994) Wood polymer composites. Wood Fiber Sci 26(1):142–151

Siau JF (1969) The swelling of basswood by vinyl monomers. Wood Sci 1(4):250–253

Siau JF, Davidson RW, Meyer JA, Skaar C (1968) A geometrical model for wood-polymer composites. Wood Sci 1(2):116–128

Siau JF, Meyer JA (1966) Comparison of the properties of heat and radiation cured wood-polymer combinations. Forest Prod J 16(8):47–56

Siau JF, Smith WB, Meyer JA (1978) Wood-polymer composites from southern hardwoods. Wood Sci 10(3):158–164

Stamm AJ (1959) Effect of polyethylene glycol on the dimensional stability of wood. Forest Prod J 9:375–381

Stamm AJ (1964a) Factors affecting the bulking and dimensional stabilization of wood with polyethylene glycols. Forest Product J 14:403–408

Stamm AJ (1964b) Wood and Cellulose Science. Ronald Press Co, New York

Su S, Wilkie CA (2003) Exfoliated poly(methyl methacrylate) and polystyrene nanocomposites occur when the clay contains a vinyl monomer. J Polym Sci, Part A: Polym Chem 41: 1124–1135

Sugiama M, Obataya E, Norimoto MC (1996) Temperature dependence of dynamic viscoelasticity for chemically treated woods. Paper presented at proceedings of the third pacific rimbiobased composites symposium, Kyoto, Japan, 2–5 December 1996

Timmons TK, Meyer JA Jr, Cote WA (1971) Polymer location in the wood- polymer composite. Wood Sci 4(1):13–24

Tunc MS, Lawoko M, Heiningen A (2010) Understanding the limitations of removal of hemicelluloses during autohydrolysis of a mixture of southern hardwoods. BioRes 5(1): 356–371

Vetter LD, Stevens M, Acker JV (2009) Fungal decay resistance and durability of organosilicon-treated wood. Int Biodeterior Biodegrad 63:130–134

Yalinkilic MK, Tsunoda K, Takahashi M, Gezer ED, Dwianto W, Nemoto H (1998) Enhancement of biological and physical properties of woof by boric acid-vinyl monomer combination treatment. Holz 52(6):667–672

Yamaguchi H (1994a) Properties of silicic acid compounds as chemical agents for impregnation and fixation of wood. Mokuzai Gakkaishi 40(8):830–837

Yamaguchi H (1994b) Preparation and physical properties of wood fixed with silicic acid compounds. Mokuzai Gakkaishi 40(8):838–845

Yan Wu, Ding GZ, Si QW, Yang Z (2009) Polypropylne composites reinforced with rice straw micro/nano fibrils isolated by high intensity ultrasonication. BioRes 4(4):1487–1497

Yildiz UC, Yildiz S, Gezer ED (2005) Mechanical properties and decay resistance of wood polymer nanocomposites prepared from fast growing species in Turkey. Biores Technol 96:1003–1011

Zollfrank C, Wegener G (2002) FTIR microscopy and ultrastructural investigation of silylated solid wood. Holz 56(1):39–42

Combined Styrene/MMA/Nanoclay Crosslinker Effect on Wood Polymer Nanocomposites (WPNCs)

M.R. Rahman and J.C.H. Lai

Abstract In the current study, to develop the compression strength, thermal stability, and surface morphology of batai wood (Paraserianthes moluccana) which was impregnated by blended of styrene, methyl methacrylate, and nanoclay. Wood polymer nanocomposites (WPNCs) were produced by the co-polymerization reaction which was occurred with cellulose in the wood cell wall by Styrene (ST) and methyl methacrylate (MMA) crosslinker and it was confirmed by Fourier Transform Infrared (FT-IR) Spectroscopy. Thermogravimetric Analysis (TGA) was employed to investigate the thermal stability. The mechanical properties of the WPNCs were expressively improved compared to the raw wood whereas WPNCs demonstrated higher thermal permanence comparative to the raw wood due to the co-polymerization reaction. The surface morphologies of the fracture surface for both the raw wood and WPNCs were recorded using Scanning Electron Microscopy (SEM). WPNCs display smoother surface and adhesion compared to that of raw wood due to the co-polymerization reaction that was seen in the SEM micrographs.

Keywords Crosslinker · Batai wood · Mechanical properties · Thermal properties

1 Introduction

Wood has great value and prominence in the world's economy due to it extraordinary properties whereas it is a regular and renewable resource which is made of cellulose, hemicellulose, and lignin. The high demand for the twenty-first-century materials will be green, environmentally friendly, sustainable, renewable, and biodegradable (Bledzki et al. 2007; Mohanty et al. 2002; Nishino 2004). Due to the excellent material properties and desirable environment characteristics, wood is appropriate to many usages and there is a plentiful availability of wood worldwide

M.R. Rahman (✉) · J.C.H. Lai
Faculty of Engineering, Universiti Malaysia Sarawak,
94300 Kota Samarahan, Sarawak, Malaysia
e-mail: rmrezaur@unimas.my

M.R. Rahman, *Wood Polymer Nanocomposites*, Engineering Materials,
DOI 10.1007/978-3-319-65735-6_3

while raw wood has contributed to great environmental benefits over other building materials because of its biodegradable properties. Plywood, wood plastic composites were produced from wood sawdust. However, the motor fuels and lubricants also can be produced from the conversion of wood sawdust. Nevertheless, wood serves as a great insulator and uses less energy to process compared to steel, concrete, aluminum, or plastic, and it shows the three times lower carbon dioxide emissions compare to the steel and metal. The most significant features of wood are its renewability due to their first growing nature. In Borneo Islands, Batai softwood is abundantly available due to their first growing nature, and it is eco-friendly; however, there are some disadvantages to softwood species. Moreover, softwood species are more vulnerable to attack by fungi and termites due to their hydrophilic nature and this species do not easily withstand drastic changes in temperature. Due to their low densities, softwood species have low dimensional stability and durability. In order to overcome the aforementioned drawbacks of softwood species, wood specimens were impregnated with ST/MMA/clay co-polymer crosslinker to produce wood polymer nanocomposites (WPNCs).

Acetylation and thermal modification are the most common processes to modify the wood species whereas this reaction process only a single-site reaction occurred in which acetic anhydride reacts with the wood hydroxyl groups (Segerholm et al. 2012). According to the Rowell (2006), acetylation and the thermal-modified wood product showed the rigidity at a swollen state and it has good dimensional stability with high resistance to decay by fungi and microorganisms. The following softwoods modified or treatment was done by few researches such as heat treatment, oil treatment, furfurylation, resin impregnation, and treatments involving silicones and silanes (Hill 2006; Hon 1996; Mai and Militz 2004).

In the present work, wood specimens were impregnated with a combination of ST/MMA/Nanoclay to produce wood polymer nanocomposites (WPNCs), and mechanical, thermal, and morphological properties were investigated in this study.

2 Experimental

2.1 Materials

The dimensions of the wood specimens were 6 cm × 2 cm × 2 cm. styrene (ST), methyl methacrylate (MMA), and benzoyl peroxide were used to produce WPNCs which are supplied by Merck, Germany. Nanoclay Nanomer was supplied by Sigma-Aldrich. The constituent of the clay was 35–45 wt% dimethyl dialkyl (C14–C18) amines with a bulk density of 200–500 kg/m^3 and average particle size around 20 µm.

2.2 Preparation of Monomers

ST/MMA/Nanoclay monomer system was prepared in the presence of benzoyl peroxide. Styrene and methyl methacrylate copolymerized in the presence of Benzoyl peroxide which showed in Fig. 1a. Benzoyl peroxide acts as an initiator to influence the reaction between styrene and methyl methacrylate. Two grams of benzoyl peroxide and 5 g of nanoclay were added into each monomer system. Aluminium foil was used to cover the mixtures and put in an autoclave for 15 min to complete the reaction. Table 1 shows the different mixture ratios of styrene, methyl methacrylate, and the nanoclay monomer system.

2.3 Impregnation of Wood Specimens/Co-polymerization Reaction with Cellulose in Wood Cell

ASTM standard was used to prepare the wood specimens and weighed out using an electronic balance. The specimens were immersed into different monomer systems in accordance with the prepared monomer and undertook a vacuum process for 1 h. After 1 h, the samples were taken out and covered with aluminum foil and put in an oven for 48 h at 70 °C to co-polymerize in wood cell wall which shown in Fig. 1b

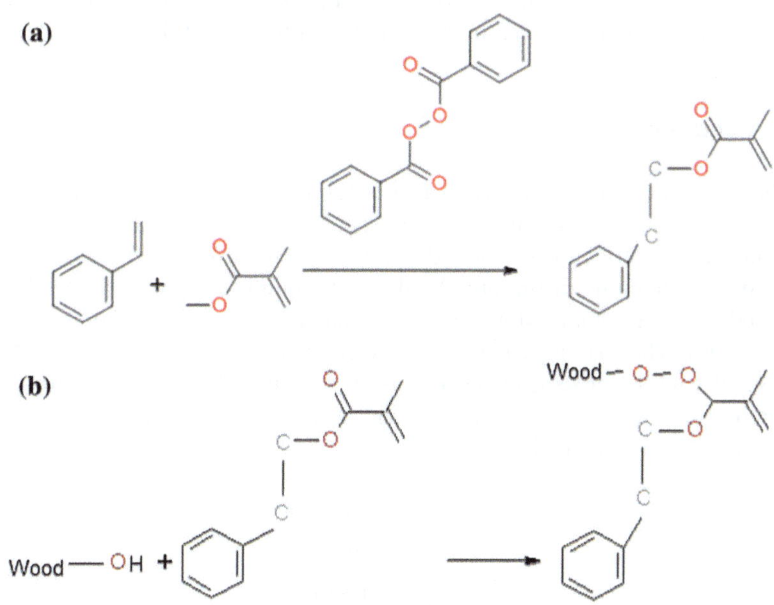

Fig. 1 **a** Reaction scheme for synthesis of ST/MMA/Nanoclay crosslinker, **b** Coupling reaction of wood with ST/MMA/nanoclay

Table 1 Preparation of monomer system at different ratio

Volume of styrene (mL)	Volume of methyl methacrylate (mL)	Amount of nanoclay (g)
20	80	5
30	70	5
40	60	5
50	50	5

and produce wood polymer nanocomposites (WPNCs). After 48 h, the WPNCs were taken out of the oven and the unreacted chemicals were removed. The weight percent gain (WPG) of the samples was then measured using Eq. (1),

$$\mathrm{WPG} = \frac{m_{\mathrm{treated}} - m_{\mathrm{untreated}}}{m_{\mathrm{untreated}}} \times 100\% \tag{1}$$

where m_{treated} and $m_{\mathrm{untreated}}$ are the over-dried weight of the untreated and treated wood samples, respectively.

2.4 Microstructural Characterizations

2.4.1 Fourier Transform Infrared Spectroscopy (FT-IR)

The infrared spectra of the raw wood and WPNCs were recorded on a Shimadzu IRAffinity-1. The transmittance range of the scan was 4000–7000 cm^{-1}.

2.4.2 Compression Test

The dimensions of the compression test specimens were 6 cm × 2 cm × 2 cm. Before the mechanical tests, the prepared specimens were kept at a temperature of 85 °C inside a cool incubator for 4 days. The mechanical properties of the specimens, such as modulus of rupture (MOR) and modulus of elasticity (MOE), were confirmed using a compression machine branded Shimadzu according to ASTM D-143 (1996). The loading capacity of the Shimadzu Universal Testing Machine was 300kN, and the crosshead speed was 10 mm/min. The MOE and MOR were measured using the compression test and were calculated using the following Eqs. (2) and (3), respectively.

$$\mathrm{MOE} = \frac{L^3 m}{4bd^2} \tag{2}$$

$$\mathrm{MOR} = \frac{1.5PL}{bh^2} \tag{3}$$

2.4.3 Thermogravimetric Analysis (TGA)

The finely ground sample was used to run TGA which was started by weighing and exposing it to a heated chamber in the presence of oxygen. The weight loss of the sample was measured by the heating the sample which was suspended on a sensitive balance. The weight change of the specimens was recorded as a function of time at a constant temperature ranging from 20 to 700 °C set in an isothermal TGA. After setting up, the rate of weight loss as a function of time and temperature could be measured with the use of a derivative computer. After the data was obtained, curve levelling and other processes were done to find the precise points of inflection. The weight loss was increased according to the increases of the temperature.

2.4.4 Scanning Electron Microscopy (SEM)

The SEM specimens were first fixed with Karnovsky's fixative. The scanning electron microscope (SEM) (JSM-6710F) was used to examine the interfacial bonding between the cell wall and monomer which supplied by JEOL Company Limited, Japan. The specimens were first fixed with Karnovsky's fixative and then taken through a graded alcohol dehydration series. Once dehydrated, the specimen was coated with a thin layer of gold before viewing on the SEM image. The micrographs were taken at a magnification of 250× and 1500×.

3 Results and Discussion

3.1 Weight Percent Gain (WPG %)

Table 2 shows the values of WPG for both raw wood and WPNCs which were measured before and after modification while the WPG for WPNCs significantly increased due to the crosslinking of wood cell OH with ST/MMA. It was noticeable that the wood's cell wall react with polymer and increasing the adhesion between wood cell wall and polymer which results in a higher WPG. According to the outcomes, 50:50 ST/MMA/clay systems were the highest WPG followed by 40:60, 30:70, and 20:80, respectively

Table 2 Average weight percent gain, WPG (%)

ST/MMA/clay	Average weight for raw (g)	Average weight for WPNCs (g)	Weight percent gain (WPG %)
20:80	9.399	11.301	20.24
30:70	8.149	10.928	34.10
40:60	7.805	11.860	51.95
50:50	7.378	11.858	60.72

3.2 Fourier Transform Infrared Spectroscopy (FT-IR)

The Fig. 2a, b represents the FT-IR spectrum of the raw wood and WPNCs which showed the ST/MMA/clay peak at certain region and proves that ST/MMA/clay successfully impregnated the wood specimens. However, the wood-ST/MMA/clay nanocomposites were produced by the crosslinking reaction with ST/MMA/clay with hydroxyl groups of the cellulose in the wood's cell wall. The absorption bands showed the significant difference in raw wood and WPNCs which was confirmed by the FT-IR spectroscopic analysis. That was due to the differences in structural and chemical composition within the raw wood and modified wood specimens (Islam et al. 2011a, b). The FT-IR spectrum of the raw wood specimens showed the stretching vibration at absorption bands in the region of 3332.99, 1028.06, and 1730.15 cm^{-1} for O–H stretching vibration, C–H stretching vibration, and C=O stretching vibration whereas the FT-IR spectrum WPNCs clearly show that there was no absorption bands due to O–H stretching vibration. It indicates that after chemical modification, the OH groups were dramatically reduced. Moreover, the absorption band at 1109.07 and 1732.08 cm^{-1} due to C–H stretching vibration and C=O stretching vibration, respectively. The FT-IR result proved that the OH groups in the wood specimens were reduced by the blocking of the OH groups for using the ST/MMA/clay combined monomer mixture (Islam et al. 2010, 2012).

3.3 Mechanical Properties Test

The ST/MMA/clay impregnation on the batai wood was evaluated, and the MOE and MOR for raw wood and WPNCs are shown in Fig. 3a, b which shows the effect of ST/MMA/clay on raw wood and WPNCs. Figure 3 shows that when combined polymer loading was higher, the WPNCs showed the better elasticity compare to the raw wood whereas the increments in MOR for WPNCs were expressively greater than that of raw wood. Polymer-impregnated wood yielded higher MOE compared to raw wood because of the co-polymerization of the wood's cell wall, which was in accordance with other research (Adams et al. 1970; Hamdan et al. 2010a, b; Yildiz et al. 2005; Pandey et al. 2010). Wood specimens coupled with monomers and filled the void spaces which led to the increase in stiffness of WPNCs. The increments of MOE and MOR for 50:50 ST/MMA/clay systems were higher followed by 40:60, 30:70, 20:80, and raw wood.

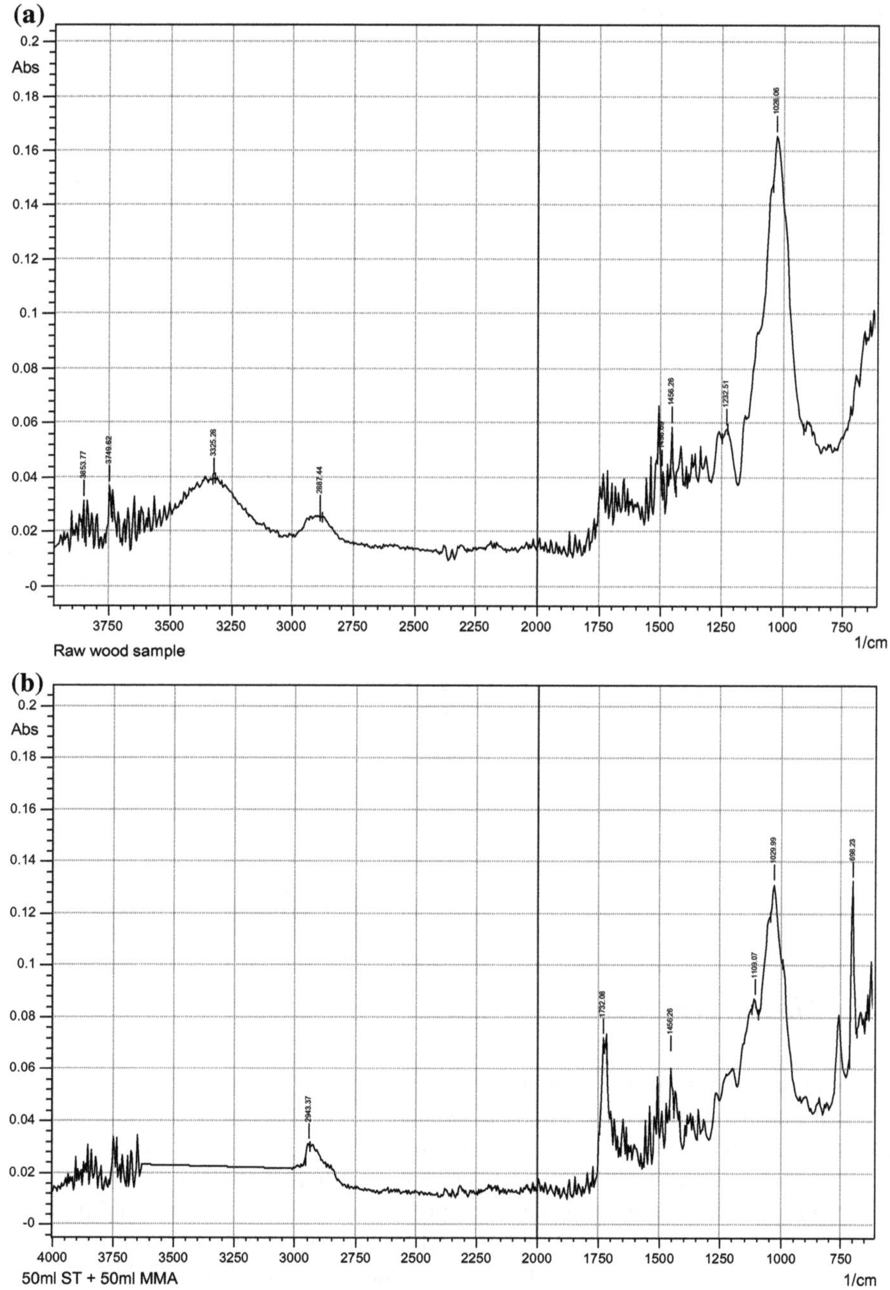

Fig. 2 **a** FT-IR spectra of raw wood specimens, **b** FT-IR spectra of WPNCs

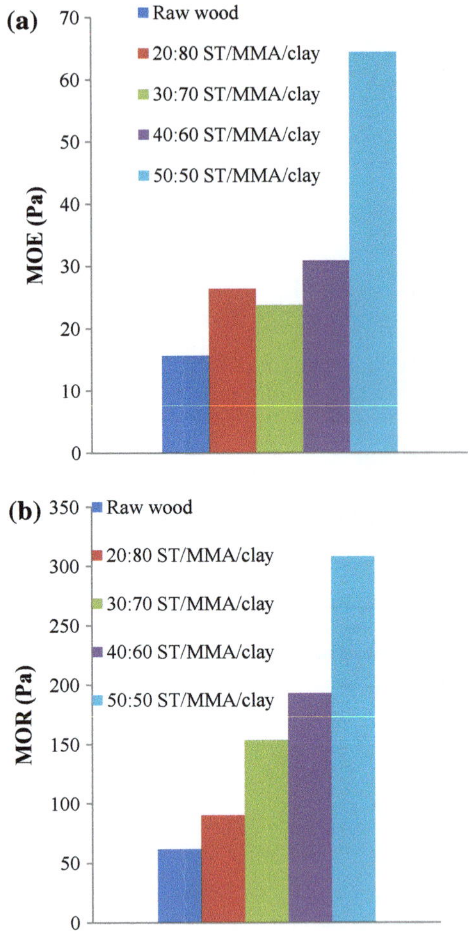

Fig. 3 **a** MOE of raw wood and WPNCs, **b** MOR of raw wood and WPNCs

3.4 Thermogravimetric Analysis (TGA)

Thermogravimetric Analysis is a useful analytical technique to investigate the thermal stability of materials. Thermal stability of wood is a very essential factor in the manufacture of wood polymer nanocomposites. Figure 4 represents the TGA curves of raw wood and WPNCs. According to Wielage et al. (1999), both raw wood and WPNCs showed no degradation up to 160 °C. Thermal stability gradually decreases where the decomposition takes place after 160 °C. From Fig. 4, the weight loss of raw wood started at 248.08 °C and ended at 379.89 °C. For WPNCs, the weight loss range was from 274.45 °C until 431.6 °C. It was clearly showed within Fig. 4 that raw wood had lower starting and ending degradation temperature compared to WPNCs. This was because WPNCs had well monomer impregnation where the residue of WPNCs ended the degradation at a higher temperature.

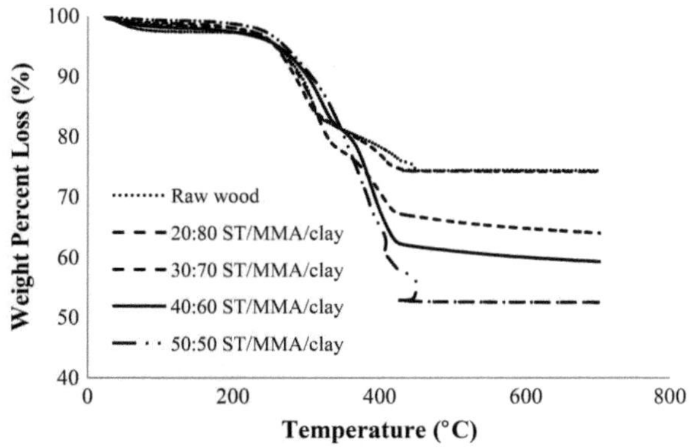

Fig. 4 TGA curves of raw wood and WPNCs

Besides, the thermal stability of WPNCs is higher due to the crosslinking reaction with cellulose OH groups. New chemical bonding between styrene and methyl methacrylate with the aid of benzoyl peroxide and nanoclay was formed that improved the chemical bond strength within the WPNCs which reduced the retardation of the WPNCs (Ramiah 1976; Wielage et al. 1999). Overall, the 50:50 ST/MMA/clay monomer systems showed better thermal stability followed by 40:60, 30:70, 20:80, and raw wood. The TGA result proved that WPNCs have better thermal stability compared to raw wood, as reflected in the mechanical properties.

3.5 *Scanning Electron Microscopy (SEM) Analysis*

The Scanning Electron Microscopy showed the morphology of the interaction of wood cell wall and polymer. The morphology of the fracture surface displays the phase information for WPNCs and raw wood as shown in Fig. 5a, b. From Fig. 5a, the surface morphology of raw wood showed no interfacial bonding between the filler and the matrix. On the other hand, WPNCs indicated enhanced interfacial adhesion between the wood cell wall and the monomer matrix. This was highly due to the ST/MMA/clay crosslinking with wood cell (Rahman et al. 2010a, b, c). Therefore, it can be concluded that the better interfacial bonding between the wood cell wall and the matrix can be observed in WPNCs as reflected in the enhancement of the mechanical properties of the WPNCs.

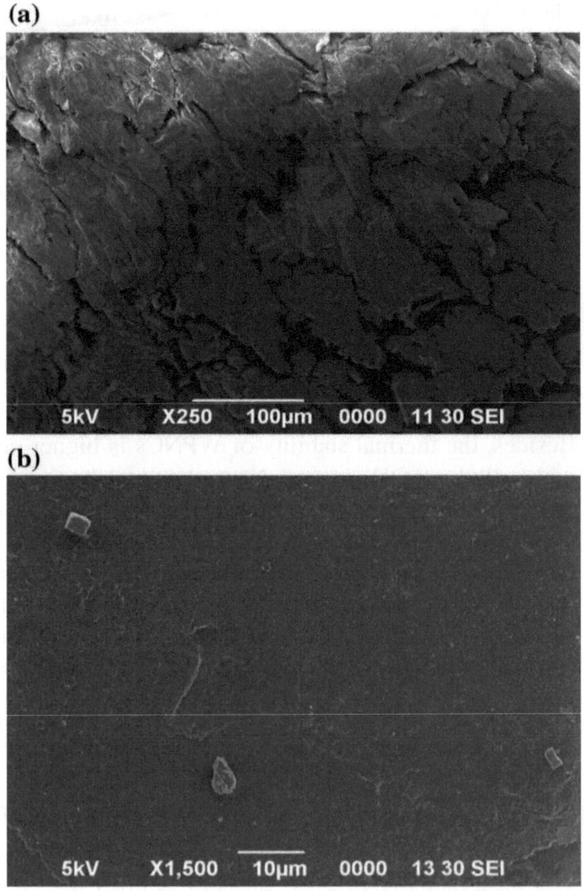

Fig. 5 **a** SEM for raw wood specimens, **b** SEM for WPNCs

4 Conclusion

In this study, ST/MMA/Nanoclay was impregnated into the raw wood to produce WPNCs. The wood specimens impregnated with ST/MMA/clay were fabricated to improve the thermal, mechanical, and morphological properties by reducing the number of hydroxyl groups in the wood cell wall. From the weight percent gain (WPG), 50:50 ST/MMA/clay WPNCs showed the highest value among all the samples fabricated. From the FT-IR analysis, it was confirmed that the coupling reaction occurred in wood with cellulose due to C–H stretching vibration and C=O stretching vibration, respectively. Both MOE and MOR were higher for WPNCs than the raw wood. Thermogravimetric Analysis (TGA) proved that WPNCs had higher decomposition temperature with exothermic peaks increased after impregnation. WPNCs showed smoother surface texture compared to raw wood through SEM analysis.

Acknowledgements The authors would like to acknowledge the financial support from Ministry of Higher Education Malaysia, for their financial support [Grant no. ERGS/02 (08)/860/2912(12)] during the research.

References

Adams DG, Choong ET, Mcllhenny RC (1970) Bending strength of radiation- produced southern pine wood-plastic combinations. Forest Prod J 20(4):25–28

Bledzki AK, Mamun AA, Faruk O (2007) Abaca fibre reinforced PP composites: comparison with jute and flax fibre composites considering fibre contents. eXPRESS Polym Lett 1(11):755–762

Hamdan S, Rahman MR, Ahmed AS, Talib ZA, Islam MS (2010a) Influence of *N,N*-dimethylacetamid on the thermal and mechanical properties of polymer-filled wood. BioRes 5 (4):2611–2624

Hamdan S, Talib ZA, Rahman MR, Saleh AA, Islam MS (2010b) Dynamic Young's modulus measurement of treated and post treated tropical wood polymer composities (WPC). BioRes 5 (1):324–342

Hill CAS (2006) Wood modification—chemical, thermal and other processes. Wiley, United Kingdom

Hon DNS (1996) Chemical modification of lignocellulosic materials. Marcel Dekker, New York, pp 159–179

Islam MS, Hamdan S, Jusoh I, Rahman MR, Ahmed AS (2010) The effect of crosslinker on mechanical and morphological properties of tropical wood material composites. J Mater Des 32 (2011):2221–2227

Islam MS, Hamdan S, Jusoh I, Rahman MR, Ahmed AS (2011a) The effect of alkali pretreatment on mechanical and morphological properties of tropical wood polymer composites. J Mater Des 33(2012):419–424

Islam MS, Hamdan S, Rahman MR, Jusoh I, Ahmed AS, Idrus M (2011b) Dynamic Young's modulus, morphological, and thermal stability of 5 tropical light hardwoods and modified by benzene diazonium salt treatment. BioRes 6(1):737–750

Islam MS, Hamdan S, Rusop M, Rahman MR, Ahmed AS, Idrus MAMM (2012) Dimensional stability and water repellent efficiency measurement of chemically modified tropical light hardwood. BioRes 7(1):1221–1231

Mai C, Militz H (2004) Modification of wood with silicon compounds. Inorganic silicon compounds and sol-gel systems: a review. Wood Sci Technol 37:339–348

Mohanty AK, Misra M, Drzal LT (2002) Sustainable bio-composite from renewable resources: opportunity and challenges in the green materials world. Polym Environ 10:19–26

Nishino T (ed) (2004) Natural fibre sources, green composites: polymer composites and the environment. Woodhead Publishing Ltd, Cambridge, England, p 49

Pandey KK, Hughes M, Vuorinen T (2010) Dimensional stability, UV resistance, and static mechanical properties of scots pine chemically modified with alkylene epoxides. BioRes 5 (2):598–615

Rahman MR, Hamdan S, Ahmed AS, Islam MS (2010a) Mechanical and biological performance of sodium metaperiodate-impregnated plasticized wood (PW). BioRes 5(2):1022–1035

Rahman MR, Hamdan S, Saleh AA, Islam MS (2010b) Mechanical and biological performance of sodium mataperiodate impregnated plasticized wood (PW). Biores 5:1022–1035

Rahman MR, Islam MN, Huque MM, Hamdan S, Ahmed AS (2010c) Effect of chemical treatment on rice husk (RH) reinforced polyethylene (PE) composites. BioRes 5(2):854–869

Ramiah MV (1976) Thermogravimetric and differential thermal analysis of cellulose, hemicellulose and lignin. J Appl Polym Sci 14:1323–1337

Rowell RM (2006) Chemical modification of wood: a short review. Wood Mater Sci Eng 1:29–33

Segerholm BK, Ibach RE, Westin M (2012) Moisture sorption, biological durability, and mechanical performance of WPC containing modified wood and polyactates. BioRes 7 (4):4575–4585

Wielage B, Lampke T, Mark G, Nestler K, Starke D (1999) Thermogravimetric and differential scanning calorimetric analysis of natural fibers and poplypropylene. Thermochim Acta 337:169–177

Yildiz CU, Yildiz S, Grezer DE (2005) Mechanical properties and decay resistance of wood polymer composites prepared from fast growing species in Turkey. Biores Technol 96:1003–1011

Oxidation of Wood Species by Sodium Metaperiodate and Impregnation with Phenyl Hydrazine

M.R. Rahman

Abstract A large variety of tropical wood species has been found in Malaysia, especially the Borneo state of Sarawak. The nominated fresh wood species, namely *Artocarpus Elasticus*, *Artocarpus Rigidus*, *Xylopia* spp., *Koompassia Malaccensis,* and *Eugenia* spp. were chemically modified with sodium metaperiodate for manufacturing them into plasticized wood (PW). Prepared plasticized wood models were analysed using Fourier Transform Infrared Spectroscopy, Scanning Electron Microscopy and mechanical testing [modulus of elasticity (MOE), modulus of rupture (MOR), static Young's modulus (E_s)], decay resistance and water absorption. MOE and MOR were calculated using the compression parallel to grain test and the natural laboratory decay test, respectively. PW showed the higher MOE, MOR, E_s and lower water content compared to raw one whereas modified wood showed higher resistance to decay exposure. *Eugenia* spp. had highest resistance compared to the others.

Keywords Mechanical properties · Decay resistance · Chemical treatment · Plasticized wood

1 Introduction

Sodium metaperiodate were used on the selected wood species, namely *Artocarpus Elasticus, Artocarpus Rigidus, Xylopia* spp., *Koompassia Malaccensis, and Eugenia* spp. for manufacturing the wood polymer nanocomposite (WPNC). Further, the phenyl hydrazine was introduced in the WPC to produce them into subordinate wood polymer nanocomposites also called post-treated WPNC (PTWPNC). The mechanical properties of the final products are influencing by the chemical treatment and post-treatment. Dynamic mechanical thermal analysis

M.R. Rahman (✉)
Faculty of Engineering, Universiti Malaysia Sarawak,
94300 Kota Samarahan, Sarawak, Malaysia
e-mail: rmrezaur@unimas.my

M.R. Rahman, *Wood Polymer Nanocomposites*, Engineering Materials,
DOI 10.1007/978-3-319-65735-6_4

(DMTA) and free-free flexural vibrational testing were carried out to measure the storage modulus (E') and the dynamic Young's modulus (E_d), respectively. The mechanical test three point bending used to calculate the MOE and MOR. Compression parallel to grain test and the natural laboratory decay test were conducted to measure the static Young's modulus (E_s) and decay resistance, respectively. According to the literature, different wood species showed the different elastic properties, i.e., stiffness (E_d), storage modulus (E'), MOE, E_s of the nanocomposite. Raw wood showed lower E' compare to the WPNC and PTWPNC, whereas the glass transition temperature (T_g) of WPNC and PTWPNC is much lower than that of raw wood. WPNC and PTWPNC showed higher MOE, MOR, and E_s compare to those of raw wood while lower water content and high resistance to decay exposure are appeared with *Eugenia* spp. which having the highest resistance compared to the others species.

From the free-free vibration and other testing, authors would like to quickly draw conclusion about the value of the nanocomposite material based on their stiffness (E_d) of the PTWPNC compared to the respective WPNC and raw woods. Fourier Transform Infrared (FT-IR) Spectroscopy and Scanning Electron Microscopy (SEM) were used to characterize the WPNC and PTWPNC. The carbonyl stretching at 1635 cm^{-1} could be observed by FT-IR spectroscopic analysis. Due to the oxidation reaction, the increased absorption band near 1718 and 1604 cm^{-1} are appeared for carbonyl aldehyde of dialdehyde cellulose and 2,3-diphenylhydrozo cellulose which represent the WPNC and PTWPNC, respectively.

The physical, mechanical, thermal, and biological properties were enhanced to using these chemicals beside the moderately minor disadvantages are found to utilize these types of chemicals, such as changes relative to the natural color. However, the wood product and other wood nanocomposites, due to wood components being enveloped in the entire the wood cell wall WPNC and PTWPNC are subjected to fungal and termite attacks. Previous studies have shown that most of the fungi penetrate the material through a check related with scratches in the wood whereas 43.6% of wood crossties were removed from track due to decay (Russel 1986). Many studies have been carried out on the decay resistance of wood and WPNCs and more efforts have been made to develop the decay resistance of WPNC with zinc borate and other chemicals (Verhey et al. 2001; Khavkine et al. 2000). The hydrophilic nature is the major disadvantage of using these species to utilize in the interior and exterior uses because of their consequent deformation of the product. In order to overcome this problem, all species were chemically modified with sodium metaperiodate and phenyl hydrazine. The objectives of this work were to compare the rate of decay of untreated wood, WPNC and PTWPNC crossties against the brown and white-rot decay fungi and to obtain improved mechanical and thermal properties.

2 Experimental

2.1 Materials

Various tropical wood species were chosen for this study, namely *Artocarpus Elasticus*, *Artocarpus Rigidus*, *Koompassia Malaccensis,* and *Eugenia* spp. Among the wood species *Xylopia spp.* was hardwood and remains species was the softwood. Chemicals used to treat the wood were H_2SO_4, $NaIO_4$, and phenyl hydrazine (Merck, Germany).

2.2 Specimen Preparation

The three bolts of 1.2 m long were cut into each tree. Each bolt was quartersawn to produce planks of 4 cm thickness and subsequently conditioned to air-dry in a room with relative humidity of 60% and ambient temperature of around 25 °C for one month prior to testing. The planks were ripped and machined to 340 mm (L) × 20 mm (T) × 10 mm (R) for free-free vibration test, 300 mm (L) × 20 mm (T) × 20 mm (R) for three point bending test, 100 mm (L) × 25 mm (T) × 25 mm (R) for compression parallel to grain test 9 mm (L) × 25 mm (T) × 25 mm (R) for decay resistant test and 10 mm (L) × 2 mm (R) × 7.5 mm (T) specimens for dynamic mechanical thermal analysis. Sodium periodate used as an oxidizing agent to oxidize the raw wood using an autoclave in order to convert them into wood polymer nanocomposites. Temperature and pressure used were 120 °C and 85 kPa, respectively. Manufactured WPNC was impregnated with phenyl hydrazine to produce PTWPNC using a vacuum chamber at 25 °C and 60 cm Hg.

2.3 Microstructural Characterizations

2.3.1 Fourier Transform Infrared Spectroscopy (FT-IR)

The aim of the analysis was to determine the functional group of the specimens. Shimadzu Fourier Transform Infrared Spectroscopy (FT-IR) 81001 Spectrophotometer was used to record the infrared spectra of the raw woods and WPNC. The transmittance range of the scan was 4000–370 cm^{-1}.

2.3.2 Scanning Electron Microscopy (SEM)

Scanning Electron Microscopy used to determine the morphology of the specimens. A graded alcohol was used for dehydration of sample series and then fixed with

Karnovsky's fixative. Once dehydrated, the specimen was coated with thin layer of gold before viewed on the SEM. Various magnification of the micrographs are taken and presented in the results and discussions section.

2.3.3 Dynamic Mechanical Thermal Analysis (DMTA)

The storage modulus (log E') and loss tangent (tan δ) of the raw woods, WPNC and PTWPNC were characterized by dynamical mechanical thermal analysis (DMTA) to evaluate the effect of temperature on raw woods, WPNC, and PTWPNC. DMTA is a mechanical test which interact woods molecules with mechanical stress while all molecular relaxation process are detected using this technique. A Perkin Elmer dynamic mechanical thermal analyzer (PE-DMTA) was used at 10 Hz frequency, ×4 strains and 2 °C min^{-1} temperature rise. The rectangular specimens with moisture content around 15% were tested using dual-cantilever bending mode on a standard bending head. Liquid nitrogen used in the chamber to thermal scan at 2 °C min^{-1} with various frequencies ranging from −100 to 200 °C. The specimen surroundings humidity was 65% RH. The graph of tan δ versus T was showed the determination of the T_g.

2.3.4 Free-Free Flexural Vibration Testing

The free-free flexural vibration testing system was developed by Hearmon in 1965. He used standard specimen which was supported by threads at its nodal point of specific mode. Electromagnet transducer and electromagnet drive were used in this study to face the iron plate whereas the electromagnet transducer was connected to a recorder and an oscilloscope. The iron plate used at one end of specimen which attracting the electromagnet driver and served with alternating current that excites the vibration. The vibration produces sound that travels through the sample and detected by a microphone. Sound level meter was used to get the response which is connected to the microphone and increases the amplitude of sound and transforms the analog signal into digital signal. Then the signal is transferred to the PICO oscilloscope and personal computer. The frequency is varied until a maximum response is detected while the maximum level is called resonant frequency. The dynamic Young's modulus (E_d) was calculated from the resonant frequency by using the Eqs. (1) and (2) (Fig. 1):

$$E' = \frac{4\pi^2 f^2 l^4 A\rho}{I(m_n)^4} \quad (1)$$

$$I = \frac{bd^3}{12} \quad (2)$$

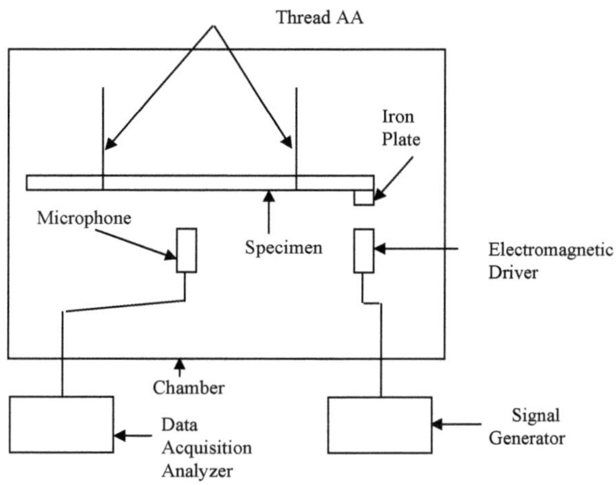

Fig. 1 Schematic diagram of free-free vibrational testing

where

d	Beam depth,
b	Beam width,
1	Beam length,
f	Natural frequency of the specimen,
ρ	Density,
A	Cross-sectional area and
$n = 1$	Is the first mode of vibration where $m_1 = 4.730$.

2.3.5 Determination of MOE and MOR Using Three Point Bending Test

MOE and MOR were carried out according to ASTM D-143 (2006) and the length, width, and thickness of the specimens were measured. A loading capacity of 300 kN was used for the three point bending test which having the crosshead speed of 2 mm/min. The machine was supplied by Shimadzu Corporation, Japan and name of the machine is "A Shimadzu Universal Testing Machine." The MOE and MOR were calculated using Eqs. (3) and (4), respectively.

$$\text{MOE} = \frac{L^3 m}{4bd^3} \tag{3}$$

$$\text{MOR} = \frac{1.5PL}{bh^2} \tag{4}$$

where

L Span length of sample, 180 mm,

b Width of sample, 20 mm,
d Thickness of sample, 20 mm,
m Slope of the tangent to the initial line of the force displacement curve,
P The maximum breaking load, and
h Depth of the beam.

2.3.6 Determination of Static Young's Modulus (E_s) Using Compression Parallel to Grain Test

According to ASTM D143-52, standard specimens were used to conduct the compression parallel to grain test. The uniaxial compression test was performed to determine the static Young's modulus of E_s. A Shimadzu Universal Testing Machine having a loading capacity of 300 kN was used for the test with the crosshead speed of 2 mm/min and static Young's modulus, E_s was calculated using Eq. (5).

$$E = \frac{\text{Stress}}{\text{Strain}}$$

$$E = \frac{F/A}{\Delta L/L}$$

where

F Force,
A Cross-sectional area,
ΔL Displacement, and
L Length of sample.

$$F = \frac{EA}{L} \times \Delta L$$

$$m = \frac{EA}{L}$$

$$\therefore\ E = \frac{EA}{L} \tag{5}$$

2.3.7 Laboratory Fungal Decay Resistance Test

According to the ASTM D2017 (1967), the standard method of accelerated laboratory test of natural decay resistance of wood was carried out by using different fungi. Based on the ASTM Standard D2017, decay resistance is classified using the scale described such as highly resistant heartwood experiences (0–10)% weight loss, resistant wood (11–24)% weight loss, moderately resistant wood (25–44%) weight loss, and non-resistant wood experiences weight loss greater than 45%.

Dark container maintained at 20 ± 1 °C and a relative humidity of 65 ± 4% were used for the air dried specimens and after conditioning to constant weight then they were weighed accurately in the laboratory. There are two types of fungi, white-rot (*polyporous versicolor* L.ex. Fr.) with ATCC No. 12679 and brown-rot (*postia placenta* (Fr.) Cke. with ATCC No. 11538 were used to evaluate the efficiency of sodium metaperiodate against the decay while the reference blocks were made of sweet gum. There were eight replications for each specimen used for the test and it is terminated after 14 weeks when the reference blocks obtained a weight loss of 60%. After 14 weeks, the mycelium was brushed off and test specimens were air dried and again conditioned to constant weight and it recorded for each specimen. The rate of decay was calculated the difference in specimens weights before and after the decay.

2.3.8 Water Uptake

Rectangular specimens were prepared the followed dimensions of 39 mm (L) × 10 mm (T) ×4 mm (R) for the measure the water absorption characteristics of the raw wood and WPNC. A Denver Instron balance was used for weight measurement of oven dried specimens at 105 °C and cooled in a desiccator which containing silica gel. The dried and weighted specimens were immersed in distilled water according to ASTM D570-99 (2002). Both hot and cold water were used for immersion and hot water immersion test was conducted only for 2 h, while the duration of the cold water immersion was 24 h. The final weight of the specimens was taken after the excess water on the surface of the specimens was removed using a soft cloth. The increase in the weight of the specimens was calculated using the following Eq. (6).

$$\text{Water Uptake } (\%) = \frac{\text{Final weight} - \text{Original weight}}{\text{Original weight}} \times 100 \qquad (6)$$

3 Results and Discussion

3.1 Fourier Transform Infrared Spectroscopy (FT-IR)

The FT-IR spectrum for the raw wood, WPNC, and PTWPNC were showed in Fig. 2a–c, respectively. However, absorption band of carbonyl stretching at 1635 cm^{-1} was observed for raw wood which shown in Fig. 2a. The increased absorption band near 1718 cm^{-1} was shown due to the WPNC and PTWPNC

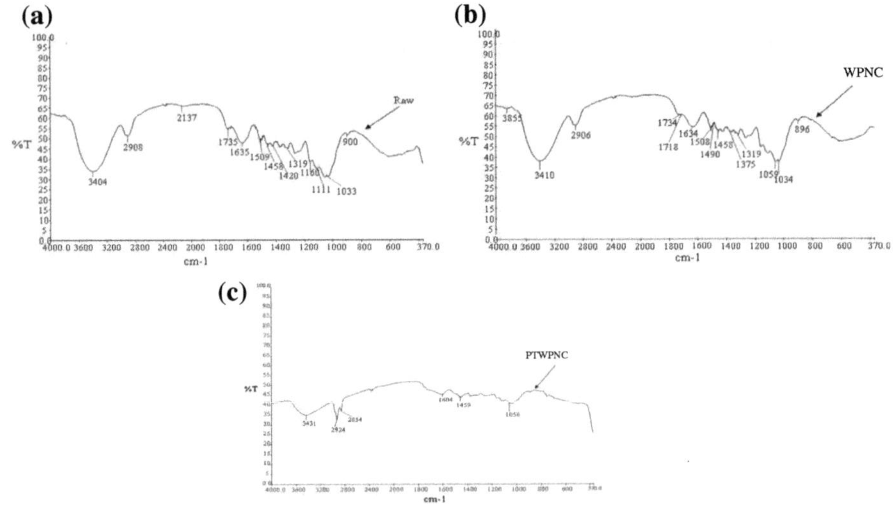

Fig. 2 **a** FT-IR spectrum for the raw, **b** FT-IR spectrum for WPNC, **c** FT-IR spectrum for PTWPNC

whereas the absorption band 1604 cm^{-1} represent the carbonyl aldehyde of dialdehyde cellulose which is formed by the oxidation of wood fiber as shown in Fig. 3. After post-treatment (PTWPNC) the changes of the microstructure are less significant as also being shown by the less drastic increase of *E*′ in PTWPNC.

3.2 *Storage Modulus (log* E′*) and Loss Tangent (tan δ) of Raw Wood, WPNC, and PTWPNC*

DMTA was applied to study the viscoelastic behavior of the wood and their nanocomposites. The viscoelastic properties of the raw wood, WPNC, and PTWPNC were shown in the Fig. 4a–e whereas the *E*′ of the WPNC and PTWPNC has been improved considerably over that of the raw wood. The increase in *E*′ values in the total temperature range is much less as compared with WPNC but the initial enhancement of *E*′ was observed for the PTWPNC as compared to the raw wood for *Artocarpus Elasticus* and *Artocarpus rigidus*. *E*′ values *Xylopia* spp. (hardwood) were higher for the PTWPNC compared with WPNC which was due to the poor interphase interaction of the hydrophilic hydroxyl group in cellulose in *Xylopia* spp. and the cured hydrophobic polymer matrix lead to poor mechanical properties of the resulting WPNC showed in Fig. 4e. The *E*′ was increased three times from the raw wood upon post-treatment with phenyl hydrazine that is almost comparable and similar to the *Eugennia* spp. which indicating that this species is

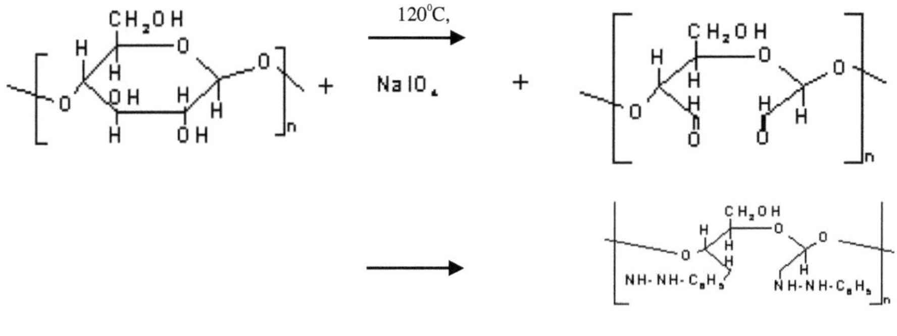

2,3-di-phenylhydrazo cellulose

Fig. 3 Oxidation of raw wood fiber specimens with sodium metaperiodate and phenyl hydrazine

considered similar to hard wood (Fig. 4d). The *E*′ values increase with the PTWPNC up to two times and eventually decrease to the same value with WPNC at 200 °C whereas the higher *E*′ values for PTWPNC could reflect the crosslinking or chain scission (increase in crystallinity) caused by the treatment as seen in the FT-IR results.

The higher tan δ value of raw wood was observed in Fig. 5a–e which is compared to the WPNC and PTWPNC (as seen in Fig. 4a, e) while the loss tangent is related to the loss of energy i.e., to viscous behavior (irreversible deformation). Due to the crosslinking reaction between wood cell and polymer, the result from the treatment showed the lower tan δ value in WPNC and PTWPNC which can probably be attributed to higher polymer content and implies higher elastic recovering.

Assuming that the T_g is the temperature where tan δ is maximum, the raw wood showed a transition in the range −50 °C. This peak was attributed to the relaxation which has properties of the glass-rubber transition. A secondary relaxation occurs 60–70 and 80–110 °C which can be described as vibration and reorientation motion within the crystal (Bikiaris et al. 1999). The relaxation occurs at broader ranges for the WPNC which had similar relaxation temperatures but it can be attributed to crosslinking and branching caused by the treatment. The post-treatment increases the polymer chain mobility to increase the secondary relaxation caused a great shift in the relaxation to a higher temperature for the post-treatment to PTWPNC.). In general, the post-treatment caused a shift of the tan δ to a higher temperature which indicating that there was an increase in the relaxation temperature but interpreted as relaxation of the constrained molecules with reduced mobility close to crystallites (Bikiaris et al. 1999). During the dynamic analysis the buckle cell walls and the amorphous polymers exert a certain restorative force in rubbery state.

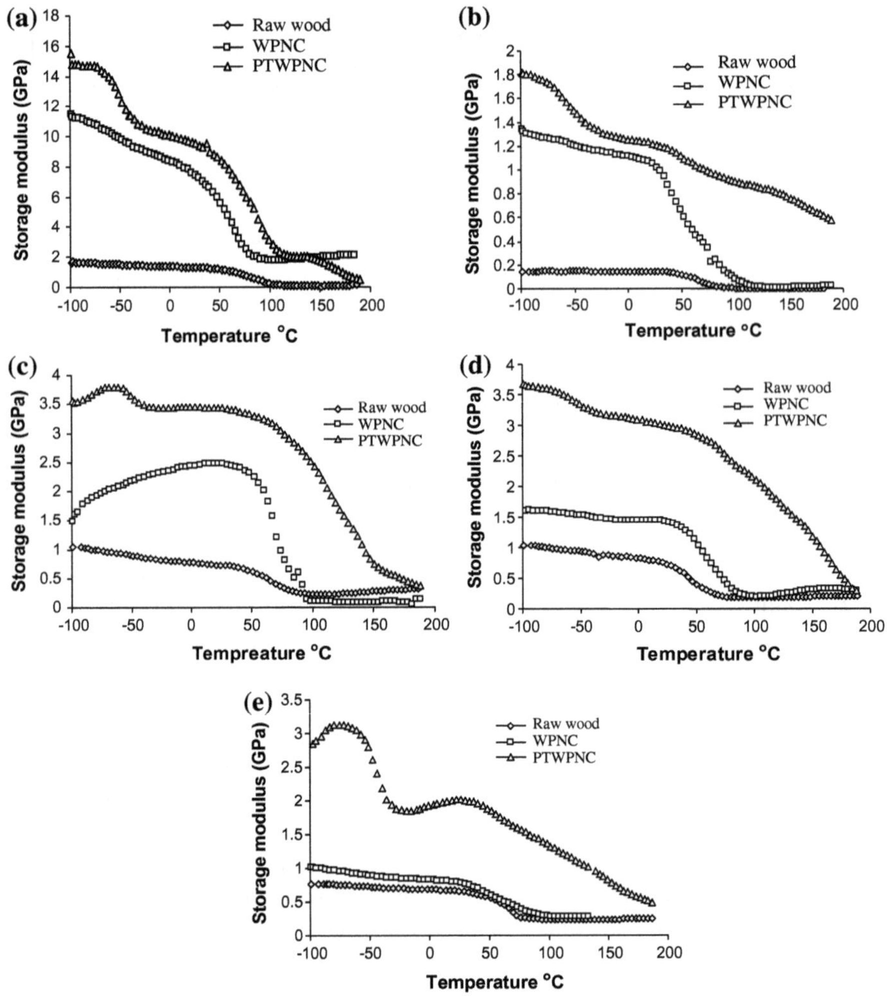

Fig. 4 **a** *E*′ versus temperature for *Artocarpus Elasticus* raw wood, WPNC and PTWPNC, **b** *E*′ versus temperature for *Artocarpus rigidus* raw wood, WPNC and PTWPNC, **c** *E*′ versus temperature for *Koompasia Malacennis* raw wood, WPNC and PTWPNC, **d** *E*′ versus temperature for *Eugennia* spp. raw wood, WPNC and PTWPNC, **e** *E*′ versus temperature for *Xylopia* spp. raw wood, WPNC and PTWPNC

3.3 Dynamic Young's Modulus of Raw Wood, WPNC, and PTWPNC

The free-free flexural vibration test was conducted to measure the stiffness of the raw wood, WPNC, and PTWPNC which are shown in Fig. 6a–e. The Young's modulus of all species moderately increases by sodium metaperiodate chemical

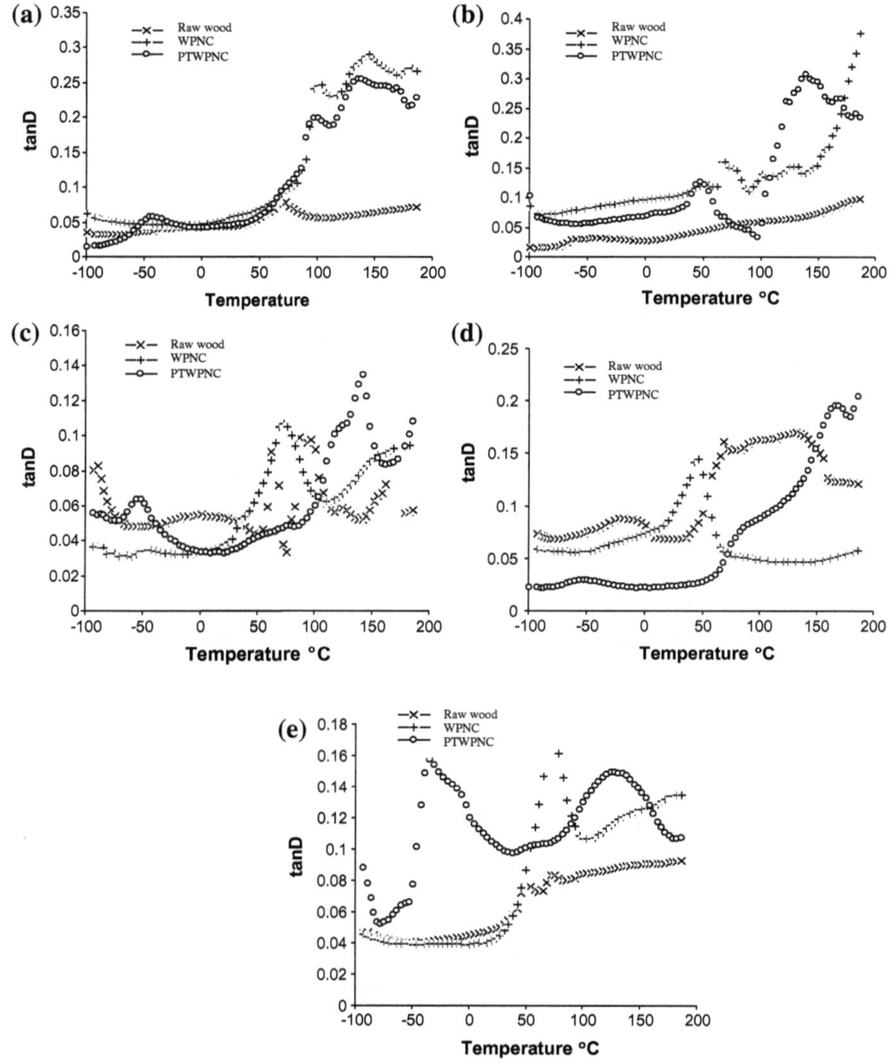

Fig. 5 **a** Tan δ versus temperature for *Artocarpus Elasticus* raw wood, WPNC and PTWPNC, **b** Tan δ versus temperature for *Artocarpus rigidus* raw wood, WPNC and PTWPNC, **c** Tan δ versus temperature for *Koompasia Malacennis* raw wood, WPNC and PTWPNC, **d** Tan δ versus temperature for *Eugennia* spp. raw wood, WPNC and PTWPNC, **e** Tan δ versus temperature for *Xylopia* spp. raw wood, WPNC and PTWPNC

modification whereas the storage modulus is significantly affected by the chemical modification (see Fig. 6a–e) but the elastic properties showed significant changes only in *Artocarpus Elasticus*. The Young's modulus for all species was enhanced upon further phenyl hydrazine treatment.

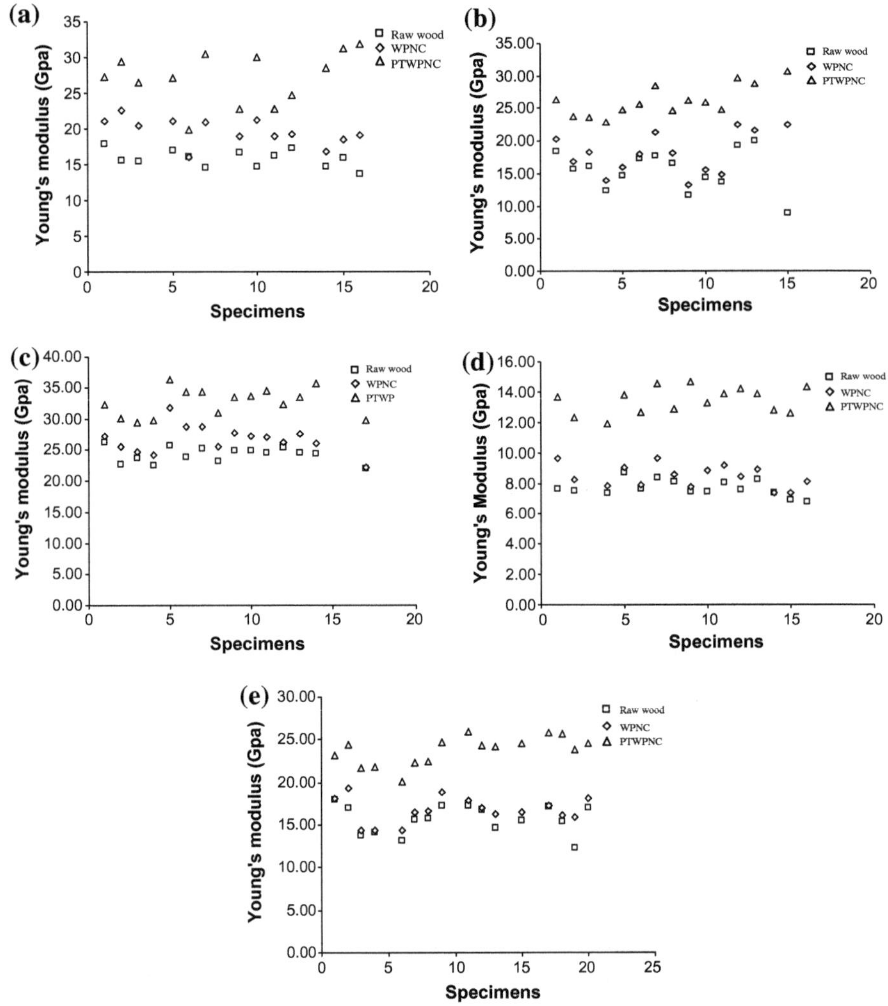

Fig. 6 **a** E_d of *Artocarpus Elasticus* raw wood, WPNC and PTWPNC, **b** E_d of *Artocarpus rigidus* raw wood, WPNC and PTWPNC, **c** E_d of *Koompasia Malacennis* raw wood, WPNC and PTWPNC, **d** E_d of *Eugennia* spp. raw wood, WPNC and PTWPNC, **e** E_d of *Xylopia* spp. raw wood, WPNC and PTWPNC

3.4 MOE and MOR Measurement

The difference of the MOE and MOR for raw wood, WPNC, and PTWPNC is shown in Tables 1 and 2 as well as Figs. 7 and 8 which is, namely *Artocarpus Elasticus, Artocarpus Rigidus, Xylopia* spp., *Koompassia Malaccensis and Eugenia* spp. respectively. The MOE and MOR of the raw wood, WPNC, and PTWPNC were investigated which was treated by the sodium metaperiodate and phenyl

Table 1 *t*-test analysis of raw wood and wood polymer nanocomposites[a]

Treatment	Modulus of elasticity	*t*-test grouping[b]
Raw wood (*Artocarpus Elasticus*)	6.48 ± 0.52	A
WPNC (*Artocarpus Elasticus*)	12.44 ± 0.99	B
Raw wood (*Artocarpus Rigidus*)	5.13 ± 0.38	C
WPNC (*Artocarpus Rigidus*)	9.56 ± 1.97	D
Raw wood (*Xylopia* spp.)	6.71 ± 0.34	E
WPNC (*Xylopia* spp.)	9.23 ± 1.81	E
Raw wood (*Koompassia Malaccensis*)	15.68 ± 1.11	F
WPNC (*Koompassia Malaccensis*)	16.75 ± 0.96	F
Raw wood (*Eugenia* spp.)	9.25 ± 0.37	G
WPNC (*Eugenia* spp.)	13.16 ± 0.41	H

[a]Each value is the average of 10 specimens
[b]The same letters are not significantly different at $\alpha = 5\%$

Table 2 *t*-test analysis of raw wood and wood polymer nanocomposites[a]

Treatment	Modulus of elasticity	*t*-test grouping[b]
Raw wood (*Artocarpus Elasticus*)	76.17 ± 3.01	A
WPNC (*Artocarpus Elasticus*)	83.92 ± 4.20	B
Raw wood (*Artocarpus Rigidus*)	33.02 ± 3.63	C
WPNC (*Artocarpus Rigidus*)	73.05 ± 9.26	D
Raw wood (*Xylopia* spp.)	45.91 ± 1.36	E
WPNC (*Xylopia* spp.)	80.35 ± 9.01	E
Raw wood (*Koompassia Malaccensis*)	122.20 ± 15.49	F
WPNC (*Koompassia Malaccensis*)	125.98 ± 6.77	F
Raw wood (*Eugenia* spp.)	46.10 ± 3.62	G
WPNC (*Eugenia* spp.)	72.8 ± 4.60	H

[a]Each value is the average of 10 specimens
[b]The same letters are not significantly different at $\alpha = 5\%$

hydrazine. *Eugenia* spp. showed highest MOE followed by *Artocarpus Rigidus, Artocarpus Elasticus, Xylopia* spp. and *Koompassia Malaccensis*, due to the crosslinking reaction between wood cell wall and the chemicals while *Koompassia Malaccensis* showed lowest increment because of its high density. Higher MOE was observed for WPNC and PTWPNC compared to the untreated wood because of the chemical modification, which is in accordance with other researchers (Adams et al. 1970; Ates et al. 2009; Mankowshi and Morrell 2000; Yildiz et al. 2005). *Eugenia* spp. showed significantly different MOE compare to raw wood which shown in the Table 1. This is due to the sodium metaperiodate reacted with the cellulose in wood cells, which converted single bonds into double bonds whereas it was stronger than single bonds and change the structure of wood cells. However, on

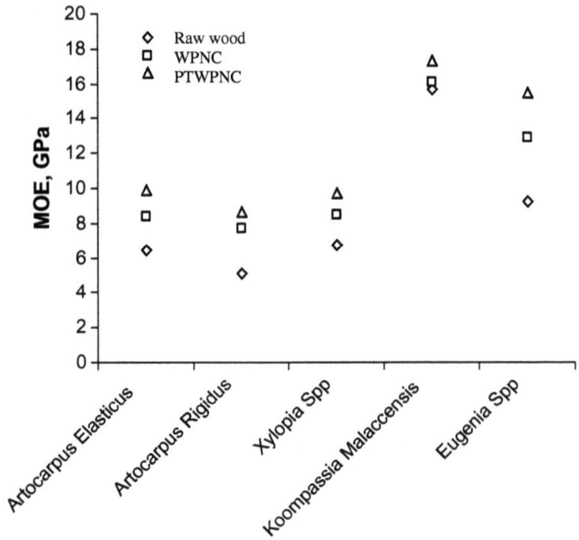

Fig. 7 MOE of untreated wood, WPNC and PTWPNC

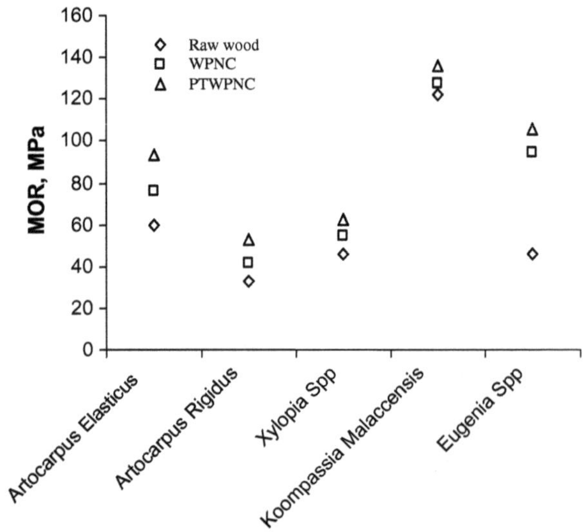

Fig. 8 MOR of untreated wood, WPNC and PTWPNC

Koompassia Malaccensis, there was no significant effect of chemical treatment because of its hardness.

According to Adams et al. (1970), the chemical treatment enhances MOR which shown in Table 2. The MOR of the WPNC and PTWPNC was significantly different from untreated wood whereas *Eugenia* spp. wood polymer nanocomposites were higher than MOR followed by *Artocarpus Rigidus, Artocarpus Elasticus, Xylopia* spp. and *Koompassia Malaccensis*, respectively. On the other hand, the chemical treatment does not reflect the WPNC and PTWPNC of *Koompassia Malaccensis,* however, MOR of WPNC and PTWPNC was slightly higher than those of untreated one.

3.5 Static Young's Modulus (E) Measurement

Static Young's modulus (*E*) measurement was measured by the compression parallel to grain test. Figure 9 and Table 3 show that summarized values of the compressive strength of 10 repetitions but there was significant difference between untreated wood, WPNC, and PTWPNC due to the impact of chemical modification. The highest E value was observed in Artocarpus Elasticus, followed by *Artocarpus Rigidus*, *Eugenia* spp., *Xylopia* spp. and *Koompassia Malaccensis* whereas increment of *E* was very minimal for and Koompassia Malaccensis because the chemical modification does not work on this species. The increase of *E* in WPNC and PTWPNC compared to untreated wood was also reported by different researchers (Ates et al. 2009; Autio and Miettinen 1970; Singha and Thakur 2009). The chemical modification of untreated wood puts a coating on the walls which thickens them, thus greatly increasing their lateral stability.

3.6 Fungal Decay Resistance Test

Decay resistance test was carried out to determine the sustainability of the WPNC and PTWPNC. Figures 10 and 11 are representing the weight loss due to fungal

Fig. 9 Static Young's modulus of untreated wood, WPNC and PTWPNC

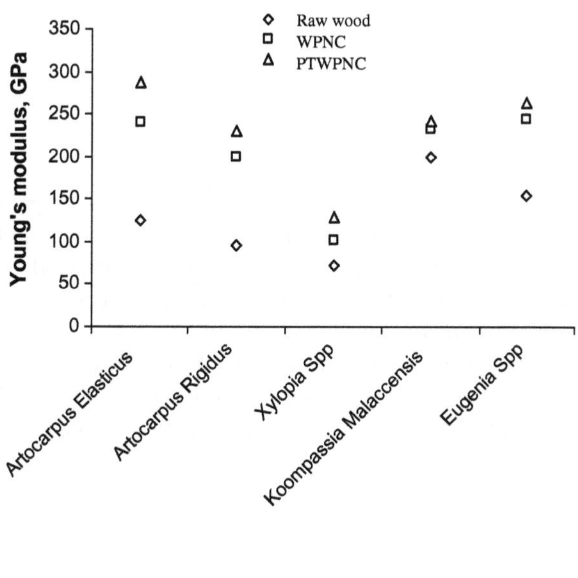

Table 3 *t*-test analysis of raw wood and wood polymer nanocomposites[a]

Treatment	Static Young's modulus	*t*-test grouping[b]
Raw wood (*Artocarpus Elasticus*)	1.36 ± 0.23	A
WPNC (*Artocarpus Elasticus*)	2.43 ± 0.51	B
Raw wood (*Artocarpus Rigidus*)	1.88 ± 0.58	C
WPNC (*Artocarpus Rigidus*)	3.14 ± 1.11	D
Raw wood (*Xylopia* spp.)	1.57 ± 0.39	E
WPNC (*Xylopia* spp.)	3.48 ± 0.37	E
Raw wood (*Koompassia Malaccensis*)	2.05 ± 0.18	F
WPNC (*Koompassia Malaccensis*)	3.19 ± 0.31	F
Raw wood (*Eugenia* spp.)	1.67 ± 0.60	G
WPNC (*Eugenia* spp.)	3.16 ± 0.75	H

[a]Each value is the average of 10 specimens
[b]The same letters are not significantly different at $\alpha = 5\%$

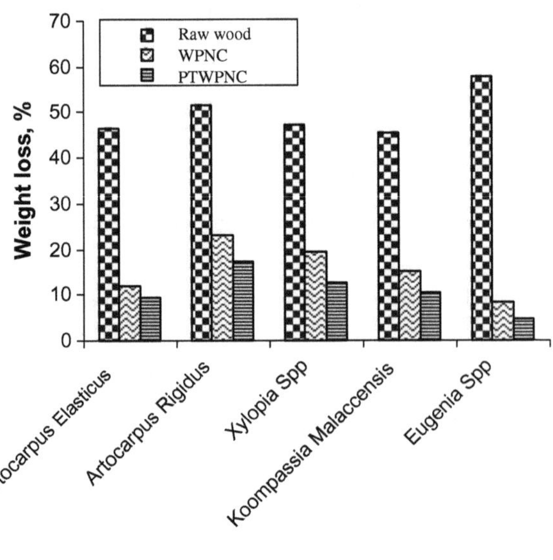

Fig. 10 Weight losses of untreated wood, WPNC and PTWPNC after exposure to the decay fungus *polyporous versicolor (White-rot fungus)* for 12 weeks

attack for raw wood and WPNC. The decay fungi, namely *polyporous versicolor* and postia placenta affect the raw wood, WPNC, and PTWPNC whereas the WPNC and PTWPNC of *Eugenia* spp. were less affected by polyporous versicolor decay fungi compared to *Artocarpus Elasticus, Koompassia Malaccensis, Xylopia* spp., and *Artocarpus Rigidus,* respectively. From the results, it also showed that raw wood species were non-resistant to decay exposure. However, the weight losses decreased both fungi for all wood species due to the sodium metaperiodate and phenyl hydrazine enhanced the decay resistance while the WPNC and PTWPNC of

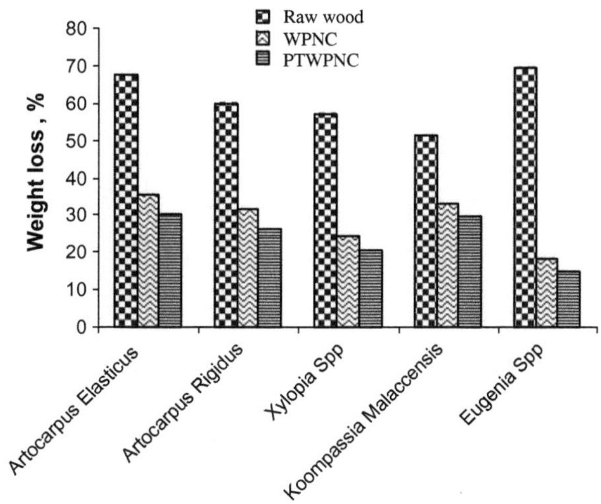

Fig. 11 Weight losses of untreated wood, WPNC and PTWPNC after exposure to the decay fungus *postia placenta* (Brown-rot fungus) for 12 weeks

Eugenia spp. were highly resistant to white-rot fungi decay exposure followed by *Artocarpus Elasticus, Koompassia Malaccensis, Xylopia* spp., and *Artocarpus Rigidus.* Moreover, Brown-rot fungi decay moderately attacks the WPNC and PTWPNC of *Eugenia* spp., *Xylopia* spp., *Artocarpus Rigidus, Koompassia Malaccensis* and *Artocarpus Elasticus,* respectively. It is recommended that selected chemicals are highly effective on the decay resistance test compare with other chemicals modifications (Yalinkilic et al. 1998).

3.7 Water Uptake

Water absorption study investigates the dimensional stability of the manufacturing raw wood, WPNC, and PTWPNC. Water absorption properties of raw wood, WPNC, and PTWPNC of cold and hot water immersion are shown in Figs. 12 and 13, respectively. The water uptake significantly decreases for WPNC and PTWPNC compared with untreated ones whereas WPNC and PTWPNC of *Eugenia* spp. had the least water absorption, followed by *Xylopia* spp., *Artocarpus Elasticus, Artocarpus Rigidus,* and *Koompassia Malaccensis,* respectively. This is due to the chemical reaction with wood cell and cellulose contains hydroxyl groups which had lower water content for WPNC and PTWPNC compared to the untreated ones. The number of hydroxyl groups in the raw wood increased the water absorption. However, the hydroxyl groups in wood specimens react with sodium metaperiodate and phenyl hydrazine and thus reduced the water absorption nature for WPNC and PTWPNC (Cai et al. 2008).

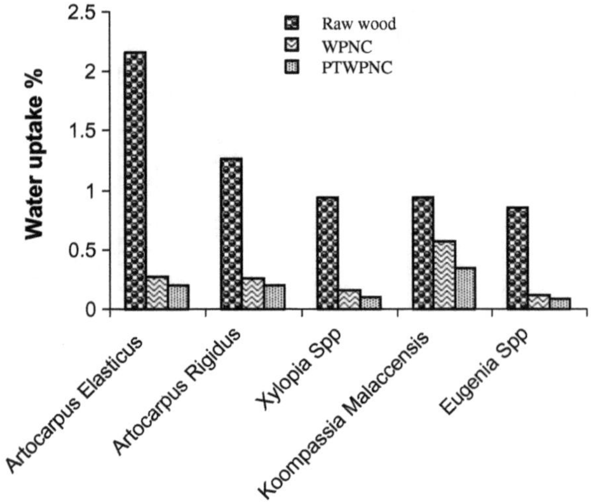

Fig. 12 Variation of water uptake for untreated wood, WPNC and PTWNC immersed in cold water

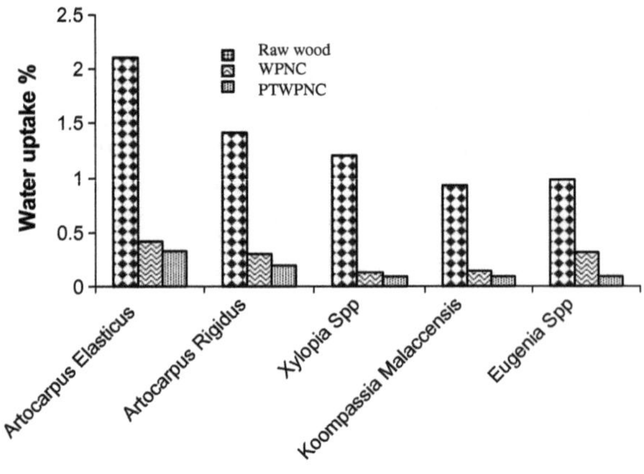

Fig. 13 Variation of water uptake for untreated wood, WPNC and PTWPNC immersed in hot water

3.8 Scanning Electron Microscopy (SEM) Analysis

The SEM micrograph shows the effect of chemical modification on wood cell wall which shown in Fig. 14a, b, respectively. According to Fig. 14a, tropical softwood is suitable to convert into the wood polymer nanocomposites because selective chemical impact on this species that show the smooth surface texture. During the chemical modification, raw hardwood (*Xylopia* spp.) is less affected due to their strong cell

Fig. 14 **a** SEM micrograph for typical softwood (*Artocarpus Rigidus*), **b** SEM micrograph for typical hardwood (*Xylopia* spp.)

wall bonded with cellulose hemicellulose and lignin. However, further post-treatment had some impact on the microstructure which drastically change the surface morphology and enhance the strength of PTWPNC as seen in the E' values (Fig. 14b). These findings confirm the small increment of E' values in WPNC.

4 Conclusion

It could be recommended and concluded that the post-treatment enhances the E′, Ed, MOE, MOR, and Es for all the selected tropical wood species. Sodium metaperiodate with nanoclay successfully convert raw softwood to WPNC which increase the

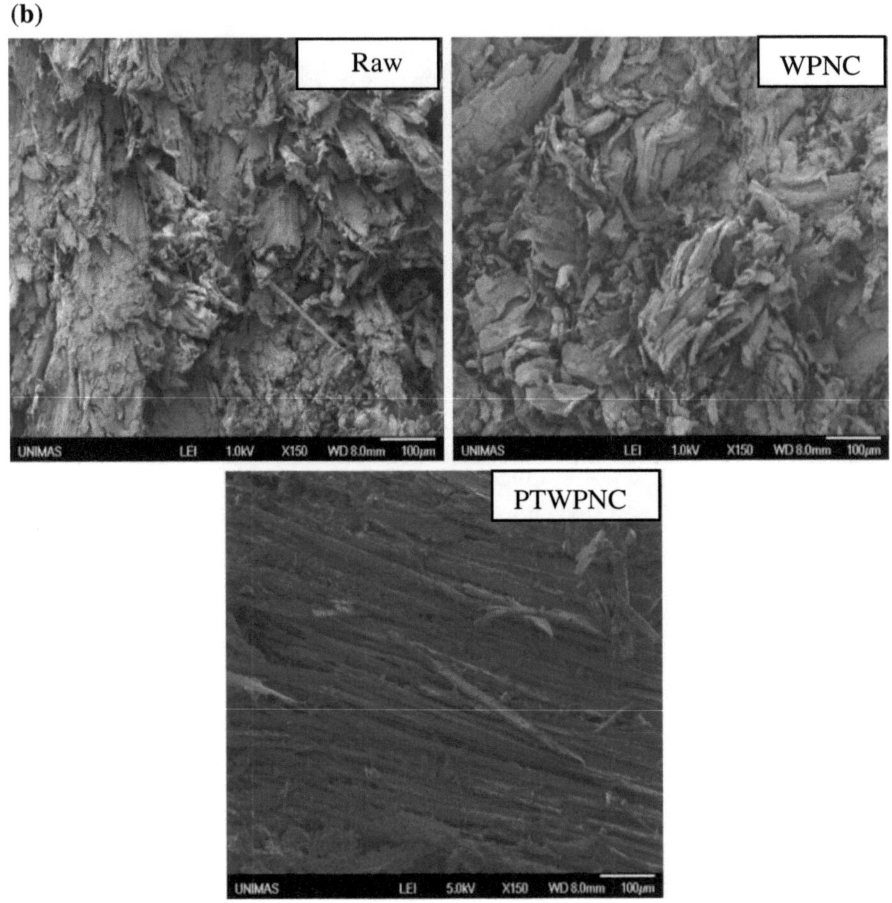

Fig. 14 (continued)

E' values of all WPNCs except *Xylopia* spp. because this species was consistently larger than raw wood. For all species, phenyl hydrazine successfully converts WPNC to PTWPNC upon post-treatments which further increase the E' values of all PTWPNCs including *Xylopia* spp. because this species react with phenyl hydrazine and further improve the E value of its WPNC. The phenyl hydrazine had significantly raised the value of the E_d for PTWPNC whereas the sodium metaperiodate slightly improve the E_d values of raw wood. However, the hardwood viscoelastic properties and stiffness are less affected by the sodium metaperiodate which confirms by similar E_d and E' for hardwood (*Xylopia* spp.). All selected tropical wood species exhibited the properties of glass (high modulus) at low temperatures and rubber (low modulus) at higher temperatures while the glass transition temperature of raw wood species ranged from −50 to 70 °C. Sodium metaperiodate modified WPNC showed the higher glass transition temperature and it was further increase when treated with and phenyl hydrazine. PTWPNC and WPNC yielded higher MOE, MOR and E_s value compared to the untreated ones. The less water absorption showed the PTWPNC compared to the WPNC and raw wood, respectively. The decay exposure of PTWPNC was less than those of WPNC and untreated wood. The authors recommended that above mentioned selected chemical modification of the *Eugenia* spp. WPNC and PTWPNC had the optimum set of decay properties in comparison with other fabricated nanocomposites.

References

Adams DG, Choong ET McIlhenny RC (1970) Bending strength of radiation-produced southern pine wood-plastic combinations. Forest Prod J 20(4):25–28

ASTM D 2017 (1967) Standard method of accelerated laboratory test of natural decay resistance of wood. Am Soc Testing Mater, Philadelphia, p 682

ASTM D-143 (2006) Standard method of testing small clear specimens of timber. Am Soc Testing Mater, Philadelphia

ASTM Standard D 570–99 (2002) Standard test methods for water absorption of plastics. Versa Tech Inc, New York, pp 1–10

Ates S, Akyildiz MH, Ozdemir H (2009) Effects of heat treatment on calabrian pine (Pinus brutia ten.) wood. BioRes 4(3):1032–1043

Autio T, Miettinen JK (1970) Experiments in Finland on properties of wood-polymer combinations. Forest Prod J 20(3):36–42

Bikiaris D, Aburto J, Alric I, Borredon E, Botev M, Betchev C, Panayiotou C (1999) Mechanical properties and biodegradibilty of LDPE blends with fatty acid esters of amylase and starch. J Appl Sci 71:1089–1100

Cai X, Riedl B, Zhang SY, Wan H (2008) The impact of the nature of nanofillers on the performance of wood polymer nanocomposites. Compos Part A 39:727–737

Hearmon RFS (1965) The assessment of wood properties by vibrations and high frequency acoustic waves. Washington State University, Pullman, pp 49–67

Khavkine M, Kazayawoko M, Law S, Balatinecz JJ (2000) Durability of wood flour-thermoplastic composites under extreme environmental conditions and fungal exposure. Int J Polym Mater 21(4):171–177

Mankowshi M, Morrell JJ (2000) Patterns of fungal attack in wood plastic composites following exposure in a soil block test. Wood Fiber Sci 32(3):340–345

Russel A (1986) The decay-underrated factor. Railway Track Struct 82(8):34–37

Singha AS, Thakur VK (2009) Study of mechanical properties of urea formaldehyde thermosets reinforced by pine needle powder. BioRes 4(1):292–308

Verhey S, Laks P, Richiter D (2001) Laboratory decay resistance of wood-fiber/thermoplastic composites. Forest Prod J 51:44–49

Yalinkilic MK, Tsunoda K, Takahashi M, Gezer ED, Dwianto W, Nemoto H (1998) Enhancement of biological and physical properties of wood by boric acid-vinyl monomer combination treatment. Holz 52(6):667–672

Yildiz CU, Yildiz S, Gezer DE (2005) Mechanical properties and decay resistance of wood polymer composites prepared from fast growing species in Turkey. Biores Technol 96: 1003–1011

Characterization of *N*,*N*-Dimethylacetamide Impregnated Wood Polymer Nanocomposites (WPNCs)

M.R. Rahman

Abstract Raw wood was impregnated with *N*,*N*-dimethylacetamide to form fabricated wood polymer nanocomposites (WPNCs). FT-IR spectra showed enhanced absorption at 1419 and then 1267 cm^{-1} which confirmed the occurrence of a modification reaction. TGA data of the fabricated WPNCs indicated a better thermal stability compared to the raw wood. The dynamic Young's modulus of the WPNCs was significantly increased compared to raw wood. Through impregnation, SEM micrographs showed porous cells of raw wood was fully filled with the polymer, which led to the better stability of WPNCs. XRD analysis indicated that the crystallinity of WPNCs increased due to the increment in the stiffness as well as the thermal stability of WPNCs.

Keywords Stiffness · Modulus of elasticity (MOE) · Modulus of rupture (MOR) · Wood polymer nanocomposites (WPNCs)

1 Introduction

In the previous research, mercerization is one of the most conventional treatment methods for natural fibers. This method helps to modify its surface properties so that it can be successfully implemented in composite formation (Gassan et al. 1999). Based on the literature on the alkali treatment toward many natural fibers, the main aim of this process is to remove the hemicelluloses. However, the rate of removal of hemicellulose is depending on the concentration of the alkali used as well as the time and temperature of treatment. On the other hand, the rate of polymer impregnation is depending on the density of wood. The introduction of the polymer into the raw wood increased the density of the wood by filling the voids. This helps to improve the thermal and mechanical strength of wood as well as the

M.R. Rahman (✉)
Faculty of Engineering, Universiti Malaysia Sarawak, 94300 Kota Samarahan, Sarawak, Malaysia
e-mail: rmrezaur@unimas.my

M.R. Rahman, *Wood Polymer Nanocomposites*, Engineering Materials,
DOI 10.1007/978-3-319-65735-6_5

chemical bonding of the basic macromolecules within the wood. The chemical bonding can be generally achieved by any reactive group on the removal of hydroxyl groups from the wood.

However, there is no work reported on tropical wood impregnated with *N,N*-dimethylacetamide polymer. This chapter presented the selected raw tropical wood species, namely *Artocarpus elasticus, Artocarpus rigidus, Xylopia* spp., *Koompassia malaccensis* and *Eugenia* spp., were impregnated with the aid of sodium hydroxide. Continuously, all sodium hydroxide mercerized raw woods were impregnated with *N,N*-dimethylacetamide after a week. This study was investigated and the outcomes were expected. The expected FT-IR showed the enhanced absorption at 1419 cm^{-1} (–C–/CH_3) and the 1267 cm^{-1} (–N–/CH_3) stretching band. This stretching band confirmed the polymerization reaction. TGA results were expected that WPNCs showed better thermal stability in comparison with the raw wood. DSC results were expected that the decomposition temperature of WPNCs provided a higher thermal stability compared to raw wood. Dynamic Young's modulus of WPNCs was expected to be significantly increased compared with the raw wood. Besides, the MOE and MOR were expected to perform better than raw wood. The Young's modulus of *Xylopia* spp., *Artocarpus rigidus,* and *Eugenia* spp. were expected to be significantly different between raw wood and WPNCs. In addition, SEM was expected to show the fully filled porous cells on the fabricated WPNCs which gave better stability of WPNCs. The increase in the stiffness and the thermal stability of WPNCs were expected to be increased due to the crystallinity of WPNCs as indicated by XRD analysis.

2 Materials and Methods

2.1 Materials

Five wood species were collected. The collected wood was divided into softwood and hardwood. The softwoods that were used, namely *Eugenia* spp., *Artocarpus rigidus*, *Artocarpus elasticus,* and *Xylopia* spp., while the hardwood chosen was *Koompassia malaccensis*. For chemicals, NaOH and *N,N*-Dimethylacetamid (Merck, Germany) were chosen to be used to treat the wood species. All the purity grade of the chemicals were 99%.

2.2 Manufacturing of Wood Polymer Nanocomposites

Raw wood samples were soaked into 1M sodium hydroxide solution using an autoclave for 5 h. The temperature and pressure used were 120 °C and 85 kPa, respectively. For WPNCs, raw wood was impregnated with *N,N*-dimethylacetamide using a vacuum chamber at 25 °C and 60 cm Hg.

2.3 Microstructural Characterizations

2.3.1 Fourier Transform Infrared Spectroscopy (FT-IR)

Both raw wood and WPNCs fabricated were undergoing FT-IR where the infrared spectra were recorded on a Shimadzu Fourier Transform Infrared Spectroscopy (FT-IR) 81001 Spectrophotometer. The transmittance range of the scan was 4000–370 cm^{-1} for all samples.

2.3.2 X-ray Diffraction (XRD)

XRD analysis for raw wood and WPNCs were performed with a Rigaku diffractometer (Cu Kα radiation, λ = 0.1546 nm). This equipment was running at 40 kV and 30 mA. An accepted experimental procedure for the determination of degree of crystallinity was applicable to this material where most of the X-ray scattering curves were resolvable into crystalline and amorphous scattering regions.

2.3.3 Thermogravimetric Analysis (TGA)

TGA measurements of raw wood and WPNCs were carried out with about 5–10 mg of samples at heating rate of 10°C/min in a nitrogen atmosphere using a Thermogravimetric Analyzer (TA Instrument SDT Q600). All the samples were subjected to TGA in high purity nitrogen under a constant flow rate of 5 ml/min. The temperature range for the thermal decomposition of each sample occurred was from 30 to 200 °C. The weight loss and temperature of all the samples were determined and analyzed to determine the thermal degradation rate (% weight loss/min), initial degradation temperature, and residual weight at 200 °C. According to the Broido (1969), the activation energy was calculated based on the Eq. (1).

$$k = Ae^{\frac{-Ea}{RT}} \tag{1}$$

2.3.4 Differential Scanning Calorimetric (DSC)

All DSC measurements were carried out on a DSC Q10 (TA instrument) thermal system, using a sealed aluminum capsule. Each test specimen was weighed to about 3–3.5 mg, and all the samples were held at a single heating rate of 10 °C/min with scanning temperature from 30 to 200 °C. Each of the data reported represented an average of five runs. Melting endotherm was determined which represented the integration of the area under the peak to assess the quantitative reproducibility. The crystallinity was determined and measured dynamically by integrating the peak area of a DSC curve.

2.3.5 Scanning Electron Microscopy (SEM)

Both raw wood and WPNCs were first fixed with Karnovsky's fixative. After that the samples were taken through a graded alcohol dehydration series. Once dehydrated, the specimen was coated with thin layer of gold before viewed on the SEM. The micrographs were taken at a magnification of 150, as presented in the results and discussions section.

2.3.6 Free-Free Flexural Vibration Testing

The free-free flexural vibration testing system was developed. The threads were used to support the specimen at its nodal point of specific mode. The iron plate was facing toward the electromagnet transducer and electromagnet drive where the electromagnet transducer was connected to a recorder and an oscilloscope. Besides, an electromagnet driver with the introduction of alternating current excited the vibration by attracting the iron plate at one end of specimen. The vibration produced sound that travels through the sample and detected by a microphone where the response of the microphone was connected to a sound level meter. The amplitude of sound was increased through the sound level meter, and the sound was transformed from analog signal to digital signal. All the detected signals were transferred to the PICO oscilloscope and personal computer. The frequency was varied until a maximum response was detected. The maximum frequency was named as resonant frequency. Dynamic Young's modulus (E_d) was calculated from the resonant frequency by using the Eqs. (2) and (3):

$$E' = \frac{4\pi^2 f^2 l^4 A\rho}{I(m_n)^4} \tag{2}$$

$$I = \frac{bd^3}{12} \tag{3}$$

where

- d beam depth,
- b beam width,
- l beam length,

f natural frequency of the specimen,
ρ density,
A cross-sectional area, and

$n = 1$ is the first mode of vibration where $m_1 = 4.730$.

2.3.7 Three-Point Bending Test for MOE and MOR

The length, width, and thickness of all raw wood and WPNCs were measured. Three-point bending test was carried out using Shimadzu Universal Testing Machine with loading capacity of 300 kN and crosshead speed of 2 mm/min. MOE and MOR were carried out according to ASTM D-143 (2006). The MOE and MOR were calculated using Eqs. (4) and (5), respectively.

$$\text{MOE} = \frac{L^3 m}{4bd^3} \tag{4}$$

$$\text{MOR} = \frac{1.5PL}{bh^2} \tag{5}$$

where

L span length of sample, 180 mm,
b width of sample, 20 mm,
d thickness of sample, 20 mm,
m slope of the tangent to the initial line of the force displacement curve,
P the maximum breaking load, and
h depth of the beam.

2.3.8 Compression Parallel to Grain Test for Static Young's Modulus (E_s)

All the samples undergo compression parallel to grain test based on the test standard, namely ASTM D143-52. Static Young's modulus, E_s, was the only test determined using the uniaxial compression test. All the testing was carried out using Shimadzu Universal Testing Machine with loading capacity of 300kN and crosshead speed of 2 mm/min. Static Young's modulus, E_s was calculated using Eq. (6).

$$E = \frac{\text{Stress}}{\text{Strain}}$$

$$E = \frac{F/A}{\Delta L/L}$$

where

F force,
A cross-sectional area,
ΔL displacement, and
L length of sample.

$$\begin{aligned} F &= \frac{EA}{L} \times \Delta L \\ m &= \frac{EA}{L} \\ \therefore E &= \frac{EA}{L} \end{aligned} \tag{6}$$

3 Results and Discussion

3.1 Fourier Transform Infrared Spectroscopy (FT-IR)

Figure 1 showed the FT-IR spectra of raw wood and WPNCs. The spectrum of the raw wood showed the basic structure of all wood samples into three regions. The first region was at 4000–3300 cm^{-1} was attributed to strong broad OH stretching while the second region was attributed to C–H stretching in methyl and methylene groups at 3000–2800 cm^{-1}. The third region located from 1000 to 1750 cm^{-1} which was attributed to a strong broad superposition with sharp and discrete absorptions in the region (Owen et al. 1989). There were few peaks to detect for hemicellulose and lignin. The peaks at 1508 and 1600 cm^{-1} were caused by lignin. For lignin, it could be detected at peak of 1734 cm^{-1}. According to Owen et al. (1989), the C=O stretch in non-conjugated ketones, carbonyls, and in ester groups was investigated. Moreover, the region between 1800 and 1100 cm^{-1} was compared with the bands assigned to the main components from wood such as cellulose, hemicelluloses, and lignin. From Fig. 1, it was clearly detected that both raw wood and WPNCs showed different infrared spectra and location. The reaction between raw wood and *N,N*-dimethylacetamide was well defined and was shown in Fig. 2.

The peak at 3424 cm^{-1} was attributed to the intensity of –OH absorption band. This peak was greatly reduced on WPNCs after the impregnation of *N,N*-dimethylacetamide into raw wood. However, the peak at 1734 cm^{-1} was clearly observed in raw wood which represented the higher xylan content in raw wood. For polymerized wood, this peak was clearly shifted to 1570 cm^{-1} due to the stronger carbonyl bands within the polymerized wood. In addition, the absorption at 1419 and 1267 cm^{-1} was attributed to –C–/CH_3 and –N–/CH_3 stretching band which confirmed the polymerization reaction.

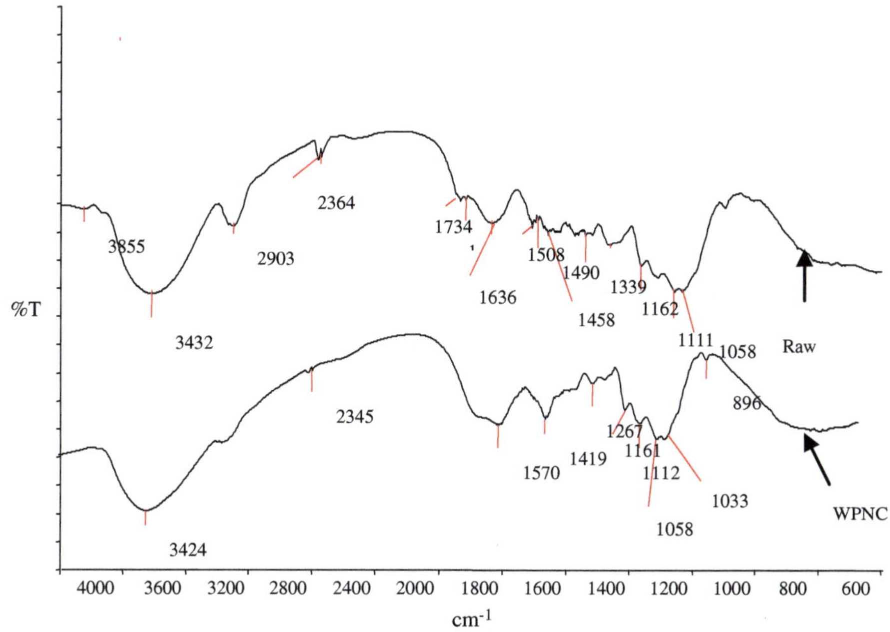

Fig. 1 IR spectrum of raw wood and WPNCs

Fig. 2 Mecerization of wood specimens with polymerization reaction

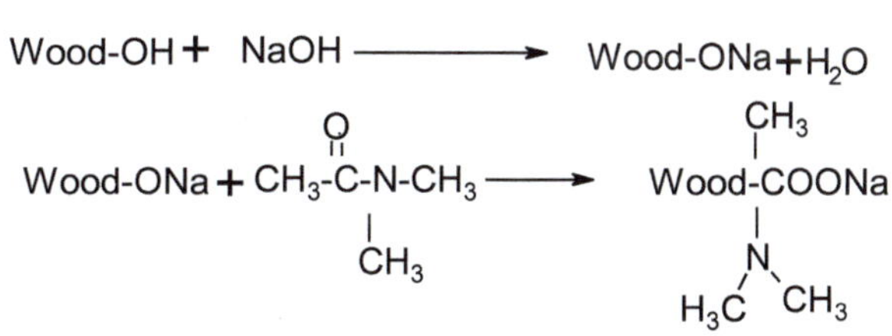

3.2 *Thermogravimetric Analysis (TGA)*

Thermogravimetric Analysis (TGA) was used to determine and investigate the thermal stability of wood. TGA is a very important parameter in the production of WPNCs. According to Wielage et al. (1999), wood acted as a filler material to improve mechanical strength in plastics that were processed at temperature below 200 °C. Through the TGA results, raw wood undergone dehydration process where there was 5–8% of absorbed water was removed. From the literature, it had been established that there was no degradation up to 160 °C (Wielage et al. 1999). Above this temperature, thermal stability of WPNCs on different species gradually decreased and the decomposition took place. Both Fig. 3a–e and Table 1 represented the TGA curves obtained by the dynamic scans for raw wood and WPNCs.

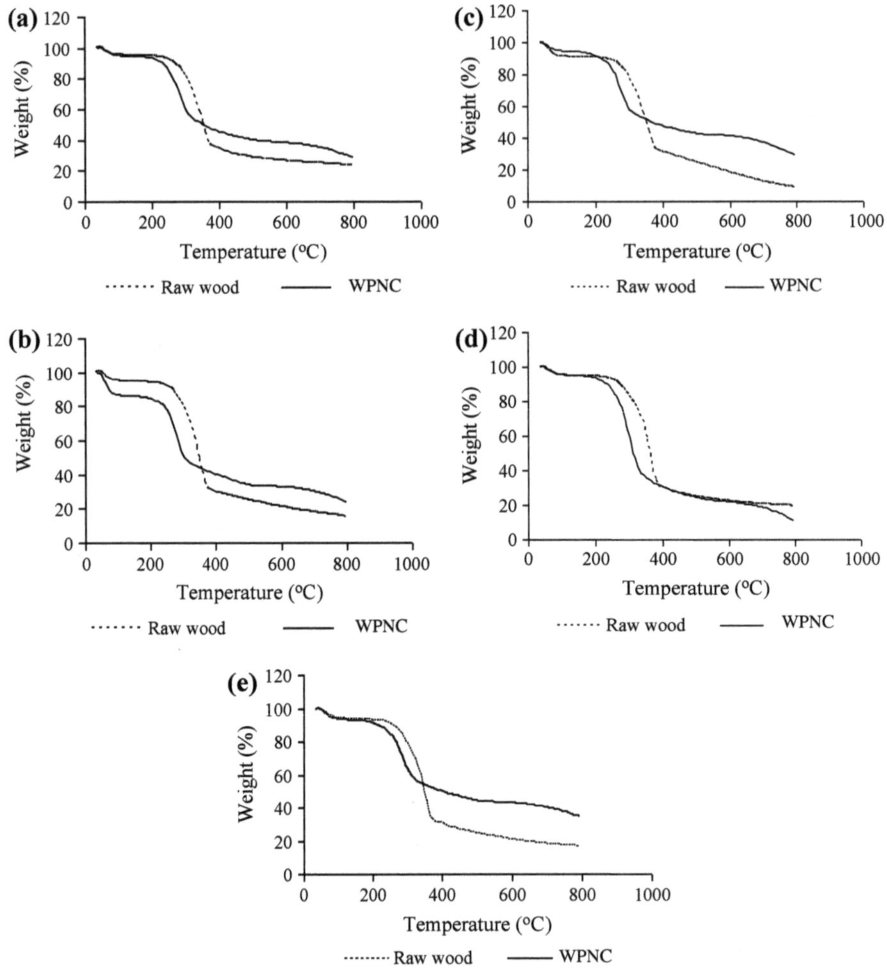

Fig. 3 **a** TGA curve of *Artocarpus elasticus* raw wood and WPNCs, **b** TGA curve of *Artocarpus rigidus* raw wood and WPNCs, **c** TGA curve of *Xylopia* spp. raw wood and WPNCs, **d** TGA curve of *Koompassia malaccensis* raw wood and WPNCs, **e** TGA curve of *Eugenia* spp. raw wood and WPNCs

The TGA curves of raw wood and WPNCs indicated that water started at 10 °C till 150 °C for both raw and WPNCs. *N,N*-dimethylacetamide was well impregnated into WPNCs to improve the thermal stability of wood. Besides, the evidence that the initial and final decomposition temperature (T_i, T_f) of all WPNCs was higher than raw wood. Therefore, the activation energy was calculated which could be helpful in reaching conclusions on the thermal stability of WPNCs. From the analysis, higher activation energy represented higher thermal stability.

Table 1 Thermal characteristics of raw wood and WPNCs

Sample		T_i (°C)	T_m (°C)	T_f (°C)	W_i (%)	W_m (%)	W_f (%)	Activation energy, E_a (J/K)
Artocarpus elasticus	Raw wood	201	332	419	98.96	64.73	32.67	33.11
	WPNCs	210	269	622	97.89	77.16	38.17	73.49
Artocarpus rigidus	Raw wood	211	340	388	99.67	56.45	30.63	25.79
	WPNCs	215	277	537	98.16	63.46	33.68	34.99
Xylopia spp.	Raw wood	212	341	420	98.78	56.15	29.85	27.04
	WPNCs	225	285	630	98.38	64.35	41.16	30.14
Koompassia malaccensis	Raw wood	212	355	403	98.99	57.91	30.00	16.12
	WPNCs	213	301	576	98.06	59.94	22.51	41.34
Eugenia spp.	Raw wood	214	340	443	99.77	60.32	27.51	19.17
	WPNCs	218	285	521	98.48	70.68	44.35	28.68

Arrhenius equation was used to measure the TGA thermograms of all raw wood and WPNCs based on Broido's research (1973) as shown in Eq. (1). The activation energy could be obtained from the plotting of ln k versus $1/T$. The linear plots were shown in Fig. 4a–e where the value for activation energy was tabulated in Table 1. The activation energy of WPNCs was higher compared with raw wood on different types of wood species.

3.3 *Differential Scanning Calorimetry (DSC)*

Thermogravimetric Analysis was carried out to determine thermal analysis as well as to determine the moisture content and volatile components presented in WPNCs. According to Herrera-France and Aguilar-Vega (1997), the properties of WPNCs were deteriorated due to the moisture content and volatile components that reduced the application of WPNCs. Through DSC analysis, raw wood and WPNCs were investigated and the results were analyzed. The thermal behavior of raw wood and WPNCs was determined and analyzed. In addition, DSC analysis also enabled the chemical activity within the raw wood and WPNCs over the temperature.

The DSC curves of raw wood and WPNCs were presented in Fig. 5a–e. Besides, Table 2 presented the enthalpy, and exotherm peaks were analyzed. Both raw wood and WPNCs showed a broad endotherm observed in the temperature range approximately 50–161 °C for all species. During this temperature range, water molecules were presented within the fibers. Based on the first decomposition temperature, WPNCs provided a higher value than the raw wood which defined the

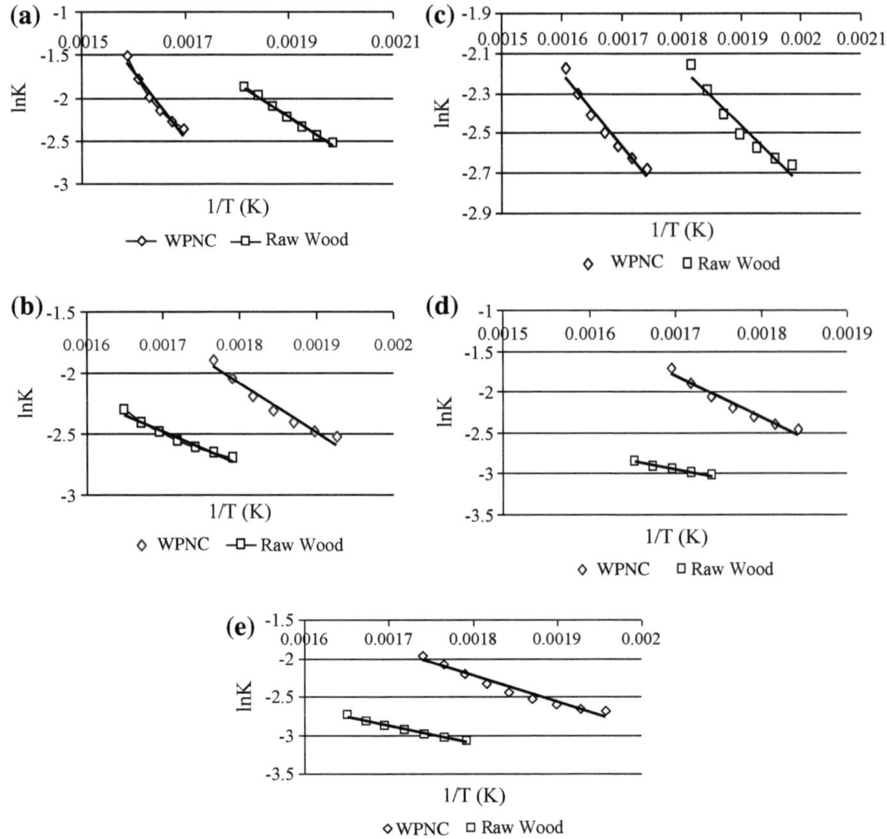

Fig. 4 **a** Temperature dependency of the decomposition rate of the *Artocarpus elasticus* raw wood and WPNCs, **b** Temperature dependency of the decomposition rate of the *Artocarpus rigidus* raw wood and WPNCs, **c** Temperature dependency of the decomposition rate of the *Xylopia* spp. raw wood and WPNCs, **d** Temperature dependency of the decomposition rate of the *Koompassia malaccensis* raw wood and WPNCs, **e** Temperature dependency of the decomposition rate of the *Eugenia* spp. raw wood and WPNCs

higher thermal stability within WPNCs. This was due to the bond between the functional groups within *N,N*-dimethylacetamide and the wood carboxyl. Based on the research work by Akita and Kase (1967), lignin degraded at a temperature around 200 °C within the cellulose fibers while the other polysaccharides such as cellulose performed thermal degradation at higher temperature. Thus, the second endothermic peaks which were higher than 200 °C, represented the decomposition temperatures of the cellulose in the wood fibers. There were two main decompositions namely the first thermal decomposition and second thermal decomposition. In addition, alkali and silane treatment increased the thermal stability of WPNCs

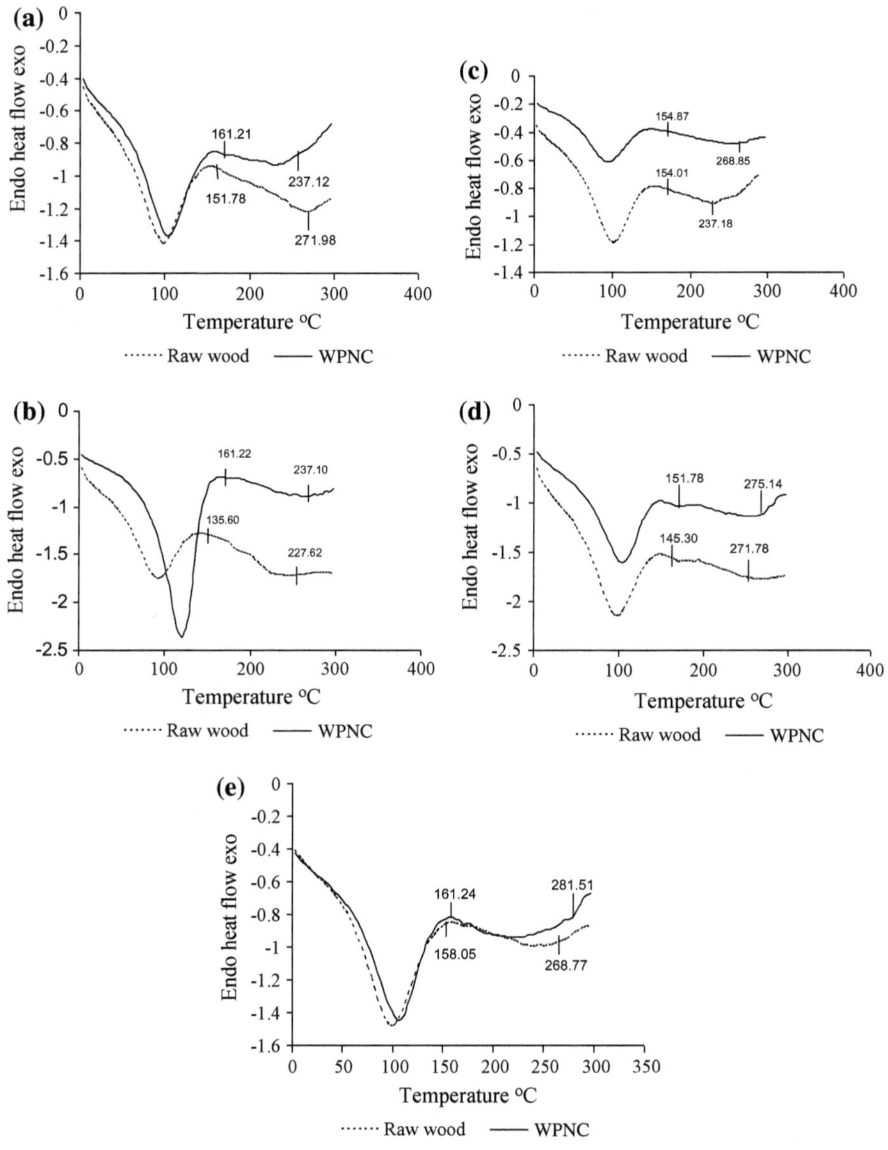

Fig. 5 **a** DSC thermographs of the *Artocarpus elasticus* raw wood and WPNCs, **b** DSC thermographs of the *Artocarpus rigidus* raw wood and WPNCs, **c** DSC thermographs of the *Koompasia malacennis* raw wood and WPNCs, **d** DSC thermographs of the *Xylopia* spp. raw wood and WPNCs, **e** DSC thermographs of the *Eugennia* spp. raw wood and WPNCs

Table 2 Crystallization enthalpy and exotherm peaks of raw wood and WPNCs

Species	Crystallization enthalpy ΔH_c (J/gm)		1st exotherm peaks (°C)	
	Raw wood	WPNCs	Raw wood	WPNCs
Artocarpus elasticus	232.10	238.64	151.78	161.21
Artocarpus rigidus	248.64	474.16	135.60	161.22
Koompasia malacennis	192.86	194.20	154.01	154.87
Xylopia spp.	252.31	323.00	145.30	151.78
Eugennia spp.	248.20	288.78	158.05	161.24

(Sreekala et al. 1997). Table 2 showed the crystallization enthalpy (ΔH_c) of all WPNCs higher than raw wood that indicated that WPNCs were more crystalline compared with raw wood, as reflected in the XRD results.

3.4 Dynamic Young's Modulus

Free-free flexural vibration testing system provided the stiffness of the raw wood and WPNCs as shown in Fig. 6. From this characterization, ten specimens were used for each wood species to ensure the accuracy of the results. The N,N-dimethylacetamide impregnation increased the Young's modulus of WPNCs on different wood species. Activation energy was significantly affected by the polymerization for WPNCs as shown in Table 1. However, only WPNCs *Artocarpus elasticus* showed significant changes in the elastic properties among all the WPNCs as well as raw wood.

Fig. 6 E_d of raw wood and WPNCs for all species

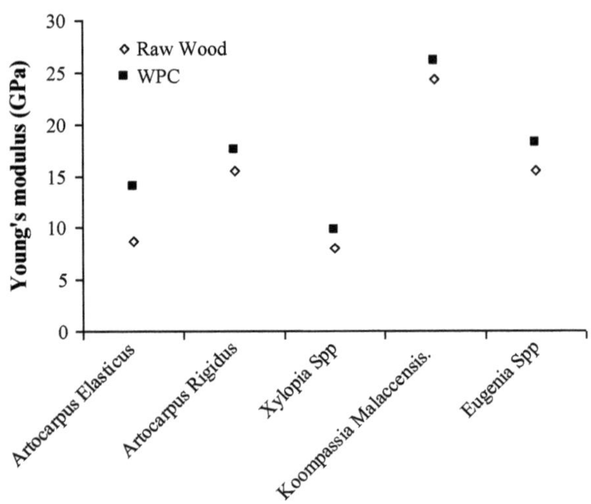

3.5 MOE and MOR Measurement

Tables 3 and 4 as well as Figs. 7 and 8 showed MOE and MOR results for raw wood and WPNCs on different types of wood species through the impregnation of *N,N*-dimethylacetamide. WPNCs on *Artocarpus elasticus* and *Artocarpus rigidus* showed higher MOE values while WPNCs on *Eugenia* spp., *Xylopia* spp., and *Koompassia malaccensis* showed moderate MOE values, respectively. From Fig. 7, it was clearly shown that WPNCs yielded higher MOE compared to the raw wood due to the *N,N*-dimethylacetamide impregnation, which was agreed by other research (Adams et al. 1970; Yildiz et al. 2005).

Table 3 showed that the increment of MOE on *Artocarpus elasticus*, *Artocarpus rigidus,* and *Eugenia* spp. were significantly higher than raw wood. However, the

Table 3 *t*-test analysis of raw wood and wood polymer nanocomposites[a]

Treatment	Modulus of elasticity	*t*-test grouping[b]
Raw wood (*Artocarpus elasticus*)	6.48 ± 0.52	A
WPNCs (*Artocarpus elasticus*)	12.44 ± 0.99	B
Raw wood (*Artocarpus rigidus*)	5.13 ± 0.38	C
WPNCs (*Artocarpus rigidus*)	9.56 ± 1.97	D
Raw wood (*Xylopia* spp.)	6.71 ± 0.34	E
WPNCs (*Xylopia* spp.)	9.23 ± 1.81	E
Raw wood (*Koompassia malaccensis*)	15.68 ± 1.11	F
WPNCs (*Koompassia malaccensis*)	16.75 ± 0.96	F
Raw wood (*Eugenia* spp.)	9.25 ± 0.37	G
WPNCs (*Eugenia* spp.)	13.16 ± 0.41	H

[a]Each value is the average of 10 specimens
[b]The same letters are not significantly different at $\alpha = 5\%$

Table 4 *t*-test analysis of raw wood and wood polymer nanocomposites[a]

Treatment	Modulus of elasticity	*t*-test grouping[b]
Raw wood (*Artocarpus elasticus*)	76.17 ± 3.01	A
WPNCs (*Artocarpus elasticus*)	83.92 ± 4.20	B
Raw wood (*Artocarpus rigidus*)	33.02 ± 3.63	C
WPNCs (*Artocarpus rigidus*)	73.05 ± 9.26	D
Raw wood (*Xylopia* spp.)	45.91 ± 1.36	E
WPNCs (*Xylopia* spp.)	80.35 ± 9.01	E
Raw wood (*Koompassia malaccensis*)	122.20 ± 15.49	F
WPNCs (*Koompassia malaccensis*)	125.98 ± 6.77	F
Raw wood (*Eugenia* spp.)	46.10 ± 3.62	G
WPNCs (*Eugenia* spp.)	72.8 ± 4.60	H

[a]Each value is the average of 10 specimens
[b]The same letters are not significantly different at $\alpha = 5\%$

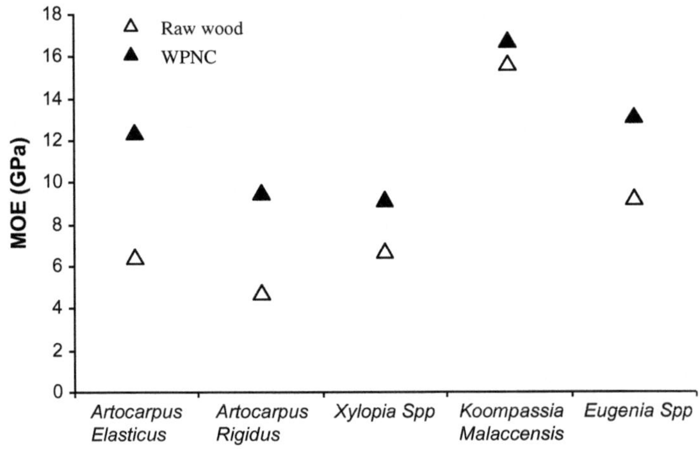

Fig. 7 MOE of raw wood and WPNCs

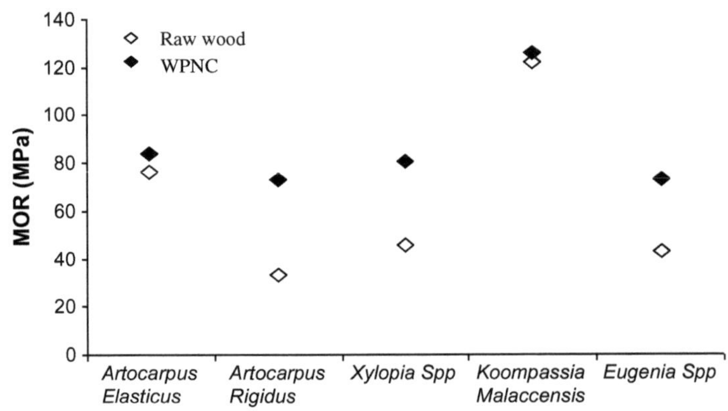

Fig. 8 MOR of raw wood and WPNCs

only hardwood *Koompassia malaccensis* as well as softwood *Xylopia* spp. showed less significant effect on the impregnation of *N*,*N*-dimethylacetamide as they had better hardness in origin form.

In this study, sodium hydroxide reacted with the cellulose in wood cells. This reaction reduced the water molecules from the raw wood to fabricate better WPNCs. *N*,*N*-Dimethylacetamid impregnation on raw wood managed to fill the void spaces completely to increase its stiffness.

MOR of WPNCs were increased with the impregnation of *N*,*N*-dimethylacetamide in agreement with previous research (Adams et al. 1970). Besides, MOR of WPNCs on *Artocarpus rigidus*, *Xylopia* spp., and *Eugenia* spp. were significantly improved as tabulated in Table 4. Figure 8 showed that MOR for *Artocarpus rigidus* was the highest while *Xylopia* spp., *Eugenia* spp., *Artocarpus elasticus, and*

Koompassia malaccensis showed moderate MOR values. Both raw wood and WPNCs hardwood *Koompassia malaccensis* showed almost similar values which indicated that *N,N*-dimethylacetamide impregnation was not effective on hardwood which proven the statement on our previous work (Rahman et al. 2010).

3.6 Static Young's Modulus (E)

Figure 9 showed the static Young's modulus was determined from 10 repetitions. WPNCs on *Xylopia* spp. showed the highest increment of *E* value, followed by *Artocarpus rigidus*, *Eugenia* spp., *Artocarpus elasticus,* and *Koompassia malaccensis*, respectively. The increment of Young's modulus for WPNCs on *Xylopia* spp., *Artocarpus rigidus,* and *Eugenia* spp. was significantly higher than raw wood. The increment of *E* in WPNCs was also reported by different researchers (Autio et al. 1970; Hamdan et al. 2010). With the introduction of *N,N*-dimethylacetamide into WPNCs, it acted as plasticizer on the cell walls through thickening which gradually improved the lateral stability (Table 5).

3.7 X-ray Diffraction (XRD)

Figures 10 and 11 showed the X-ray diffraction patterns of raw wood and WPNCs. The patterns of raw wood exhibited only one well-defined 2θ peak at 22.0° as shown in Fig. 10, which was due to cellulose (Yi et al. 2010). On the other hand, WPNCs exhibited four broad 2θ peaks at 42.08°, 43.68°, 49.23°, and 72.67°. All these peaks were observed due to the introduction of polymer into raw wood that caused polymerization. According to the report of Yi et al. (2010), chemical

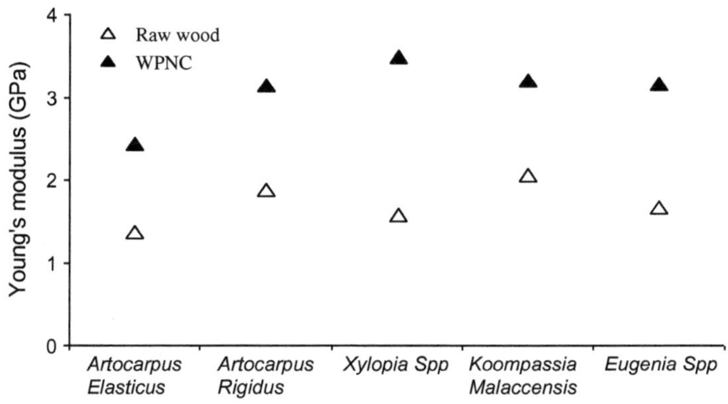

Fig. 9 Static Young's modulus of raw wood and WPNCs

Table 5 *t*-test analysis of raw wood and wood polymer nanocomposites[a]

Treatment	Static young's modulus	*t*-test grouping[b]
Raw wood (*Artocarpus elasticus*)	1.36 ± 0.23	A
WPNCs (*Artocarpus elasticus*)	2.43 ± 0.51	B
Raw wood (*Artocarpus rigidus*)	1.88 ± 0.58	C
WPNCs (*Artocarpus rigidus*)	3.14 ± 1.11	D
Raw wood (*Xylopia* spp.)	1.57 ± 0.39	E
WPNCs (*Xylopia* spp.)	3.48 ± 0.37	E
Raw wood (*Koompassia malaccensis*)	2.05 ± 0.18	F
WPNCs (*Koompassia malaccensis*)	3.19 ± 0.31	F
Raw wood (*Eugenia* spp.)	1.67 ± 0.60	G
WPNCs (*Eugenia* spp.)	3.16 ± 0.75	H

[a]Each value is the average of 10 specimens
[b]The same letters are not significantly different at $\alpha = 5\%$

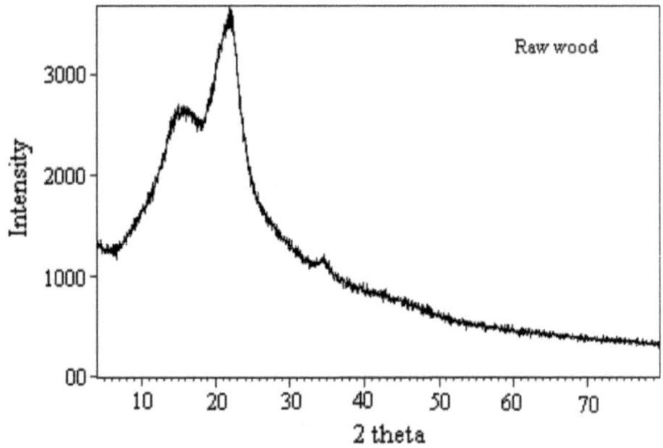

Fig. 10 X-ray diffraction of raw wood fiber

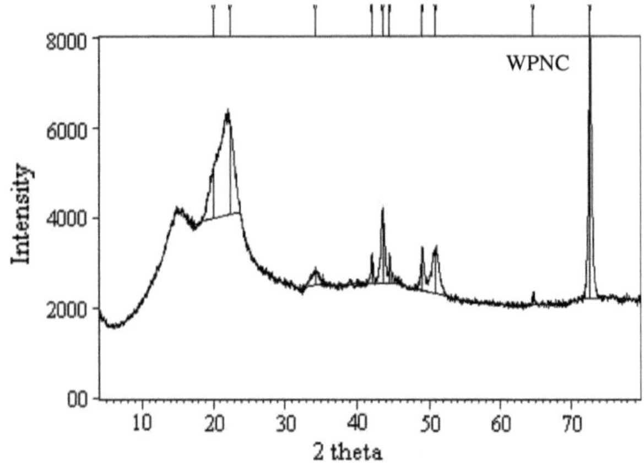

Fig. 11 X-ray diffraction of WPNCs

modification of lignocellulosic materials greatly affected their crystallinity. The impregnation process started with the reagent reacted with the chain ended on the surface of crystallites. This occurred as the polymers could not diffuse into the crystalline region. This resulted in the opening of some of the hydrogen-bonded cellulose chains. Through this, it was significantly proved that the impregnation of *N,N*-dimethylacetamide into raw wood helped to increase the crystallinity of WPNCs, by an drastic removal of the fibers' amorphous constituents.

3.8 Scanning Electron Microscopy (SEM)

SEM micrograph for the raw wood and WPNCs were clearly presented in Fig. 12. Figure 12a showed the SEM micrograph of raw wood on *Artocarpus rigidus* while Fig. 12b showed the SEM micrograph of WPNCs. The changes between Fig. 12a,

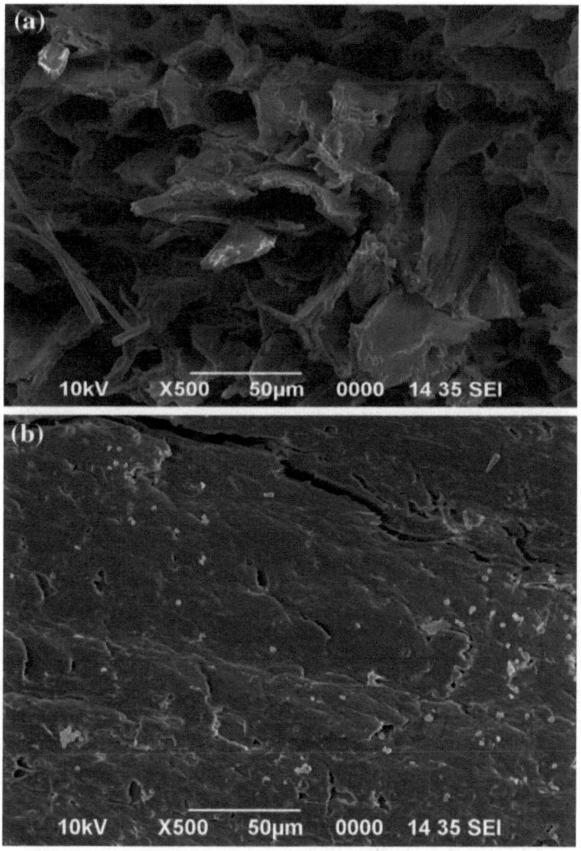

Fig. 12 SEM micrograph of raw wood (**a**) and WPNCs (**b**)

b were dramatic which proved the changes after the impregnation process. The dramatic changes on the surface morphology would led to the drastic increase of activation energy as well as the final decomposition temperature of WPNC after polymerization as shown in Fig. 3a–e and Table 1. The surface morphology proved that raw wood consisted of many void spaces within the wood cell wall itself. However, all the void spaces were fully covered by *N,N*-diemthylacetamide through impregnation process. At the same time, waxy substances were removed from fabricated WPNCs which was confirmed by our previous work (Rahman et al. 2010).

4 Conclusion

The present study was to investigate the introduction of *N,N*-dimethylacetamide into raw wood. FT-IR spectra indicated the reaction between organic *N,N*-Dimethylacetamid and raw wood through impregnation. The absorption band at 1419 cm^{-1} which represented –C–/CH_3 and 1267 cm^{-1} was attributed to –N–/CH_3 stretching bands confirmed the polymerization reaction. This polymerization of *N*, *N*-dimethylacetamide with raw wood improved WPNCs thermally over certain temperature range compared to raw wood. Therefore, WPNCs showed higher thermal stability among all the samples prepared. Through mechanical testing, the stiffness, MOE, and MOR of the WPNCs was significantly increased compared with raw wood on *Artocarpus elasticus, Artocarpus rigidus,* and *Eugenia* spp., respectively. In addition, the Young's modulus of *Xylopia* spp., *Artocarpus rigidus,* and *Eugenia* spp. was significantly different compared to raw wood and other WPNCs. X-ray diffraction patterns indicated WPNCs were more crystalline compared with raw wood.

Acknowledgements The authors would like to acknowledge the financial support from Ministry of Higher Education Malaysia, for their financial support [Grant no. FRGS/02(05)/655/2007(20)] during the research.

References

Adams DG, Choong ET Mcllhenny RC (1970) Bending strength of radiation-produced southern pine wood-plastic combinations. Forest Prod J 20(4):25–28

Akita K, Kase M (1967) Determination of kinetic parameters for pyrolysis of cellulose and cellulose treated with ammonium phosphate by differential thermal analysis and thermal gravimetric analysis. J Polym Sci 1(5):833–848

Autio T, Miettinen JK (1970) Experiments in Finland on properties of wood-polymer combinations. Forest Prod J 20(3):36–42

Broido A (1969) A simple, sensitive graphical method of treating thermogravimeric analysis data. J Polym Sci 7(10):1773

Gassan J, Bledzki AK (1999) Possibilities for improving the mechanical properties of jute/epoxy composites by alkali treatment of fibers. Compos Sci Technol 59(9):1303–1309

Hamdan S, Talib ZA, Rahman MR, Ahmed AS, Islam MS (2010) Dynamic Young's modulus measurement of treated and post-treated tropical wood polymer composites (WPC). BioRes 5(1):324–342

Herrera-France P, Aguilar-Vega M (1997) Effect of fiber treatment on the mechanical properties of LDPE-henequen cellulosic fibre composites. J Appl Polym Sci 10:197–207

Owen NL, Thomas DW (1989) Infrared studies of hard and soft woods. Appl Spec 43:451–455

Rahman MR, Hamdan S, Saleh AA, Islam MS (2010) Mechanical and biological performance of sodium mataperiodate impregnated plasticized wood (PW). BioRes 5(2):1022–1035

Sreekala M, Kumaran M, Thomas S (1997) Oil plam fibres: morphology, chemical composition, surface modification and mechanical properties. J Appl Polym Sci 66:821–835

Wielage B, Lampke Th, Mark G, Nestler K, Starke D (1999) Thermogravimetric and differential scanning calorimetric analysis of natural fibers and polypropylene. Thermochim Acta 337: 169–177

Yi C, Tain L, Tang F, Wang L, Zou H, Xu W (2010) Crystalline transition behavior of sisal in cycle process. Polym Compos 31(5):933–938

Yildiz CU, Yildiz S, Gezer DE (2005) Mechanical properties and decay resistance of wood polymer composites prepared from fast growing species in Turkey. Biores Technol 96: 1003–1011

Mechanical and Thermal Characterization of Urea-Formaldehyde Impregnated Wood Polymer Nanocomposites (WPNCs)

M.R. Rahman

Abstract In this study, urea-formaldehyde resin wood polymer nanocomposites (WPNCs) were investigated. All the WPNCs undergo characterizations. The FT-IR spectra confirmed the impregnation of organic urea-formaldehyde into the raw wood. Besides, WPNCs were generally more thermally stable over temperature compared to the raw wood due to the introduction of urea-formaldehyde into the raw wood. From mechanical testing, WPNCs showed higher MOE and MOR for *Eugenia* spp. and *Xylopia* spp., respectively. Besides, WPNCs on *Eugenia* spp. showed higher Young's modulus compared to raw wood and other WPNCs. From the X-ray diffraction patterns, the crystallinity of WPNCs increased with the introduction of urea-formaldehyde resin into raw wood. The SEM micrograph of WPNCs clearly proved that the void space was fully filled with urea-formaldehyde resins and most of the waxy substance was removed. Therefore, urea-formaldehyde-impregnated WPNCs showed significantly effective on *Eugenia* spp., continued by *Xylopia* spp. and *Artocarpus elasticus* wood species.

Keywords Differential scanning calorimetry · Thermal stability · Impregnation · Wood polymer nanocomposites (WPNCs)

1 Introduction

With the advantage of the fibrous nature of wood, wood has been one of the most appropriate and versatile raw materials for various applications. However, the application of wood can be widely applied with some modifications. Polymeric material is one of the materials especially those obtained from renewable resources, namely natural fibers. This material has gained much attention and attraction from

M.R. Rahman (✉)
Faculty of Engineering, Universiti Malaysia Sarawak, 94300 Kota Samarahan, Sarawak, Malaysia
e-mail: rmrezaur@unimas.my

M.R. Rahman, *Wood Polymer Nanocomposites*, Engineering Materials,
DOI 10.1007/978-3-319-65735-6_6

both researchers and industrialists during the last few years due to environment concerns. Chemical compositions of the wood are highly affected by the properties of the wood fibers. However, all these chemical compositions of the fibers depend on different factors. Few factors, namely geographic location, climate, plant part, and soil conditions are significantly affected by the chemical compositions of wood. In addition, the carbohydrate portion of fiber consists of cellulose and hemicelluloses, and lignocellulosic fibers. From this wood fiber, the percentage of cellulose is the highest which makes them the most abundant natural polymer. Cellulose is usually responsible for the strength of natural fibers. This was due to the specific properties such as high degree of polymerization and linear orientation. Since the past decade, various industries such as the automotive, construction, and packaging industries have shown their great interest in the development of new biocomposites materials.

Borneo Island is one of the regions that is full of natural rainforest. However, this precious wealth of nature is not being exploited for better end products. From the various types of natural wood species, *Eugenia* spp., *Artocarpus rigidus*, *Artocarpus elasticus* and *Xylopia* spp., and the *Koompassia malaccensis* are selected as they have high potential to be used as a reinforcing material in WPNCs.

Therefore, this study was carried out to fabricate WPNCs through the impregnation of urea-formaldehyde resin into five selective wood species as mentioned above. All the WPNCs undergo various characterizations. The expected outcome of this study was the enhanced absorption peak at 1648 cm^{-1} (O=N–), C–H absorption at 1319 cm^{-1} and –C–O– stretching band at 1262 cm^{-1} through FT-IR confirmed the impregnation. Besides, the investigation on TGA and DSC was expected to show that WPNCs had better thermal stability compare to raw wood. Dynamic Young's modulus was expected to prove that WPNCs had higher MOE and MOR value as well as the static Young's modulus. XRD analysis was expected to show higher crystallinity of WPNCs. Moreover, expected outcome from SEM analysis was to show the porous cells of raw wood filled by the urea-formaldehyde as reflected in thermal and mechanical properties.

2 Materials and Methods

2.1 Materials

There were five wood species were collected for this study which was classified as softwood and hardwood. The wood species were used in this study, namely *Eugenia spp., Artocarpus rigidus, Artocarpus elasticus, Xylopia spp.*, and *Koompassia malaccensis* respectively. For impregnation, chemicals, namely sodium hydroxide and urea-formaldehyde (Marck, Germany), were used as received. The purity grade of the chemicals was 99%.

2.2 *Manufacturing of Wood Polymer Nanocomposites*

All the wood samples were oven dried for 24 h and continuously soaked in sodium hydroxide solution for 5 h. The soaking temperature used was 100 °C. Oven-dried wood samples were impregnated with urea-formaldehyde resin to form WPNCs by undergoing a vacuum chamber at 25 °C and 60 cm Hg.

2.3 *Microstructural Characterizations*

The fabricated WPNCs were characterization using the Fourier Transform Infrared Spectroscopy (FT-IR), X-ray diffraction (XRD), Thermogravimetric Analysis (TGA), differential scanning calorimetric testing (DSC), Scanning Electron Microscopy analysis (SEM), free-free flexural vibration testing, three-point bending test, and compression parallel to grain testing.

3 Result and Discussion

3.1 *FT-IR*

Fourier Transform Infrared Spectroscopy (FT-IR) spectra of raw wood and WPNCs were shown in Fig. 1. The basic structure for most of the wood samples was OH stretching at 4000–3300 cm^{-1} and C–H stretching in methyl and methylene groups at 3000–2800 cm^{-1}. Besides, Owen and Thomas (1989) showed that strong broad superposition with sharp and discrete absorptions could be observed in the region from 1750 to 1000 cm^{-1}. The absorption band at 1508 cm^{-1} was detected due to the degradation of lignin while the absorption located at 1734 cm^{-1} was caused by hemicelluloses. C=O stretch in non-conjugated ketones, carbonyls, and ester groups were clearly proven through Fig. 1. In addition, the region between 1800 and 1100 cm^{-1} was assigned to the main components from wood, namely cellulose, hemicelluloses, and lignin.

Both absorbance values and shapes of the bands as well as their location showed clear difference as shown in Fig. 1. The less xylan content in softwood was evidenced by a carbonyl band at 1734 cm^{-1} especially for chemically modified wood. The peak at 1734 cm^{-1} was clearly this being shifted to a lower wave number value at 1648 cm^{-1}. Through the enhanced absorption peak at 1648 cm^{-1} that was representing O=N– groups, C–H absorption at 1319 cm^{-1} and –C–O– stretching band at 1262 cm^{-1}, this confirmed the impregnation of urea-formaldehyde into raw wood to improve the properties of WPNCs.

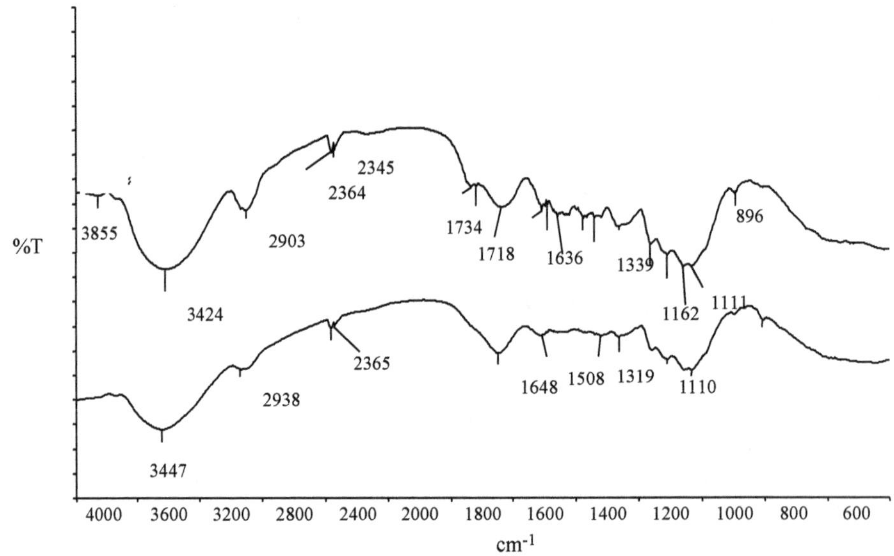

Fig. 1 IR spectrum of **a** raw wood and **b** WPNCs

3.2 TGA

Through Thermogravimetric Analysis (TGA), one of the common behaviors that the raw wood sample undergo was dehydration process, where approximately 5–8% of adsorbed water was removed. Based on the research work of Wielage et al. (1999), there was no degradation up to 160 °C. However, the thermal stability above this temperature gradually decreased as the thermal decomposition took place. The thermogravimetric curves for raw wood and WPNCs were obtained by dynamic scans and presented in Fig. 2a–e.

To ensure all the fabricated WPNCs performed well in many applications, thermal stability is a very important parameter to be investigated. There were few major chemical components, namely cellulose, hemicellulose, lignin, and extractives where all these components degraded at different temperatures. This was because different wood composition degraded at different temperatures, and thus, different degradation profiles were performed. Cellulose, as major percentage in the wood materials, was highly crystalline that created cellulose to become more thermally stable. On the other hand, the minor percentage, namely hemicelluloses and lignin were amorphous and these components started to degrade before cellulose (Autio and Miettinen 1970). This proved that hemicelluloses and lignin were the least thermally stable wood components, due to the presence of acetyl groups (Bourgois et al. 1989).

From Table 1, it is showed that WPNCs had higher decomposition temperature with lower weight loss rate compared to raw wood. In addition, WPNCs had higher activation energy compared to raw wood. It could be summarized that WPNCs were more thermally stable than raw wood in all selective wood species.

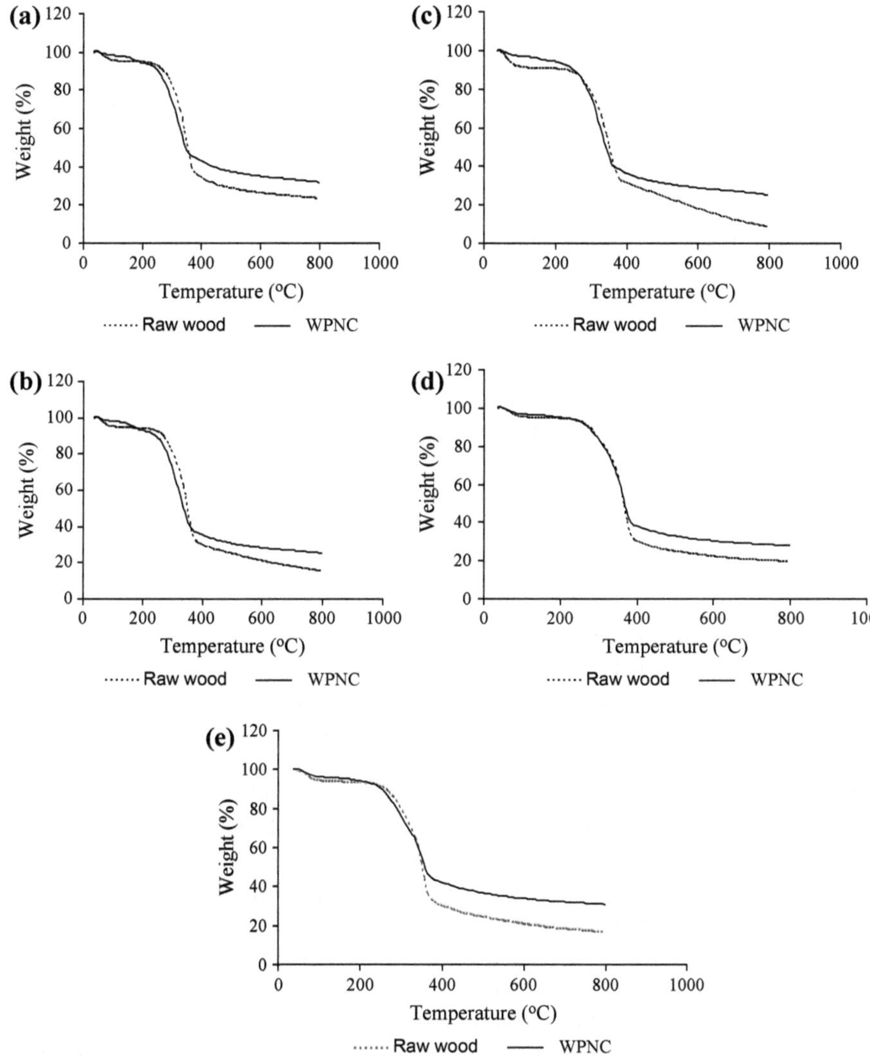

Fig. 2 **a** TGA curve of *Artocarpus elasticus* raw wood and WPNCs, **b** TGA curve of *Artocarpus rigidus* raw wood and WPNCs, **c** TGA curve of *Xylopia* spp. raw wood and WPNCs, **d** TGA curve of *Koompassia malaccensis* raw wood and WPNCs, **e** TGA curve of *Eugenia* spp. raw wood and WPNCs

3.3 DSC

Moisture content and volatile components contained within the raw wood and WPNCs were determined and investigated through differential scanning calorimetry (DSC). These two factors were important as the raw wood would deteriorate and

Table 1 Thermal characteristics of raw wood and WPNCs

Sample		T_i (°C)	T_m (°C)	T_f (°C)	W_{Ti} (%)	W_{Tm} (%)	W_{Tf} (%)	Activation Energy, E_a (J/K)
Artocarpus elasticus	Raw	201	300	419	98.96	64.73	32.67	33.11
	WPNCs	214	356	616	93.89	47.16	34.81	36.08
Artocarpus rigidus	Raw	211	310	388	99.67	56.45	30.63	25.79
	WPNCs	237	332	553	90.61	52.46	31.18	32.28
Xylopia spp.	Raw	212	321	420	98.78	56.15	25.85	27.04
	WPNCs	245	332	632	90.87	55.56	28.16	31.98
Koompassia malaccensis	Raw	212	355	403	98.99	57.91	30.00	16.12
	WPNCs	237	363	545	93.28	52.08	31.51	20.23
Eugenia spp.	Raw	214	340	443	99.77	60.32	27.51	19.17
	WPNCs	237	348	521	92.08	55.68	35.15	22.56

T_m Temperature corresponding to the maximum rate of mass loss
T_i Temperature corresponding to the beginning of the decomposition
T_f Temperature corresponding to the ending of the decomposition
W_{Ti}, W_{Tm}, and W_{Tf} Mass loss at T_i, T_m, and T_f

affect the great properties achieved (Herrera et al. 1997). DSC was carried out to determine the thermal behavior of the raw wood and fabricated WPNCs. In addition, DSC analysis also enabled the chemical impregnation to be clearly identified which could be carried out in the raw wood and WPNCs with the increment in temperature.

Figure 3a–e shows the DSC curves of raw wood and WPNCs. Besides, the enthalpy and exotherm peaks were shown in Table 2. From Fig. 3, there was a broad endotherm in the temperature range around 40–150 °C in both raw wood and WPNCs for the five selective wood species-indicated water molecules were presented within the wood fibers. Cellulose fibers especially lignin degraded rapidly at the temperature about 200 °C while the other polysaccharides especially on cellulose degraded at higher temperature (Akita and Kase 1967). Moreover, the temperature at 200 °C and above produced tar with scientific name 1, 6 anhydro-β-D-glucopyranose combined with polymeric materials (Peters and Still 1979). Therefore, the second endothermic peaks of all raw wood and WPNCs increased more than 200 °C temperature. This was due to the decomposition temperatures of the cellulose in the wood fibers. From Fig. 3, it was clearly observed that WPNCs showed a higher endothermic heat flow compared to the raw wood with exceptional on *Koompasia malacennis* and *Eugennia* spp. based on the first decomposition temperature. Sreekala et al. (1997) reported that thermal stability of WPNCs could be greatly enhanced due to the alkali and silane treatment. Table 2 showed that the crystallization enthalpy (ΔH_c) of WPNCs was two times higher compared with raw wood, which could be reflected in the XRD results. From this, it could be concluded that WPNCs were thermally more stable than raw wood on some of the selective wood species.

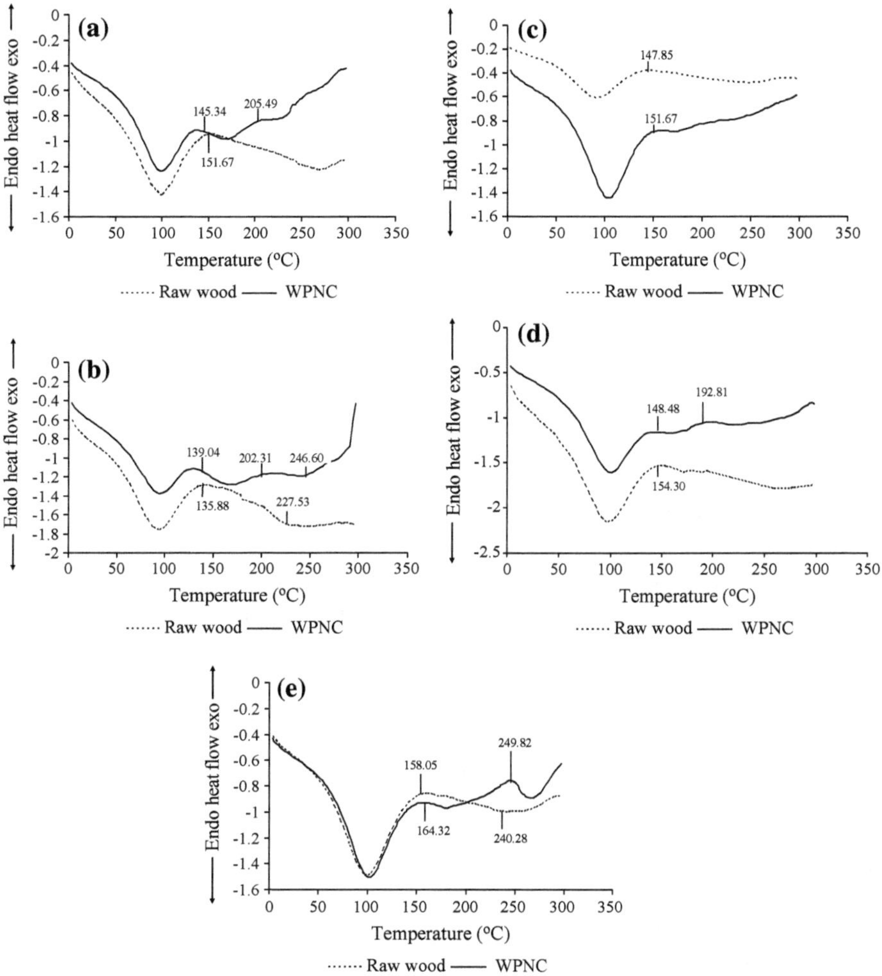

Fig. 3 **a** DSC thermographs of the *Artocarpus elasticus* raw wood and WPNCs, **b** DSC thermographs of the *Artocarpus rigidus* raw wood and WPNCs, **c** DSC thermographs of the *Koompasia malacennis* raw wood and WPNCs, **d** DSC thermographs of the *Xylopia* spp. raw wood and WPNCs, **e** DSC thermographs of the *Eugennia* spp. raw wood and WPNCs

Table 2 Crystallization enthalpy and exotherm peaks of raw wood and WPNC

Species	Crystallization enthalpy ΔH_c (J/gm)		1st Exotherm peaks (°C)	
	Raw wood	WPNC	Raw wood	WPNC
Artocarpus elasticus	126.39	233.10	151.78	145.35
Artocarpus rigidus	124.91	248.64	135.60	139.04
Koompasia malacennis	192.86	224.60	147.85	151.67
Xylopia spp.	192.29	323.00	145.30	151.78
Eugennia spp.	231.27	290.78	158.05	164.32

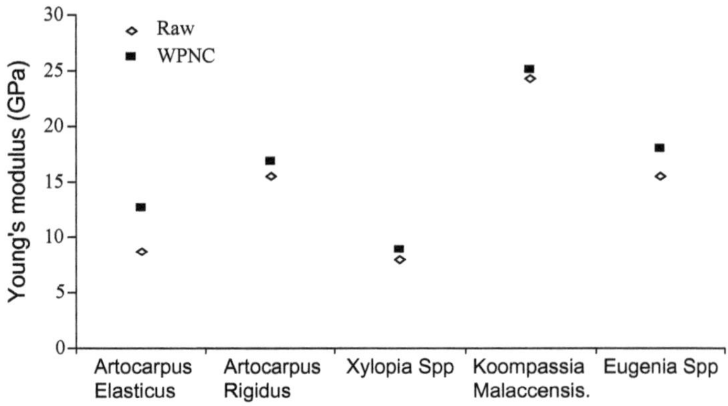

Fig. 4 E_d of raw wood, WPNCs for all species

3.4 Dynamic Young's Modulus Measurement

Figure 4 presented the dynamic Young's modulus of raw wood and WPNC through free-free flexural vibration testing system. All the results were based on ten specimens that were used for each wood species prepared. Through urea-formaldehyde impregnation, Young's modulus of WPNCs was increased which was according to the other researcher (Hamdan et al. 2010). Besides, Young's modulus of WPNCs on *Artocarpus elasticus*, *Artocarpus rigidus,* and *Eugenia* spp. was significantly higher than raw wood. On the other hand, Young's modulus of *Koompassia malaccensis* and *Xylopia* spp. was slightly higher among all the wood species due to their hardness. From Table 1, the activation energy was greatly increased due to the urea-formaldehyde impregnation. However, only WPNCs on *Artocarpus elasticus*, *Artocarpus rigidus,* and *Eugenia* spp., respectively, showed significant changes in elastic properties.

3.5 MOE and MOR Measurement

Both modulus of elasticity (MOE) and modulus of rupture (MOR) of raw wood and WPNCs were shown in Tables 3 and 4 as well as Figs. 5 and 6, respectively. MOE and MOR were investigated on the impregnation of urea-formaldehyde resin on the selective wood species. WPNCs on *Eugenia* spp. and *Artocarpus elasticus* were highest followed by, *Artocarpus rigidus, Xylopia* spp., and *Koompassia malaccensis*, respectively. WPNCs yielded higher MOE mainly due to the impregnation of resin into raw wood, which was in accordance with other researchers (Adams et al. 1970; Yildiz et al. 2005).

Table 3 summarized the MOE of raw wood and WPNCs. The result showed that WPNCs on *Artocarpus elasticus, Eugenia* spp., and *Xylopia* spp. were significantly

Table 3 *t*-test analysis of raw wood and wood polymer nanocomposites[a]

Treatment	Modulus of elasticity	*t*-test grouping[b]
Raw wood (*Artocarpus elasticus*)	6.48 ± 0.52	A
WPNCs (*Artocarpus elasticus*)	11.66 ± 0.60	B
Raw wood (*Artocarpus rigidus*)	5.13 ± 0.38	C
WPNCs (*Artocarpus rigidus*)	10.53 ± 2.60	D
Raw wood (*Xylopia* spp.)	6.71 ± 0.34	E
WPNCs (*Xylopia* spp.)	11.83 ± 0.67	E
Raw wood (*Koompassia malaccensis*)	15.68 ± 1.11	F
WPNCs (*Koompassia malaccensis*)	19.58 ± 2.58	F
Raw wood (*Eugenia* spp.)	9.25 ± 0.37	G
WPNCs (*Eugenia* spp.)	20.70 ± 0.83	H

[a]Each value is the average of 10 specimens
[b]The same letters are not significantly different at $\alpha = 5\%$

higher compared to raw wood. However, WPNCs on *Koompassia malaccensis* (hardwood) showed no significant effect due to its hardness.

Sodium hydroxide reacted with the cellulose in wood cells by reducing the water molecules from the raw wood. Wood cell walls plasticized with urea-formaldehyde resin to fill the void space in the raw wood to increase its stiffness during the impregnation process. With this, the MOE of all WPNCs was higher than raw wood as shown in Fig. 5.

Besides MOE, MOR of WPNCs especially on *Eugenia* spp. was significantly increased after the impregnation of urea-formaldehyde resin. Figure 6 shows that all the WPNCs showed higher MOR compared to raw wood which agreed with previous research (Adams et al. 1970). Table 4 indicates that the MOR of raw wood and WPNCs on *Xylopia* spp. and *Eugenia* spp. were significantly different. The

Table 4 *t*-test analysis of raw wood and wood polymer nanocomposites[a]

Treatment	Modulus of elasticity	*t*-test grouping[b]
Raw wood (*Artocarpus elasticus*)	76.17 ± 3.01	A
WPNCs (*Artocarpus elasticus*)	85.09 ± 3.44	A
Raw wood (*Artocarpus rigidus*)	33.02 ± 3.63	B
WPNCs (*Artocarpus rigidus*)	53.42 ± 11.72	B
Raw wood (*Xylopia* spp.)	45.91 ± 1.36	C
WPNCs (*Xylopia* spp.)	88.11 ± 3.78	D
Raw wood (*Koompassia malaccensis*)	122.20 ± 15.49	E
WPNCs (*Koompassia malaccensis*)	125.43 ± 24.36	E
Raw wood (*Eugenia* spp.)	46.10 ± 3.62	F
WPNCs (*Eugenia* spp.)	107.78 ± 13.44	G

[a]Each value is the average of 10 specimens
[b]The same letters are not significantly different at $\alpha = 5\%$

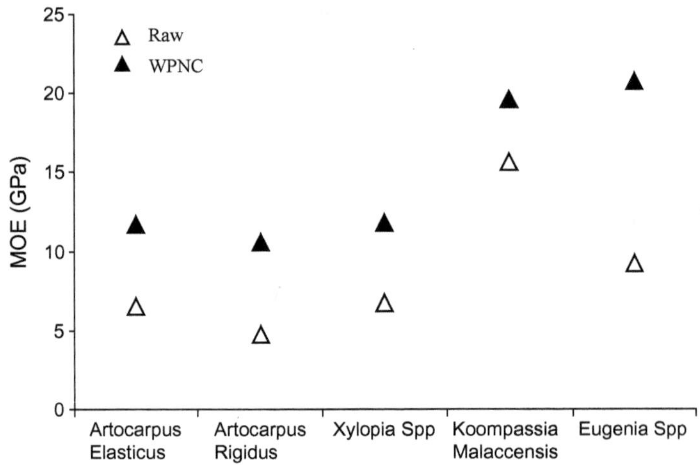

Fig. 5 MOE of raw wood and WPNCs

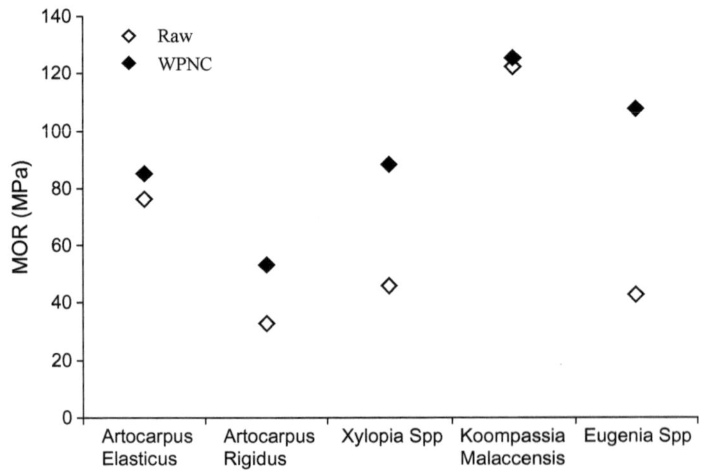

Fig. 6 MOR of raw wood and WPNCs

increment of MOR for WPNCs on *Eugenia* spp. was the highest followed by *Xylopia* spp., *Artocarpus rigidus*, *Artocarpus elasticus,* and *Koompassia malaccensis*, respectively. However, the value on raw wood and WPNCs for *Koompassia malaccensis* (hardwood) was almost similar which proved that the impregnation of urea-formaldehyde resin into hardwood was not effective as confirmed by our previous work (Rahman et al. 2010).

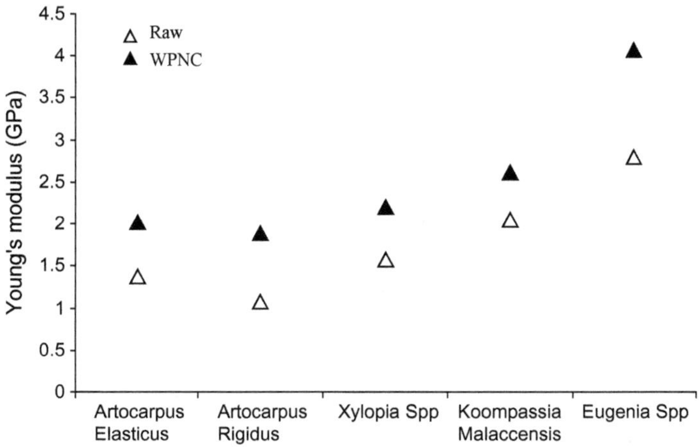

Fig. 7 Static Young's modulus of raw wood and WPNCs

3.6 *Static Young's Modulus* (E) *Measurement*

Figure 7 shows the static Young's modulus of raw wood and WPNCs which were determined from 10 repetitions. WPNCs of *Eugenia* spp. followed by *Artocarpus rigidus*, *Artocarpus elasticus*, *Xylopia* spp., *and Koompassia malaccensis*, respectively showed the highest increment of E value. Besides, Table 5 shows Young's modulus of WPNCs for *Eugenia* spp. was significantly different from raw wood. Rahman et al. (2010) proved that WPNCs showed higher increment of *E* in WPNCs compared to raw wood. This characterization confirmed that the introduction of urea-formaldehyde resin plasticized on the wood cell walls as well as greatly increased their lateral stability.

Table 5 *t*-test analysis of raw wood and wood polymer nanocomposites[a]

Treatment	Static Young's modulus	*t*-test grouping[b]
Raw wood (*Artocarpus elasticus*)	1.36 ± 0.23	A
WPNCs (*Artocarpus elasticus*)	2.01 ± 0.37	A
Raw wood (*Artocarpus rigidus*)	1.88 ± 0.58	B
WPNCs (*Artocarpus rigidus*)	1.07 ± 0.07	B
Raw wood (*Xylopia* spp.)	1.57 ± 0.39	C
WPNCs (*Xylopia* spp.)	2.19 ± 0.25	C
Raw wood (*Koompassia malaccensis*)	2.05 ± 0.18	D
WPNCs (*Koompassia malaccensis*)	2.60 ± 0.49	D
Raw wood (*Eugenia* spp.)	1.67 ± 0.60	E
WPNCs (*Eugenia* spp.)	4.06 ± 0.79	E

[a]Each value is the average of 10 specimens
[b]The same letters are not significantly different at $\alpha = 5\%$

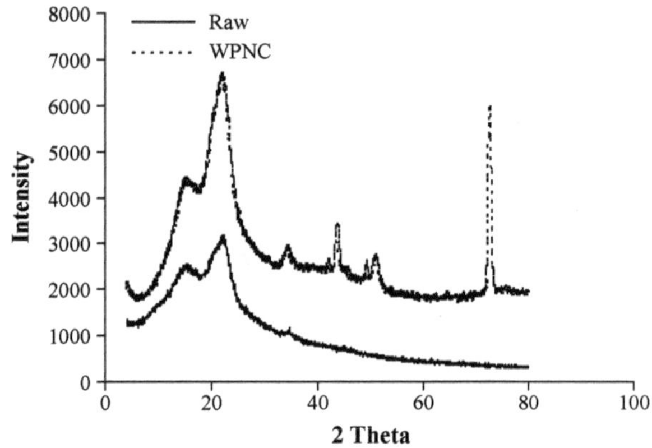

Fig. 8 X-ray diffraction of raw wood and WPNCs

3.7 XRD Analysis

Figure 8 shows the X-ray diffraction patterns of raw wood and WPNCs. From Fig. 8, raw wood only showed one well-defined 2θ peak at 22.0° which was corresponded to cellulose (Stamm 1964). On the other hand, WPNCs exhibited five broad 2θ peaks at 43.67°, 49.14°, 50.97°, 72.59°, and 75.73°. All these peaks were due to the incorporation of urea-formaldehyde resin into raw wood through impregnation. According to Stamm (1964), the crystallinity of WPNCs was greatly affected by the chemical modification of lignocellulosic materials. The urea-formaldehyde resin firstly reacted with the chain ended on the surface of crystallites. This occurred because the resins could not diffuse into the crystalline region which resulted in the opening of some of the hydrogen-bonded cellulose chains. From this, it showed that the mercerization removed the raw wood fiber's amorphous constituents upon impregnation of resin which increased the crystallinity of WPNCs.

3.8 SEM

SEM micrographs of raw wood and WPNCs were analyzed as shown in Fig. 9. Figure 9a shows the surface morphology of raw wood which consisted of many small void space. From Fig. 9b, the introduction of urea-formaldehyde resin into raw wood filled all the void space. This showed that the impregnation of urea-formaldehyde resin into raw wood increased the activation energy, crystallization enthalpy, and final decomposition temperature. Besides, WPNCs showed smooth surfaces with the waxy substance removed from the raw wood which was confirmed by our previous work (Rahman et al. 2010).

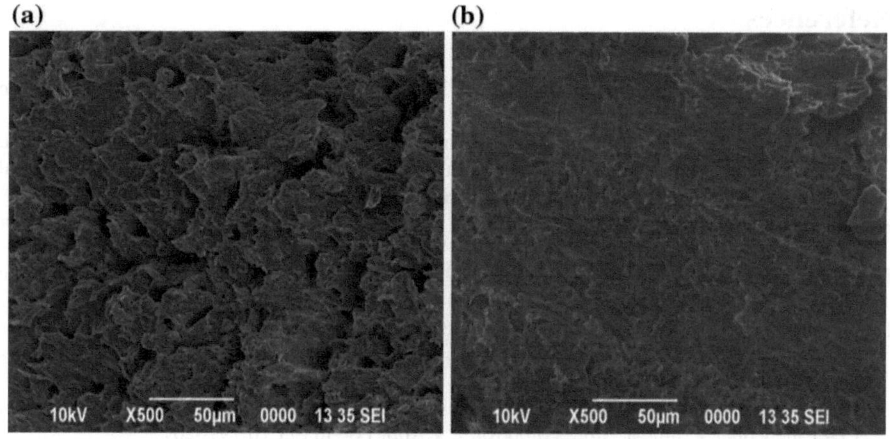

Fig. 9 Scanning Electron Microscopy of **a** raw wood and **b** WPNC

4 Conclusion

In this study, urea-formaldehyde resin WPNCs were fabricated and investigated through different characterizations. From FT-IR spectra, the impregnation of organic urea-formaldehyde resin into raw wood enhanced few peaks at 1648 cm^{-1} (O=N–), C–H absorption at 1319 cm^{-1} and –C–O– stretching band at 1262 cm^{-1}. These peaks confirmed the impregnation of WPNCs. Besides, WPNCs were more thermally stable over temperature range compared to raw wood through the well impregnation of urea-formaldehyde resin WPNCs. In addition, WPNCs showed higher MOE, MOR and stiffness of the WPNC were significantly increased compared to raw wood on *Eugenia* spp. and *Xylopia* spp., respectively. Besides, Young's modulus of *Eugenia* spp. was significantly higher for WPNCs. The X-ray diffraction patterns confirmed that the crystallinity of WPNCs was increased with polymer resin loading. From SEM micrograph, the surface morphology of WPNCs was smooth as the void space was filled by urea-formaldehyde resin to remove the waxy substance. It could be proven that urea-formaldehyde resin was significantly effective on *Eugenia* spp., continued with *Xylopia* spp. and *Artocarpus elasticus* wood species, respectively.

Acknowledgements The authors would like to acknowledge the financial support from Ministry of Higher Education Malaysia, for their financial support [Grant no. FRGS/02(05)/655/2007(20)] during the research.

References

Adams DG, Choong ET Mcllhenny RC (1970) Bending strength of radiation-produced southern pine wood-plastic combinations. Forest Prod J 20(4):25–28

Akita K, Kase M (1967) Determination of kinetic parameters for pyrolysis of cellulose and cellulose treated with ammonium phosphate by differential thermal analysis and thermal gravimetric analysis. J Polym Sci 1(5):833–848

Autio T, Miettinen JK (1970) Experiments in Finland on properties of wood-polymer combinations. Forest Prod J 20(3):36–42

Bourgois J, Bartholin MC, Guyonnet R (1989) Thermal Treatment of wood: analysis of the obtained product. Wood Sci Technol 23:303–310

Hamdan S, Talib ZA, Rahman MR, Ahmed AS, Islam MS (2010) Dynamic Young's modulus measurement of treated and post-treated tropical wood polymer composites (WPC). BioRes 5(1):324–342

Herrera-France P, Aguilar-Vega M (1997) Effect of fiber treatment on the mechanical properties of LDPE-henequen cellulosic fibre composites. J Appl Polym Sci 10:197–207

Owen NL, Thomas DW (1989) Infrared studies of hard and soft woods. Appl Spec 43:451–455

Peters R, Still R (1979) Some aspects of the degradation of polymers used in textile applications. In: Happey F (ed) Applied Fiber Science, 2nd edn. Academic Press, London, pp 321–420

Rahman MR, Hamdan S, Saleh AA, Islam MS (2010) Mechanical and biological performance of sodium mataperiodate impregnated plasticized wood (PW). BioRes 5(2):1022–1035

Sreekala M, Kumaran M, Thomas S (1997) Oil plam fibres: morphology, chemical composition, surface modification and mechanical properties. J Appl Polym Sci 66:821–835

Stamm AJ (1964) Factors affecting the bulking and dimensional stabilization of wood with polyethylene glycols. Forest Prod J 14:403–408

Wielage B, Lampke Th, Mark G, Nestler K, Starke D (1999) Thermogravimetric and differential scanning calorimetric analysis of natural fibers and polypropylene. Thermochim Acta 337: 169–177

Yildiz CU, Yildiz S, Gezer DE (2005) Mechanical properties and decay resistance of wood polymer composites prepared from fast growing species in Turkey. Biores Technol 96: 1003–1011

Characterization of Epoxy/Nanoclay Wood Polymer Nanocomposites (WPNCs)

M.R. Rahman

Abstract From the present work, the fabrication of epoxy/MMT wood polymer nanocomposites (WPNCs) was investigated. From FT-IR characterization, it confirmed the C–O stretch of C–O–H in starch at 1232 and 1182 cm^{-1} as well as the C–O stretch of C–O–C in starch at 1029 cm^{-1} with the decreasing wave number. This proved that raw wood was well impregnated by epoxy/MMT. In addition, Thermogravimetric Analysis (TGA) proved that WPNCs were more thermally stable over temperature compared to raw wood due to the high impact of montmorillonite (MMT) on wood. The stiffness, modulus of elasticity (MOE), and modulus of rupture (MOR) were significantly increased on WPNCs of *Eugenia* spp., *Xylopia* spp., *Artocarpus Rigidus,* and *Artocarpus Elasticus* compared with raw wood. From X-ray diffraction patterns, the addition of epoxy/MMT improved the crystallinity of WPNCs at the amorphous region. SEM analysis showed that the void space in raw wood was fully filled with epoxy/MMT, and the waxy substances were removed. It could be concluded that epoxy/MMT was significantly effective on *Eugenia* spp., followed by *Xylopia* spp., *Artocarpus Rigidus,* and *Artocarpus Elasticus,* respectively.

Keywords Mechanical properties · XRD · SEM · Wood polymer nanocomposites (WPNCs)

1 Introduction

One of the common polymers that are widely applied in the material yield is the epoxy resins. According to the previous research, this resin can be used as impregnating materials, adhesives, or matrices for nanocomposites (Alexandre and Dubois 2000; Basara et al. 2005; Becker et al. 2002; Chen and Yang 2002; Chen et al. 2004; Evans and Canfer 2000; Kong and Park 2003; Nigam et al. 2004; Ueki

M.R. Rahman (✉)
Faculty of Engineering, Universiti Malaysia Sarawak,
94300 Kota Samarahan, Sarawak, Malaysia
e-mail: rmrezaur@unimas.my

M.R. Rahman, *Wood Polymer Nanocomposites*, Engineering Materials,
DOI 10.1007/978-3-319-65735-6_7

et al. 2005). Epoxy resins are highly used as they are good electric insulators, having good chemical resistance but low shrinkage during cure. Besides, they have good thermal characteristics and simple processing stage. However, one major problem with the introduction of epoxy resins on engineering applications due to their low stiffness and strength when compared with metals (Basara et al. 2005). To ensure the drawback of pure epoxy resin to be offset, a good method to overcome this drawback is to incorporate reinforcing fillers. One of the common reinforcing fillers is montmorillonite (MMT). MMT has large aspect ratio and multiscale structure which can be improve properties of epoxy resins especially on gas barrier and thermomechanical properties (Becker et al. 2002; Chen and Yang 2002; Lebaron et al. 1999).

Based on the previous research work, MMT was introduced to effectively improve the properties of epoxy resins (Chen et al. 2004; Park and Jana 2003; Ray and Okamoto 2003). Besides, Nigam et al. (2004) investigated the epoxy related WPNCs were prepared and the tensile strength could be increased by 20%. Moreover, the addition of 0.5 wt% cloisite 30B improved the impact strength of an epoxy resin up to 137% (Basara et al. 2005). Through the improvement on epoxy resins, all the epoxy related WPNCs were developed in space industry and superconductive technologies (Evans and Canfer 2000). From the research carried out by Lam et al. (2005), the study proved that epoxy related WPNCs were successfully applied that greatly depended on their superior performances at extreme low temperature (such as liquid nitrogen and liquid helium temperature), particularly for reusable launch vehicles. In addition, MMT could be added to effectively improved the properties of epoxy resins at room temperature (Chen and Yang 2002; Chen et al. 2004; Evans and Canfer 2000; Lam et al. 2005; Nigam et al. 2004; Ray and Okamoto 2003).

It was clearly showed that the introduction of epoxy/MMT into raw wood helped to improve the properties of WPNCs. These WPNCs were undergo different characterizations to ensure the safety of the applications at the elevated temperature especially on the thermal and mechanical properties. Based on the previous literature, no work has been reported on the physical, thermal, and mechanical properties of epoxy/MMT impregnated WPNCs. In this present study, the aim is to investigate the physical, mechanical, and thermal properties of epoxy/MMT impregnated WPNCs as well as the effect same polymer resin with nanofiller into different types of wood species.

2 Materials and Methods

2.1 Materials

In this present study, there were five wood species were collected and used. There were two types of wood used namely softwood and hardwood. The softwoods namely *Eugenia* spp., *Artocarpus Rigidus*, *Artocarpus Elasticus* and *Xylopia* spp. while the hardwood namely *Koompassia Malaccensis* was used. For chemicals, sodium hydroxide (NaOH) and *N,N*-Dimethylacetamid (Merck, Germany) were used to treat the raw wood. The purity grade of the chemicals were 99%.

2.2 Preparation of Solution Through Impregnation

In this study, clay powders were used as received and the average particle size was 8 μm. 1% layered aluminosilicate nanofiller was introduced into the epoxy resin at a mixing speed of 2050 rpm for 20 min to form impregnation solutions to be well impregnated the different raw wood species.

2.3 Manufacturing of Wood Polymer Nanocomposites

All the raw woods were oven dried at 103 °C for 24 h until it reached constant weight. They were then placed into an impregnation chamber with no contact between samples and fully covered by epoxy/MMT resin mixtures. The mixture was evacuated for 30 min up to 60 mmHg where compressed air was applied to the system and maintained for 30 min at 0.52 MPa then released. The excess chemicals were wiped off from the samples. All the prepared specimens were weighted after dried under air circulation for 24 h, continued with oven drying at 90 °C for 24 h. The excess polymer was then removed from the surface of dried WPNCs samples. The resulting WPNCs were characterized through Fourier Transform Infrared Spectroscopy (FT-IR) analysis, X-ray diffraction (XRD) analysis, Thermogravimetric Analysis (TGA), Scanning Electron Microscopy analysis (SEM), free-free flexural vibration testing, three-point bending test, and compression parallel to grain testing.

3 Result and Discussion

3.1 Fourier Transform Infrared Spectroscopy (FT-IR) Analysis

Figure 1 showed the FT-IR spectra of raw wood and WPNCs on five different wood species. There were two main parts of the spectra namely the OH stretching vibrations in the range of 4000–2700 cm^{-1} and fingerprint region in the range of 1800–400 cm^{-1}. The peak at 4000–3300 cm^{-1} represented the strong band OH stretching while C–H stretching in methyl and methylene groups was observed in

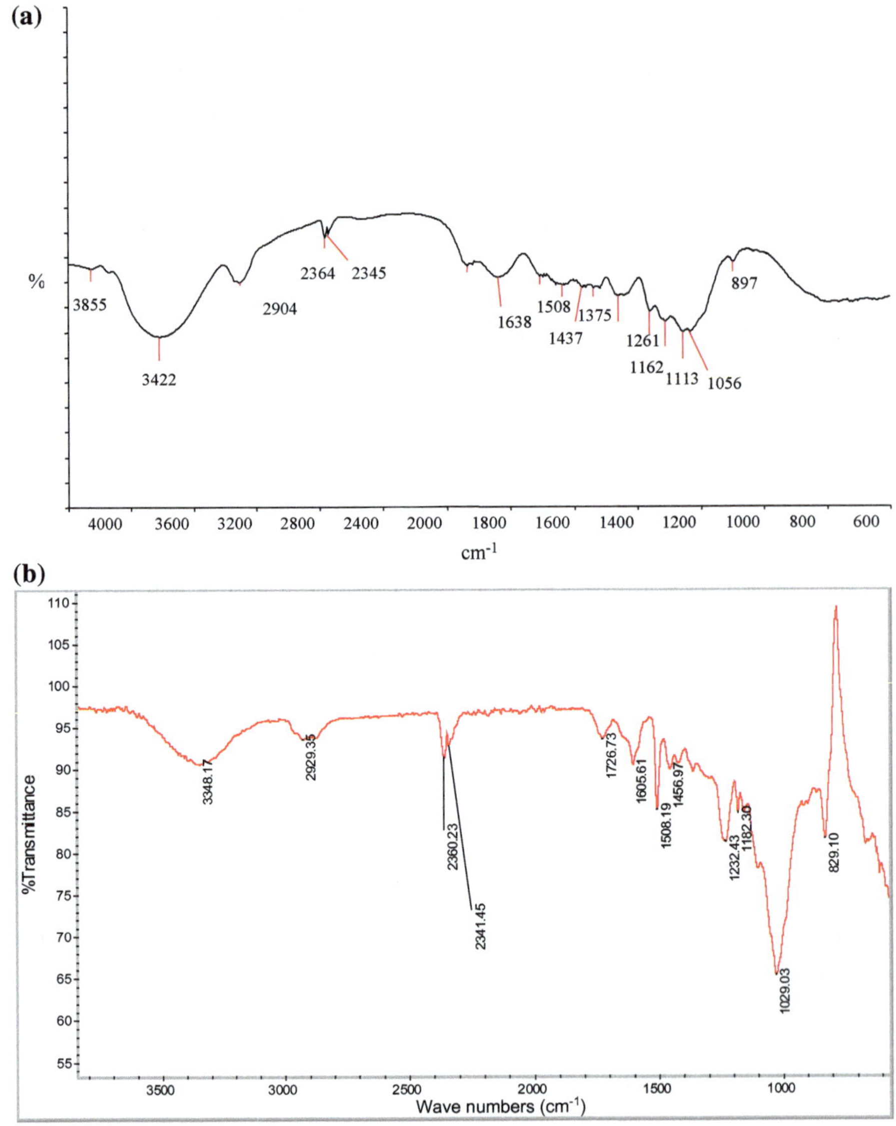

Fig. 1 **a** IR spectrum of raw wood, **b** IR spectrum of WPNCs

the range of 3000–2800 cm^{-1}. In addition, within the peak range of 1750–1000 cm^{-1}, a strong broad superposition with sharp and discrete absorptions were investigated (Owen and Thomas 1989). The absorption band at 1508 cm^{-1} was due to lignin degradation while the absorption band located at 1735 cm^{-1} was caused by hemicelluloses. Based on Owen and Thomas (1989), the C=O stretch occurred in non-conjugated ketones, carbonyls, and ester groups. For the peak region between 1800 and 1100 cm^{-1}, this was attributed to the main components from wood which were cellulose, hemicelluloses, and lignin. Less xylan content in softwood was proven by a carbonyl band at 1735 cm^{-1}. The peak for epoxy/MMT modified wood was shifted to a lower wave number value which was 1508 cm^{-1}. On the other hand, the introduction of epoxy/MMT decreased wave number at 1232 and 1182 cm^{-1} that attributed to C–O stretch of C–O–H in starch and at 1029 cm^{-1} that attributed to C–O stretch of C–O–C in starch which confirmed the impregnation.

3.2 *Thermogravimetric Analysis (TGA)*

Figure 2a–e showed the TGA curves of both raw wood and WPNCs. The general thermal decomposition of cellulose and hemicellulose started at 180 °C. In addition, the maximum value of mass loss rate was obtained at 280 °C which increased significantly over the temperature. Figure 2a–e and Table 1 showed that the raw wood started the thermal decomposition on *Artocarpus Elasticus, Artocarpus Rigidus, Xylopia* spp., *Koompassia Malaccensis* and *Eugenia* spp., respectively, at 201, 211, 212, 212, and 214 °C. However, WPNCs on A*rtocarpus Elasticus* spp. started the decomposition the earliest at 216 °C, continued with WPNCs on *Artocarpus Rigidus* at 240 °C, *Xylopia* spp. at 247 °C, *Koompassia Malaccensis* at 237 °C, and *Eugenia* spp. at 244 °C, respectively. At the end of the thermal degradation, all the WPNCs showed higher decomposition temperature compared to the raw wood. Besides, all the WPNCs showed higher activation energy compared to raw wood, which was well reflected on the mechanical properties and this behavior had been reported by Varhegyi et al. (1997). From TGA result, it was clearly summarized that WPNCs performed better in thermal stability than raw wood.

3.3 *Dynamic Young's Modulus Measurement*

Dynamic Young's modulus of raw wood and WPNCs were investigated through free-free flexural vibration testing system as shown in Fig. 3. This analysis was carried out based on ten replicates for each species. It was clearly proven that the introduction of epoxy/MMT improved the Young's modulus, which satisfied the statement from our others previous work (Hamdan et al. 2010). From Fig. 3, Young's modulus of WPNCs on *Eugenia* spp., *Xylopia* spp., *Artocarpus Rigidus,* and *Artocarpus Elasticus*, was significantly higher compared to raw wood.

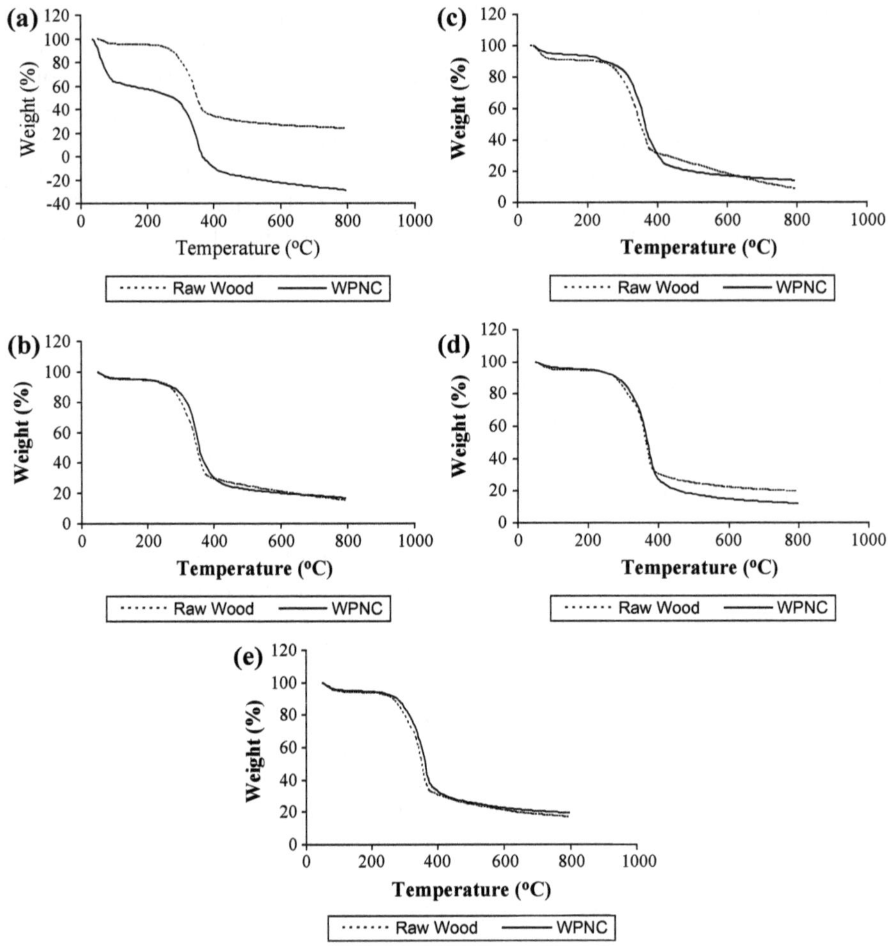

Fig. 2 **a** TGA curve of *Artocarpus Elasticus* raw wood and WPNCs, **b** TGA curve of *Artocarpus Rigidus* raw wood and WPNCs, **c** TGA curve of *Xylopia* spp. raw wood and WPNCs, **d** TGA curve of *Koompassia Malaccensis* raw wood and WPNCs, **e** TGA curve of *Eugenia* spp. raw wood and WPNCs

However, Young's modulus of WPNCs on *Koompassia Malaccensis* was slightly higher among all the wood species due to their hardness cell wall density. From this analysis, the activation energy was greatly affected by the introduction of nanofiller MMT through impregnation as shown in Table 1. This concluded that WPNCs on *Eugenia* spp., *Xylopia* spp., *Artocarpus Rigidus,* and *Artocarpus Elasticus,* respectively, showed significant changes in elastic properties.

Table 1 Thermal characteristics of raw wood and WPNCs

Sample		T_i (°C)	T_m (°C)	T_f (°C)	W_{Ti} (%)	W_{Tm} (%)	W_{Tf} (%)	Activation energy, E_a (J/K)
Artocarpus Elasticus	Raw	201	300	419	98.96	64.73	32.67	33.11
	WPNCs	216	352	624	94.87	45.26	35.81	37.05
Artocarpus Rigidus	Raw	211	310	388	99.67	56.45	30.63	25.79
	WPNCs	240	326	557	91.65	51.36	32.18	33.56
Xylopia spp.	Raw	212	321	420	98.78	56.15	25.85	27.04
	WPNCs	247	328	636	93.89	54.46	29.16	33.89
Koompassia Malaccensis	Raw	212	355	403	98.99	57.91	30.00	16.12
	WPNCs	237	359	545	96.24	53.08	33.51	22.16
Eugenia spp.	Raw	214	340	443	99.77	60.32	27.51	19.17
	WPNCs	244	340	526	95.18	53.58	36.15	22.87

T_m temperature corresponding to the maximum rate of mass loss
T_i temperature corresponding to the beginning of the decomposition
T_f temperature corresponding to the ending of the decomposition
W_{Ti}, W_{Tm}, and W_{Tf} mass loss at T_i, T_m, and T_f

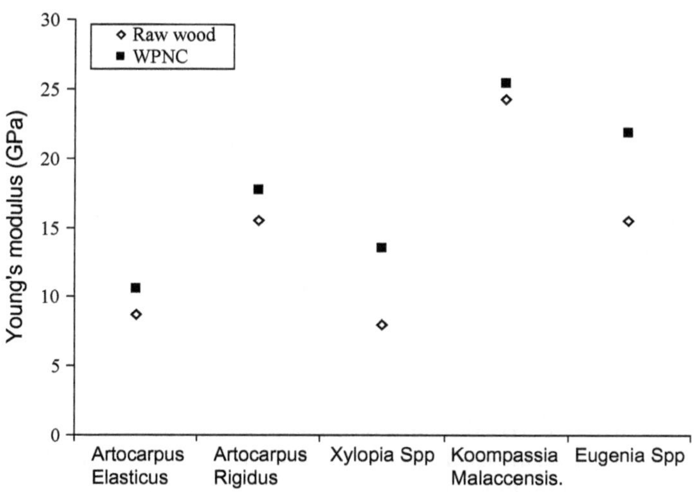

Fig. 3 E_d of raw wood, WPNCs for all species

3.4 Modulus of Elasticity (MOE) and Modulus of Rupture (MOR) Measurement

MOE and MOR for both raw wood and WPNCs were shown in Tables 2 and 3 as well as Figs. 4 and 5, respectively. Through this test, the impregnation of epoxy/MMT on five selective wood species was investigated. From Fig. 2, it was

Table 2 *t*-test analysis of raw wood and wood polymer nanocomposites[a]

Treatment	Modulus of elasticity	*t*-test grouping[b]
Raw wood (*Artocarpus Elasticus*)	6.48 ± 0.52	A
WPNCs (*Artocarpus Elasticus*)	11.78 ± 0.52	B
Raw wood (*Artocarpus Rigidus*)	5.13 ± 0.38	C
WPNCs (*Artocarpus Rigidus*)	11.51 ± 2.24	D
Raw wood (*Xylopia* spp.)	6.71 ± 0.34	E
WPNCs (*Xylopia* spp.)	13.34 ± 0.62	F
Raw wood (*Koompassia Malaccensis*)	15.68 ± 1.11	G
WPNCs (*Koompassia Malaccensis*)	19.45 ± 2.60	G
Raw wood (*Eugenia* spp.)	9.25 ± 0.37	H
WPNCs (*Eugenia* spp.)	23.83 ± 0.78	I

[a]Each value is the average of 10 specimens
[b]The same letters are not significantly different at $\alpha = 5\%$

Table 3 *t*-test analysis of raw wood and wood polymer nanocomposites[a]

Treatment	Modulus of rupture	*t*-test grouping[b]
Raw wood (*Artocarpus Elasticus*)	76.17 ± 3.01	A
WPNCs (*Artocarpus Elasticus*)	87.53 ± 2.43	B
Raw wood (*Artocarpus Rigidus*)	33.02 ± 3.63	C
WPNCs (*Artocarpus Rigidus*)	68.63 ± 9.27	D
Raw wood (*Xylopia* spp.)	45.91 ± 1.36	E
WPNCs (*Xylopia* spp.)	90.46 ± 2.31	F
Raw wood (*Koompassia Malaccensis*)	122.20 ± 15.49	G
WPNCs (*Koompassia Malaccensis*)	128.32 ± 21.71	G
Raw wood (*Eugenia* spp.)	46.10 ± 3.62	H
WPNCs (*Eugenia* spp.)	108.15 ± 11.61	I

[a]Each value is the average of 10 specimens
[b]The same letters are not significantly different at $\alpha = 5\%$

significant that MOE of WPNCs on *Eugenia* spp. and *Xylopia* spp. was the highest. For WPNCs on *Artocarpus Rigidus, Artocarpus Elasticus, and Koompassia Malaccensis* spp., the MOE values were slightly lower. Cai et al. (2008) proved that WPNCs performed higher MOE than raw wood as MMT was well reacted with the wood cell wall.

MOE of WPNCs on *Eugenia* spp., *Xylopia* spp., *Artocarpus Elasticus,* and *Artocarpus Rigidus* was significantly higher compared with raw wood. On the other hand, WPNCs on *Koompassia Malaccensis* (hardwood) showed no significant effect of epoxy/MMT impregnation. However, the MOE was slightly higher than the value of our previous work (Rahman et al. 2010).

Epoxy/MMT plasticized on the cellulose and hemicellulose in wood cells which managed to reduce the water molecules from the wood specimens. This was due to the impregnation of MMT that reacted as a binder which filled the void space in the

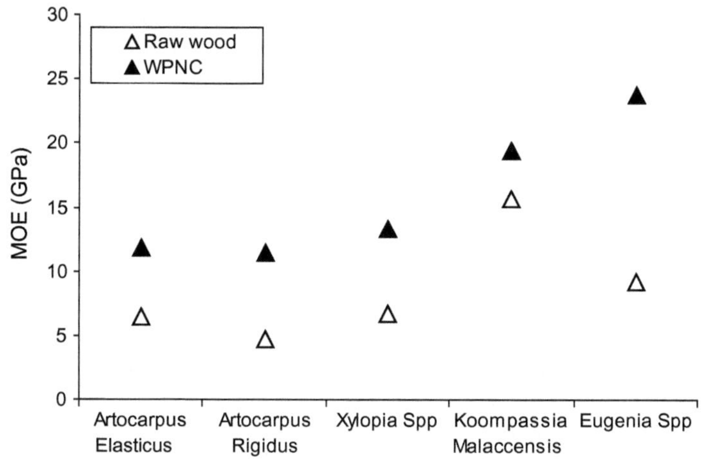

Fig. 4 MOE of raw wood and WPNCs

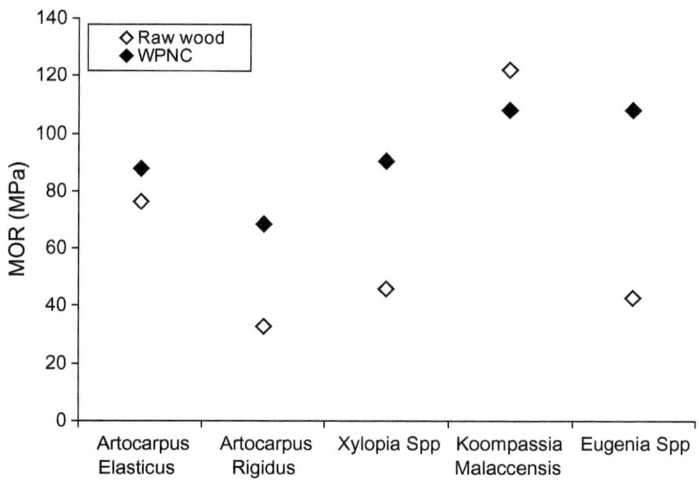

Fig. 5 MOR of raw wood and WPNCs

wood specimens to increase it stiffness. Therefore, MOE of all WPNCs was higher than that of raw wood as shown in Fig. 4.

Besides MOE, WPNCs showed higher MOR on *Eugenia* spp. due to the well impregnation of epoxy/MMT impregnation as shown in Fig. 5. Based on the previous research, MOR of WPNCs improved with the incorporation of suitable filler into the wood cell wall (Cai et al. 2007, 2008). All the MOR values with standard deviation was well tabulated in Table 3. Both raw wood and WPNCs on *Xylopia* spp., *Artocarpus Rigidus, Artocarpus Elasticus,* and *Eugenia* spp. showed different values due to the presence of epoxy/MMT. WPNCs on *Xylopia* spp., *Artocarpus*

Rigidus, Artocarpus Elasticus, and Koompassia Malaccensis, respectively, performed better than raw wood. However, WPNCs on *Koompassia Malaccensis* (hardwood) was the highest among all the samples prepared.

3.5 Static Young's Modulus (E) *Measurement*

This measurement was carried out based on the 10 replicates and the result was presented in Fig. 6. *Eugenia* spp. WPNCs showed the highest increment on static Young's modulus (*E*) continued by *Artocarpus Rigidus, Artocarpus Elasticus, Xylopia* spp., *and Koompassia Malaccensis* WPNCs, respectively. Besides, the significant difference of static Young's modulus on raw wood was clearly summarized in Table 4. The increment of E in WPNCs was supported by the research work of Cai et al. 2007, 2008 and Rahman et al. 2010. Epoxy/MMT was introduced not only plasticized on the wood cell wall, it also helped to improve their lateral stability.

3.6 X-ray Diffraction (XRD) Analysis

X-ray diffraction results for both raw wood and WPNCs were presented in Fig. 7. Both raw wood and WPNCs showed diffraction peaks. The 2θ angles of raw wood were detected at 17°, 22.5°, and 35°. According to Mulinari et al. (2010), all these 2θ angles were corresponded to the cellulose crystallographic planes I101, I002, and I040, respectively. From Fig. 7, it was clearly detected the position of these peaks did not change as the structure of cellulose did not change compared to raw wood. On the other hand, six broad 2θ peaks were observed at 42.09°, 43.70°, 49.16°, 50.86°, 72.82°, and 75.72°, respectively, which confirmed the crystallinity of WPNCs in amorphous region.

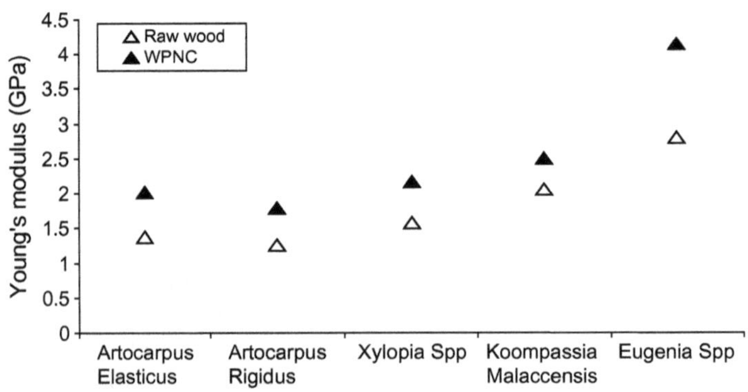

Fig. 6 Static Young's modulus of raw wood and WPNCs

Table 4 *t*-test analysis of raw wood and wood polymer nanocomposites[a]

Treatment	Static Young's modulus	*t*-test grouping[b]
Raw wood (*Artocarpus Elasticus*)	1.36 ± 0.23	A
WPNCs (*Artocarpus Elasticus*)	2.02 ± 0.35	A
Raw wood (*Artocarpus Rigidus*)	1.26 ± 0.04	B
WPNCs (*Artocarpus Rigidus*)	1.79 ± 0.56	B
Raw wood (*Xylopia* spp.)	1.57 ± 0.39	C
WPNCs (*Xylopia* spp.)	2.17 ± 0.26	C
Raw wood (*Koompassia Malaccensis*)	2.05 ± 0.18	D
WPNCs (*Koompassia Malaccensis*)	2.50 ± 0.46	D
Raw wood (*Eugenia* spp.)	1.67 ± 0.60	E
WPNCs (*Eugenia* spp.)	4.13 ± 0.78	F

[a]Each value is the average of 10 specimens
[b]The same letters are not significantly different at $\alpha = 5\%$

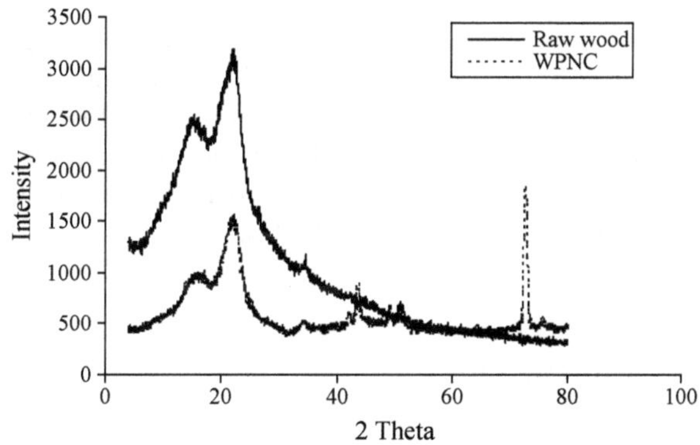

Fig. 7 X-ray diffraction of raw wood and WPNCs

All the peaks detected from Fig. 7 were attributed to modification on the raw wood with the incorporation of epoxy/MMT polymer. Polymerization occurred when epoxy/MMT polymer was well introduced into the lumen of wood cell. This formed strong crosslinking bonds between the groups of polymer and the surface hydroxyl (OH) groups of wood species.

3.7 Scanning Electron Microscopy (SEM) Analysis

Figure 8 showed the surface morphology of raw wood as well as the WPNCs were observed through SEM. Besides, the fracture surfaces of the notched bending samples were clearly investigated through SEM images as shown in Fig. 8a. This showed that raw wood was not homogeneous aggregated and this prevented the surface morphology to be rough and uneven. On the other hand, for WPNCs, the

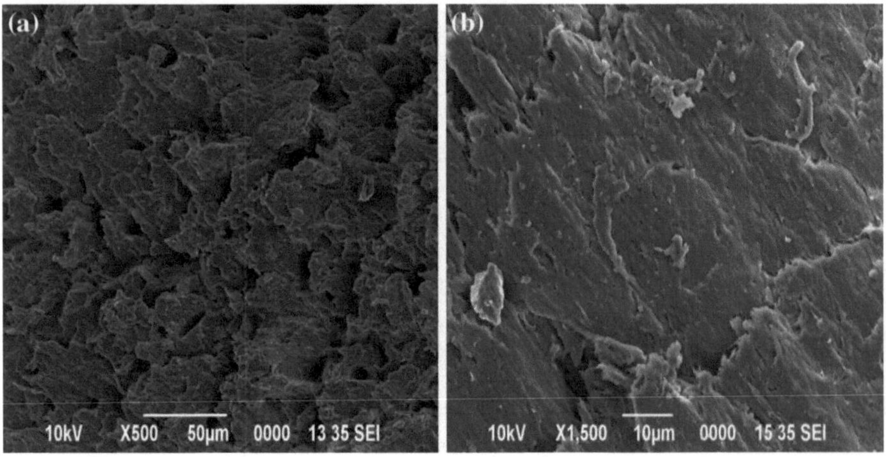

Fig. 8 Scanning Electron Microscopy of **a** Raw wood and **b** WPNCs

surface was smooth and compacted with homogenous materials as shown in Fig. 8b. This was due to the well impregnation of epoxy/MMT into the wood cell and created the plasticization on the cell wall (Rahman et al. 2010).

This analysis summarized that the introduction of epoxy and MMT aided to reduce the aggregation of MMT particles within the cell wall that created high impact on the stiffness of WPNCs.

4 Conclusion

The present work investigated epoxy/MMT WPNCs. All the samples undergo different characterizations to ensure the WPNCs fabricated performed better than raw wood. From FT-IR spectra, the decrement of wave number at 1232 and 1182 cm^{-1} was ascribed to C–O stretch of C–O–H in starch while the peak at 1029 cm^{-1} was ascribed to C–O stretch of C–O–C in starch. Through these peaks, it confirmed the impregnation of epoxy/MMT into raw wood sample. Besides, WPNCs showed higher thermal stability compared to raw wood through Thermogravimetric Analysis (TGA). The MOE, MOR, Young's modulus, and stiffness of WPNCs were significantly increased for all the five types of wood species namely *Eugenia* spp., *Xylopia* spp., *Artocarpus Rigidus* and *Artocarpus Elasticus,* respectively, compared to raw wood. The crystallinity of WPNCs were significantly increased at the amorphous region through X-ray diffraction patterns due to epoxy/MMT loading. From SEM analysis, WPNCs showed smooth surface as the void was fully filled by the epoxy/MMT and at the same time, the waxy substace was removed. This study summarized that the introduction of epoxy/MMT into raw wood improved its properties especially on *Eugenia* spp. followed by *Xylopia* spp., *Artocarpus Rigidus* and *Artocarpus Elasticus* wood species, respectively.

Acknowledgements The authors would like to acknowledge the financial support from Ministry of Higher Education Malaysia, for their financial support.

References

Alexandre M, Dubois P (2000) Polymer-layered silicate nanocomposites: preparation, properties and uses of a new class of materials. Mater Sci Eng 28:1–63

Basara C, Yilmazer U, Bayram G (2005) Synthesis and characterization of epoxy based nanocomposites. J Appl Polym Sci 98:1081–1086

Becker O, Varley R, Simona G (2002) Morphology, thermal relaxations and mechanical properties of layered silicate nanocomposites based upon high-functionality epoxy resins. Polym 43:4365–4373

Cai XB, Riedl SY, Zhang Wan H (2007) Formation and properties of nanocomposites made up from solid aspen wood, melamine-urea-formaldehyde, and clay. Holz 61:148–154

Cai X, Riedl B, Zhang SY, Wan H (2008) The impact of the nature of nanofillers on the performance of wood polymer nanocomposites. Compos Part A 39:727–737

Chen KH, Yang SM (2002) Synthesis of epoxy–montmorillonite nanocomposite. J Appl Polym Sci 86:414–421

Chen B, Liu J, Chen H, Wu JS (2004) Synthesis of disordered and highly exfoliated epoxy/clay nanocomposites using organoclay with catalytic function via acetone-clay slurry method. Chem Mater 16:4864–4866

Evans D, Canfer SJ (2000) Radiation stable, low viscosity impregnating resin systems for cryogenic applications. Adv Cryo Eng Mater 46:361–368

Hamdan S, Talib ZA, Rahman MR, Ahmed AS, Islam MS (2010) Dynamic Young's modulus measurement of treated and post-treated tropical wood polymer composites (WPC). BioRes 5(1):324–342

Kong D, Park CE (2003) Real time exfoliation behavior of clay layers in epoxy-clay nanocomposites. Chem Mater 15:419–424

Lam CK, Cheung HY, Lau KT (2005) Cluster size effect in hardness of nanoclay/epoxy composites. Compos Part B 36:263–269

Lebaron PC, Wang Z, Pinnavaia TJ (1999) Polymer-layered silicate nanocomposites: an overview. Appl Clay Sci 15:11–29

Mulinari DR, Voorwald HJC, Cioffi MOH, Rocha GJ, Pinto Da Silva MLC (2010) Surface modification of sugarcane bagasse cellulose and its effect on mechanical and water absorption properties of sugarcane bagasse cellulose/HDPE composites. BioRes 5(2):661–671

Nigam V, Setua DK, Mathur GN (2004) Epoxy–montmorillonite clay nanocomposites: synthesis and characterization. J Appl Polym Sci 93:2201–2210

Owen NL, Thomas DW (1989) Infrared studies of hard and soft woods. Appl Spec 43:451–455

Park J, Jana SC (2003) Effect of plasticization of epoxy networks by organic modifier on exfoliation of nanoclay. Macromol 36:8391–8397

Rahman MR, Hamdan S, Saleh AA, Islam MS (2010) Mechanical and biological performance of sodium mataperiodate impregnated plasticized wood (PW). BioRes 5(2):1022–1035

Ray SS, Okamoto M (2003) Polymer/layered silicate nanocomposites: a review from preparation to processing. Prog Polym Sci 28:1539–1541

Ueki T, Nishijima S, Izumi Y (2005) Designing of epoxy resin systems for cryogenic use. Cryo 45:141–148

Varhegyi G, Antal MJJ, Jakab E (1997) Kinetics modeling of biomass pyrolysis. J Anal Appl Pyrol 42:73–87

Influence of Nanoclay/Phenol Formaldehyde Resin on Wood Polymer Nanocomposites

M.R. Rahman

Abstract In this study, the introduction of nanofiller into phenol formaldehyde matrix formed wood polymer nanocomposites (WPNCs). FT-IR results showed that the addition of nanoclay into phenol formaldehyde (PF) formed H-bonding interaction with hydroxyl groups by reducing the wave number of the peak. The Thermogravimetric Analysis (TGA) results showed that WPNCs were thermally stable compared to the raw ones. The MOE and MOR of WPNCs were significantly improved for *Eugenia spp., Xylopia spp., Artocarpus Rigidus,* and *Artocarpus Elasticus* respectively. The Young's modulus of WPNCs on *Eugenia* spp. was significantly higher compared to raw wood. From X-ray diffraction results, WPNCs showed improved crystallinity at the amorphous region due to the polymer loading. SEM micrograph of WPNCs showed that void space was filled with the polymer, and the waxy substance was removed. All the nanofiller/phenol formaldehyde was significantly effective on *Eugenia* spp. followed by *Xylopia* spp., *Artocarpus Rigidus,* and *Artocarpus Elasticus* wood species, respectively.

Keywords Mechanical properties · Thermal stability · Chemical treatment · WPNCs

1 Introduction

In the past, most of the organic or inorganic fillers have been used as reinforcements in polymers. In the recent years, nanoparticles are widely used that acted as reinforcements and gain most of the attention in manufacturing polymeric nanocomposites. Through experimental results, polymeric nanocomposites show better variations in their properties based on the research works carried out by Adebhar et al. (2001), Kinloch et al. (2003), and Yasmin et al. (2003). However, there are

M.R. Rahman (✉)
Faculty of Engineering, Universiti Malaysia Sarawak,
94300 Kota Samarahan, Sarawak, Malaysia
e-mail: rmrezaur@unimas.my

M.R. Rahman, *Wood Polymer Nanocomposites*, Engineering Materials,
DOI 10.1007/978-3-319-65735-6_8

few reports on the unfavorable effect due to nanoparticle inclusions (Haggenmueller et al. 2006; Zilg et al. 1999, 2000). Nanofillers are well known, unique, and well dispersing particles that can be widely applied in many applications. However, this filler is still lacking especially the large surface of nanoparticles that caused agglomeration of particles that reduced the interactions among particles.

The nanocomposites were produced through different processing methods such as shear mixing (Yasmin et al. 2003), mechanical mixing (Shah et al. 2004), in situ polymerization (Haggenmueller et al. 2006), and impregnation. Through the research works, the mechanical properties of WPNCs were fabricated via the impregnated method. However, some exceptions such as the properties of the resulting nanocomposites tended to go down after reaching the maximum at a particle loading when its nanofiller reached about 3–5 wt% or less (Choi et al. 2005; Rodgers et al. 2005; Zheng et al. 2005).

There are several studies that have been reported on silica/epoxy nanocomposites fabricated with the introduction of organosilicasol (colloidal silica in organic solvent) (Adebhar et al. 2001; Kinloch et al. 2003; Uddin and Sun 2008). Besides, Uddin and Sun obtained 40% improvement in matrix compressive modulus through the incorporation of 15wt% silica nanoparticle loading. Glass fiber nanocomposites gained 60–80% in longitudinal compressive strength for different fiber volume fractions using this nanoparticle-enhanced matrix. In addition, fracture behavior of an epoxy adhesive with the inclusion of both silica nanoparticles and rubber nanocomposites was greatly improved (Kinloch et al. 2003). The present study was to fabricate WPNCs with the addition of nanoclay/phenol formaldehyde resin, which functioned as a reinforcement matrix.

2 Materials and Methods

2.1 Materials

For this study, five wood species were collected which varied of softwoods and hardwoods. The softwoods used, namely *Eugenia* spp., *Artocarpus Rigidus*, *Artocarpus Elasticus* and *Xylopia* spp., while the hardwood, namely *Koompassia Malaccensis,* were used. Besides, chemicals that used throughout the study for impregnation of all the wood species were Nanoclays Nanomer® 1.30E, montmorillonite (MMT) (Southern Clay Products, Inc. USA), and phenol formaldehyde resin (PF) (Marck, Germany). The purity grade of the chemicals was 99%.

2.2 Impregnation Solutions Preparation

The clay powder that used throughout the research work was used as received and the average particle size was 8 μm. 1% layered aluminosilicate nanofiller was introduced into the low viscosity phenol formaldehyde resin to form impregnation solutions at a mixing speed of 2050 rpm for 20 min.

2.3 *Fabrication of Wood Polymer Nanocomposites (WPNCs)*

All the samples were oven dried to constant weight at 103 °C for 24 h. After dried, all the samples were placed into an impregnation chamber where the samples had no contact between samples. Besides, all the samples were covered completely by nanoclay/phenol formaldehyde mixtures. The system was evacuated to 60 mmHg for 30 min. Continuously, compressed air was applied to the system and maintained at 0.52 MPa for 30 min then gradually release the compressed air. All the excess chemicals were wiped off the samples. After that, the samples were weighted and dried for 24 h by air circulation, continued with oven drying at 90 °C for 24 h. Any excess polymer on the surface of the resulting WPNCs was then removed. All the fabricated WPNCs were characterization using the FT-IR spectroscopy analysis, X-ray diffraction (XRD) analysis, Thermogravimetric Analysis (TGA), Scanning Electron Microscopy analysis (SEM), free-free flexural vibration testing, three-point bending test, and compression parallel to grain testing.

3 Result and Discussion

3.1 *Fourier Transform Infrared Spectroscopy Analysis*

Figure 1 showed the FT-IR spectra of raw wood and WPNCs. It was clearly detected that the spectra were separated in two regions, namely the –OH stretching vibrations in the range of 4000–2700 cm^{-1} region as well as fingerprint region in the range of 1800–400 cm^{-1} region. From Fig. 1, the strong band OH stretching was detected at the peak range of 4000–3300 cm^{-1} while C–H stretching in methyl and methylene groups was detected at the peak range of 3000–2800 cm^{-1}. Besides, according to Owen et al. (1989), strong broad superposition with sharp and discrete absorptions was detected in the region from 1750 to 1000 cm^{-1}. Lignin peak could be clearly detected at 1508 cm^{-1}, and the absorption located at 1735 cm^{-1} was caused by hemicelluloses. All the peaks detected previous indicated the C=O stretch in non-conjugated ketones, carbonyls, and ester groups (Owen et al. 1989). In addition, the main components from wood: cellulose, hemicelluloses, and lignin could be detected in the region between 1800 and 1100 cm^{-1}. It was clear to show the difference in the infrared spectra with different absorbance values and shapes of the bands together with their location. Moreover, the peak at 1735 cm^{-1} was shifted to 1591 cm^{-1} for modified wood which confirmed less xylan content in softwood. The impregnation of nanoclay and phenol formaldehyde into raw wood could be proven through C–O stretch of C–O–H in starch at 1317 cm^{-1} and 1222 cm^{-1} and C–O stretch of C–O–C in starch at 1027 cm^{-1}. This was because nanoclay/phenol formaldehyde acted as plasticizer, which formed H-bonding interaction with the hydroxyl groups.

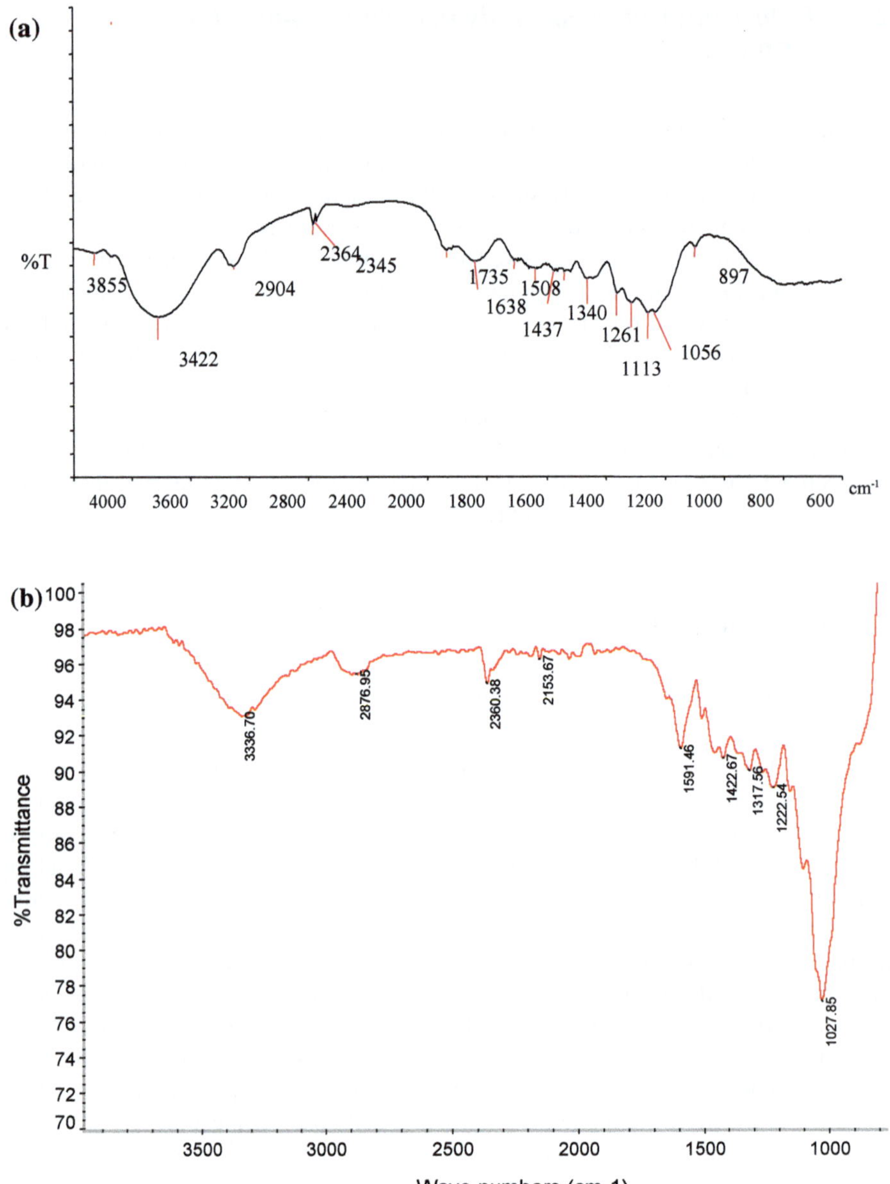

Fig. 1 **a** IR spectrum of raw wood, **b** IR spectrum of WPNCs

3.2 Thermogravimetric Analysis

Figure 2a–e showed the thermogravimetric curves of raw wood and WPNCs Fig. 2 presented the general thermal decomposition of cellulose and hemicellulose that started at 180 °C. The increment of mass loss rate corresponded to the temperature and the maximum value could be obtained at about 280 °C. The thermal decomposition started at 201, 211, 212, 212, and 214 °C for raw wood species, namely *Artocarpus Elasticus*, *Artocarpus Rigidus*, *Xylopia* spp., *Koompassia Malaccensis,* and *Eugenia* spp., respectively, through Fig. 2a–e and Table 1. The beginning of

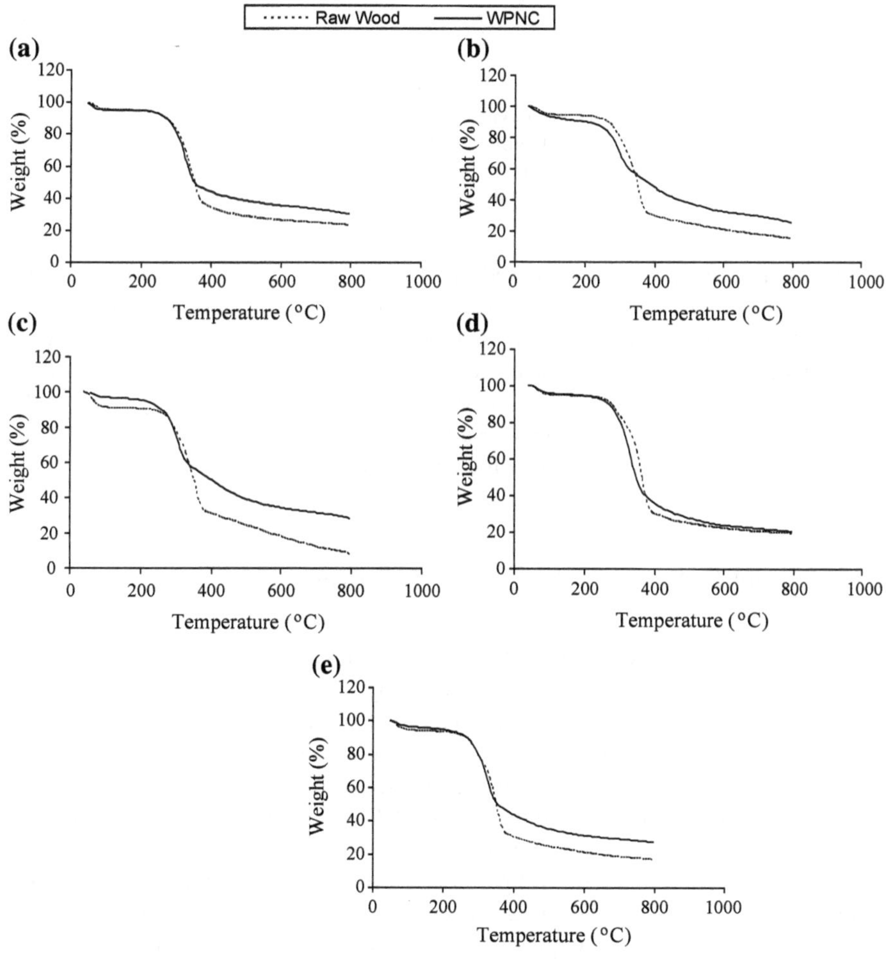

Fig. 2 **a** TGA curve of *Artocarpus Elasticus* raw wood and WPNCs, **b** TGA curve of *Artocarpus Rigidus* raw wood and WPNCs, **c** TGA curve of *Xylopia* spp. raw wood and WPNCs, **d** TGA curve of *Koompassia Malaccensis* raw wood and WPNCs, **e** TGA curve of *Eugenia* spp. raw wood and WPNCs

Table 1 Thermal characteristics of raw wood and WPNCs

Sample		T_i (°C)	T_m (°C)	T_f (°C)	W_{Ti} (%)	W_{Tm} (%)	W_{Tf} (%)	Activation Energy, E_a (J/°K)
Artocarpus Elasticus	Raw	201	300	419	98.96	64.73	32.67	33.11
	WPNCs	218	353	626	93.89	47.16	34.81	38.05
Artocarpus Rigidus	Raw	211	310	388	99.67	56.45	30.63	25.79
	WPNCs	242	328	559	90.61	52.46	31.18	34.16
Xylopia spp.	Raw	212	321	420	98.78	56.15	25.85	27.04
	WPNCs	249	329	638	90.87	55.56	28.16	34.19
Koompassia Malaccensis	Raw	212	355	403	98.99	57.91	30.00	16.12
	WPNCs	239	360	547	93.28	52.08	31.51	21.26
Eugenia spp.	Raw	214	340	443	99.77	60.32	27.51	19.17
	WPNCs	246	341	527	92.08	55.68	35.15	23.67

T_m Temperature corresponding to the maximum rate of mass loss
T_i Temperature corresponding to the beginning of the decomposition
T_f Temperature corresponding to the ending of the decomposition
W_{Ti}, W_{Tm} and W_{Tf} Mass loss at T_i, T_m and T_f

decomposition of the WPNCs started at 218, 242, 249, 239, and 246 °C, respectively. From the Fig. 2, the initial mass loss and the end of thermal decomposition of WPNCs were higher compared to raw wood. Through the calculation using Arrhenius equation, activation energy that was calculated determined was higher on WPNCs compared to raw wood. This behavior had been reported by other author (Varhegyi et al. 1997). From TGA results, it was clearly showed that the WPNCs were higher thermally stable compared to raw wood.

3.3 Dynamic Young's Modulus Measurement

Figure 3 showed that the dynamic Young's modulus of the raw wood and WPNCs tested through free-free flexural vibration testing system. This test was carried out with ten specimens for all the species. The incorporation of nanofiller with phenol formaldehyde increased the Young's modulus for all the wood species, in accordance with other researchers' previous work (Hamdan et al. 2010). After monomer impregnation, the Young's modulus of WPNCs on *Eugenia* spp., *Xylopia* spp., *Artocarpus Rigidus,* and *Artocarpus Elasticus* was significantly higher compared to raw one. On the other hand, the WPNCs on *Koompassia Malaccensis* showed higher Young's modulus due to their hardness cell wall density. All the WPNCs showed increment in activation energy due to the nanofiller effect on the all wood species that were greatly affected by the monomer impregnation as shown in Table 1. In addition, elastic properties on WPNCs showed significant changes in *Eugenia* spp., *Xylopia* spp., *Artocarpus Rigidus,* and *Artocarpus Elasticus,* respectively.

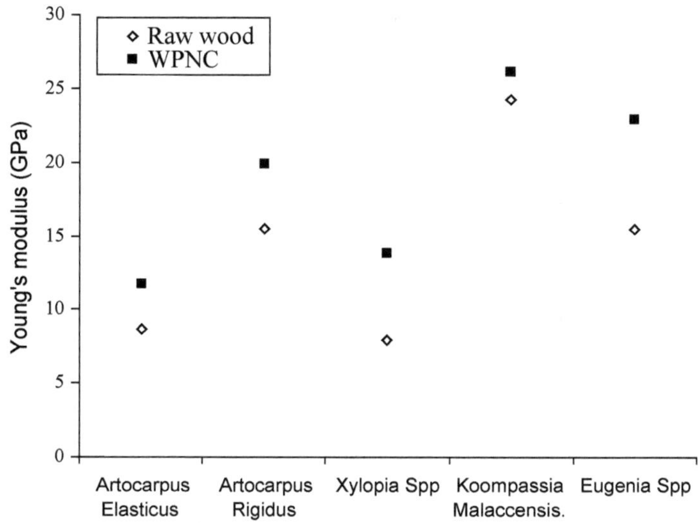

Fig. 3 E_d of raw wood, WPNCs for all species

3.4 *Modulus of Elasticity and Modulus of Rupture Measurement*

The modulus of elasticity (MOE) and modulus of rupture (MOR) of raw wood and WPNCs are shown in Tables 2 and 3 as well as Figs. 4 and 5, respectively. The impregnation of nanofiller/phenol formaldehyde on the raw wood was investigated. Higher MOE of WPNCs on *Eugenia* spp. and *Xylopia* spp. was highest continued by *Artocarpus Rigidus, Artocarpus Elasticus,* and *Koompassia Malaccensis,*

Table 2 *t*-test analysis of raw wood and wood polymer nanocomposites[a]

Treatment	Modulus of elasticity	*t*-test grouping[b]
Raw wood (*Artocarpus Elasticus*)	6.48 ± 0.52	A
WPNCs (*Artocarpus Elasticus*)	12.87 ± 0.55	B
Raw wood (*Artocarpus Rigidus*)	5.13 ± 0.38	C
WPNCs (*Artocarpus Rigidus*)	12.00 ± 2.34	D
Raw wood (*Xylopia* spp.)	6.71 ± 0.34	E
WPNCs (*Xylopia* spp.)	14.44 ± 0.51	F
Raw wood (*Koompassia Malaccensis*)	15.68 ± 1.11	G
WPNCs (*Koompassia Malaccensis*)	20.77 ± 2.30	G
Raw wood (*Eugenia* spp.)	9.25 ± 0.37	H
WPNCs (*Eugenia* spp.)	24.93 ± 0.88	I

[a]Each value is the average of 10 specimens
[b]The same letters are not significantly different at $\alpha = 5\%$

Table 3 *t*-test analysis of raw wood and wood polymer nanocomposites[a]

Treatment	Modulus of rupture	*t*-test grouping[b]
Raw wood (*Artocarpus Elasticus*)	76.17 ± 3.01	A
WPNCs (*Artocarpus Elasticus*)	88.53 ± 2.86	B
Raw wood (*Artocarpus Rigidus*)	33.02 ± 3.63	C
WPNCs (*Artocarpus Rigidus*)	69.42 ± 9.67	D
Raw wood (*Xylopia* spp.)	45.91 ± 1.36	E
WPNCs (*Xylopia* spp.)	91.36 ± 2.41	F
Raw wood (*Koompassia Malaccensis*)	122.20 ± 15.49	G
WPNCs (*Koompassia Malaccensis*)	130.22 ± 24.91	G
Raw wood (*Eugenia* spp.)	46.10 ± 3.62	H
WPNCs (*Eugenia* spp.)	110.11 ± 12.91	I

[a]Each value is the average of 10 specimens
[b]The same letters are not significantly different at $\alpha = 5\%$

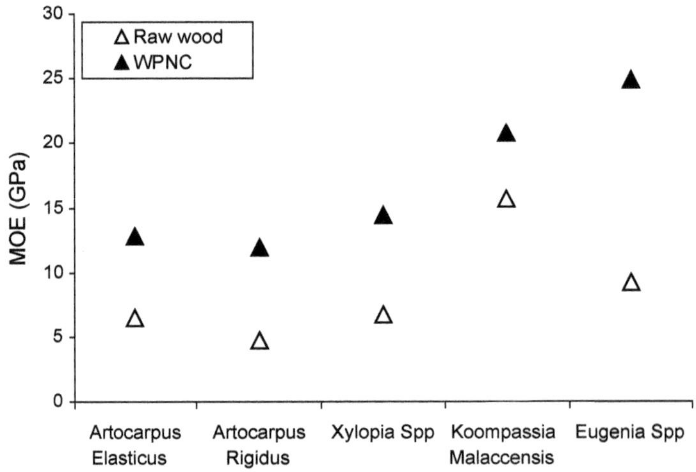

Fig. 4 MOE of raw wood and WPNCs

respectively. From this test, WPNCs yielded higher MOE compared to the raw wood as nanofiller had high impact on wood species which was in accordance with other researchers (Cai et al. 2007, 2008).

Table 2 showed that the MOE of the *Eugenia* spp., *Xylopia* spp., *Artocarpus Elasticus,* and *Artocarpus Rigidus* was significantly higher compared to raw wood. On the other hand, *Koompassia Malaccensis* (hardwood) WPNCs showed slight increment on MOE compared to our previous chapter although there was no significant effect of polymer impregnation.

The incorporation of phenol formaldehyde aided to plasticize on the cellulose and hemicellulose in wood cells by reducing the water molecules from the wood specimens. Figure 4 showed that the combination of nanoclay and phenol

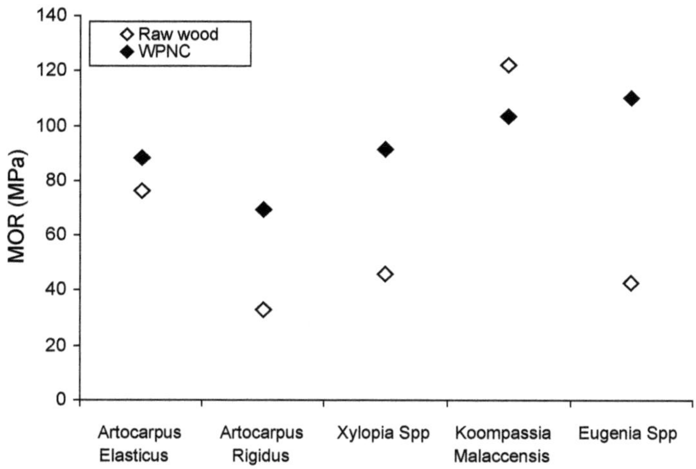

Fig. 5 MOR of raw wood and WPNCs

formaldehyde acted as a binder which filled the void space in the wood specimens and gradually improved its stiffness. Due to this factor, the MOE of all WPNCs was higher than that of raw wood.

Introduction of nanofiller with phenol formaldehyde through impregnation into the wood improved the MOR of WPNCs on *Eugenia* spp. This improvement was clearly investigated and proven by previous researchers (Cai et al. 2007, 2008). Besides, MOR was significantly varied on *Xylopia* spp., *Artocarpus Rigidus, Artocarpus Elasticus,* and *Eugenia* spp. for both raw wood and WPNCs. From Table 3, it was clearly detected that the growth of MOR for *Eugenia* spp. was highest, continued by *Xylopia* spp., *Artocarpus Rigidus, Artocarpus Elasticus, and Koompassia Malaccensis*, respectively. The *Koompassia Malaccensis* (hardwood) of WPNCs was higher than compared to raw one.

3.5 Static Young's Modulus Measurement

Figure 6 showed the static Young's modulus which was determined from 10 repetitions. The highest increment of E value was observed in *Eugenia* spp. continued by *Artocarpus Rigidus, Artocarpus Elasticus, Xylopia* spp., and *Koompassia Malaccensis*, respectively. WPNCs showed higher increment of E for *Eugenia* spp. compared to raw wood as shown in Table 4 (Cai et al. 2007, 2008; Rahman et al. 2010). From this test, it was clearly proven that the nanofiller/phenol formaldehyde not only plasticized on the wood cell walls, but it was also deeply increasing their lateral stability.

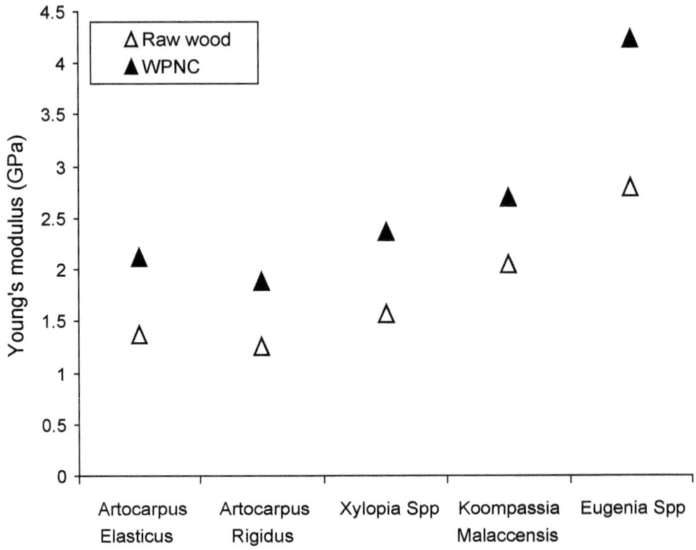

Fig. 6 Static Young's modulus of raw wood and WPNCs

Table 4 *t*-test analysis of raw wood and wood polymer nanocomposites[a]

Treatment	Static Young's modulus	*t*-test grouping[b]
Raw wood (*Artocarpus Elasticus*)	1.36 ± 0.23	A
WPNCs (*Artocarpus Elasticus*)	2.12 ± 0.37	A
Raw wood (*Artocarpus Rigidus*)	1.26 ± 0.04	B
WPNCs (*Artocarpus Rigidus*)	1.88 ± 0.58	B
Raw wood (*Xylopia* spp.)	1.57 ± 0.39	C
WPNCs (*Xylopia* spp.)	2.37 ± 0.24	C
Raw wood (*Koompassia Malaccensis*)	2.05 ± 0.18	D
WPNCs (*Koompassia Malaccensis*)	2.70 ± 0.48	D
Raw wood (*Eugenia* spp.)	1.67 ± 0.60	E
WPNCs (*Eugenia* spp.)	4.23 ± 0.79	F

[a]Each value is the average of 10 specimens
[b]The same letters are not significantly different at $\alpha = 5\%$

3.6 X-ray Diffraction Analysis

X-ray diffraction tests for raw wood and WPNCs are shown in Fig. 7. Raw wood and WPNCs showed diffraction peaks at 17°, 22.5°, and 35°. According to Mulinari et al. (2010), the different angles detected were corresponded to the cellulose crystallographic planes I101, I002, and I040, respectively. The structure of cellulose

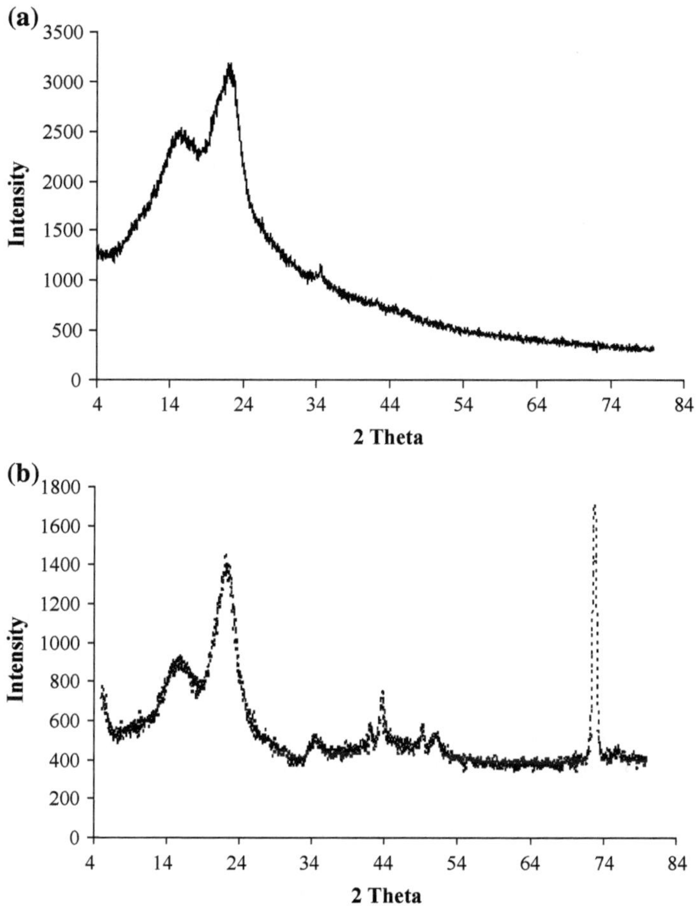

Fig. 7 **a** X-ray diffraction of raw wood, **b** X-ray diffraction of WPNCs

did not change when compared to raw wood due to the position of these peaks did not change. From this analysis, WPNCs exhibited another five broad 2θ peaks at 43.74°, 49.21°, 51.13°, 55.65°, and 72.87° due to polymerization that increased in amorphous region.

Nanofiller was introduced into the raw wood and modified with phenol formaldehyde polymer due to the well impregnation within the wood cell lumen. The surface hydroxyl groups of woods were well attached to the end of polymer groups through crosslinking reaction. However, according to Wu et al. (2004), from X-ray diffraction, a quasi-crystalline form was generated which may not represent an increase in the crystallinity of cellulose.

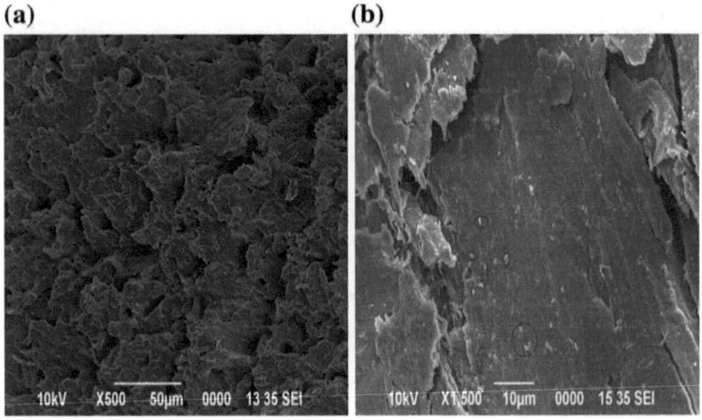

Fig. 8 Scanning Electron Microscopy of **a** Raw Wood and **b** WPNCs

3.7 Scanning Electron Microscopy Analysis

Figure 8 showed the scanning electron micrographs (SEM) of raw wood and WPNCs. From Fig. 8a, SEM micrographs showed that raw wood surfaces were uneven layer with many voids. On the other hand, treated wood surfaces were smooth as shown in Fig. 8b. The smooth surface of treated wood was shown which due to the good penetration of the polymers into the cell wall as well as the vessels of the wood (Rahman et al. 2010). Under pulse-dipping pressure, the chemical could be immerged in the wood fiber and other vertical cells. Continuously, these chemical particles were distributed discontinuously in the wood cell. From this analysis, it was clearly indicated that the polymer was impregnated into cell wall and cell cavities of wood to improve the surface morphology of the treated wood.

4 Conclusion

This study investigated the synthesis of nanofiller/phenol formaldehyde WPNCs. From FT-IR results, the peak at 1317 and 1027 cm^{-1} was significantly decreased which were attributed to C–O stretch of C–O–H as well as C–O stretch of C–O–C in starch which proved the impregnation of nanoclay/PF wood sample with more H-bonding interaction with the hydroxyl groups. Besides, WPNCs were more thermally stable over temperature compared to the raw ones due to the introduction of nanofiller on wood species. The stiffness of the WPNCs was significantly increased. WPNCs showed higher MOE and MOR *Eugenia* spp., *Xylopia* spp., *Artocarpus Rigidus,* and *Artocarpus Elasticus,* respectively, among all the samples. The Young's modulus of *Eugenia* spp. of WPNCs was significantly different with raw wood. The crystallinity of WPNCs increased at the amorphous region due to the monomer loading through X-ray diffraction test. From SEM micrographs, the void space was filled by the polymer as well as the removal of the waxy substance on WPNCs. This proved that nanofiller/phenol formaldehyde was significantly effective on *Eugenia* spp. followed by *Xylopia* spp., *Artocarpus Rigidus,* and *Artocarpus Elasticus* wood species, respectively.

Acknowledgements The authors would like to acknowledge the financial support from Ministry of Higher Education Malaysia, for their financial support [Grant no. ERGS/02 (08)/860/2912(12)] during the research.

References

Adebhar T, Roscher C, Adam J (2001) Reinforcing nanoparticles in reactive resins. Eur Coat J 4:144

Cai XB, Riedl SY, Zhang Wan H (2007) Formation and properties of nanocomposites made up from solid aspen wood, melamine-urea-formaldehyde, and clay. Holz 61:148–154

Cai X, Riedl B, Zhang SY, Wan H (2008) The impact of the nature of nanofillers on the performance of wood polymer nanocomposites. Compos Part A 39:727–737

Choi YK, Sugimoto KI, Song SM, Gotoh Y, Ohkoshi Y, Endo M (2005) Mechanical and physical properties of epoxy composites reinforced by vapor grown carbon nanofibers. Carbon 43: 208–2199

Haggenmueller R, Du F, Fischer JE, Winey KI (2006) Interfacial in situ polymerization of single wall carbon nanotube/nylon 66 nanocomposites. Polym 47:2381–2388

Hamdan S, Talib ZA, Rahman MR, Ahmed AS, Islam MS (2010) Dynamic Young's modulus measurement of treated and post-treated tropical wood polymer composites (WPC). BioRes 5(1):324–342

Kinloch AJ, Lee JH, Taylor AC, Sprenger S, Eger C, Egan D (2003) Toughening structural adhesives via nano- and micro-phase inclusions. J Adhes 79:73–867

Mulinari DR, Voorwald HJC, Cioffi MOH, Rocha GJ, Pinto Da Silva MLC (2010) Surface modification of sugarcane bagasse cellulose and its effect on mechanical and water absorption properties of sugarcane bagasse cellulose/HDPE composites. BioRes 5(2):661–671

Owen NL, Thomas DW (1989) Infrared studies of hard and soft woods. Appl Spec 43:451–455

Rahman MR, Hamdan S, Saleh AA, Islam MS (2010) Mechanical and biological performance of sodium mataperiodate impregnated plasticized wood (PW). BioRes 5(2):1022–1035

Rodgers RM, Mahfuz H, Rangari VK, Chisholm N, Jeelani S (2005) Infusion of SiC nanoparticles into SC-15 epoxy: an investigation of thermal and mechanical response. Macromol Mater Eng 290:423–429

Shah RK, Paul DR (2004) Nylon 6 nanocomposites prepared by a melt mixing masterbatch process. Polym 45:2991–3000

Uddin M, Sun CT (2008) Strength of unidirectional glass/epoxy composite with silica nanoparticle-enhanced matrix. Compos Sci Technol 68:1637–1643

Varhegyi G, Antal MJJ, Jakab E (1997) Kinetics modeling of biomass pyrolysis. J Anal Appl Pyrol 42:73–87

Wu CL, Li XP, Qing SL (2004) Cellulose solvent: current research status and its application prospect. Trans China Pulp Paper 19(2):171–175

Yasmin A, Abot JL, Daniel IM (2003) Processing of clay/epoxy nanocomposites by shear mixing. Scripta Mater 49:6–81

Zheng H, Ning R, Zheng Y (2005) Study of SiO_2 nanoparticles on the improved performance of epoxy and fiber composites. J Reinf Plastic Compos 24:223–233
Zilg C, Mulhaupt R, Finter J (1999) Morphology and toughness/stiffness balance of nanocomposites based upon anhydride-cured epoxy resins and layered silicates. Macromol Chem Phys 200:661
Zilg C, Thomman T, Finter J, Mulhaupt R (2000) The influence of silicate modification and compatibilizers on mechanical properties and morphology of anhydridecured epoxy nanocomposites. Macromol Mater Eng 280:41

Clay Dispersed Styrene-*co*-glycidyl Methacrylate Impregnated Kumpang Wood Polymer Nanocomposites: Impact on Mechanical and Morphological Properties

M.R. Rahman, S. Hamdan and J.C.H. Lai

Abstract In this study, we evaluate the physical, mechanical, and morphological properties of a clay dispersed styrene-*co*-glycidal methacrylate (ST-*co*-GMA) impregnated wood polymer nanocomposite (WPNC). The WPNC was characterized by Fourier Transform Infrared Spectroscopy (FT-IR), Scanning Electron Microscopy (SEM), 3-point bending, and free-vibration testing. The FT-IR results showed that the absorbance at $1730\ \text{cm}^{-1}$ was increased for ST-*co*-GMA-clay-WPNC compared with other nanocomposites and the raw material. The SEM results showed that ST-*co*-GMA-clay-WPNC had a smoother surface than other nanocomposites and raw wood. The modulus of elasticity (MOE), modulus of rupture (MOR), and dynamic Young's moduli (E_d) of WPNCs were considerably increased compared to wood polymer nanocomposites (WPNCs) and raw wood. The raw wood exhibited a higher water uptake (WU) than WPNCs.

Keywords ST-*co*-GMA · Nanoclay · Mechanical properties · Morphological properties

1 Introduction

Wood has a remarkable value and importance in the world's economy because of its mechanical, renewable, sustainable, and eco-friendly characteristics (Zabel and Morrell 1992). Wood has attracted attention by material researchers for replacing synthetic plastic materials, which are not renewable or eco-friendly (Bledzki et al. 2007; Mohanty et al. 2002; Rahman et al. 2015). Wood is also used as a material for musical instruments because of its wide-ranging colors and acoustic properties. The natural frequency of a wood beam can determine the cracks present in the wood,

M.R. Rahman (✉) · S. Hamdan · J.C.H. Lai
Faculty of Engineering, Universiti Malaysia Sarawak,
94300 Kota Samarahan, Sarawak, Malaysia
e-mail: rmrezaur@unimas.my

M.R. Rahman, *Wood Polymer Nanocomposites*, Engineering Materials,
DOI 10.1007/978-3-319-65735-6_9

which affects the dynamic behavior of wood (Kam and Lee 1992). Wood is a natural composite that is composed of cellulose, hemicellulose, and lignin. These constituents are basically a series of tubular fibers or cells cemented together. Cellulose is responsible for wood strength because of its linear orientation and high degree of polymerization. Hemicelluloses and lignin are also closely associated with cellulose to increase the packing density of the cell wall. Hemicelluloses act as a coupling agent to bind non-crystalline hydrophilic cellulose and amorphous hydrophobic lignin. Cellulose, hemicellulose, and lignin are responsible for the mechanical properties of wood. The macromolecule of cellulose is formed by covalent bonding, which is resistant to tensile stress. The hydrogen bonds within the cellulose provide rigidity by transferring stress. The mechanical properties of wood also depend on temperature, pressure, humidity, pH, chemical adsorption from the environment, UV radiation, fire, and biological degradation. At high humidity, wood is dimensionally unstable and susceptible to termite attack because of its hydrophilic and capillarity properties (Rowell 2006). In low-humidity environments, wood desorbs water and undergoes shrinkage. Hygroscopic and dimensionally unstable properties restrict the use of wood. In outdoor applications, wood suffers from the photodegradation process as well (Mattos et al. 2014). Wood also discolors when used indoor as a result of the oxidation of lignin, albeit with lower intensity (Feist and Hon 1984).

When wood plastic composites (WPCs) first entered the market, the wood particles were presumed to be completely encapsulated in the plastic, thereby protecting those from both moisture and microorganism attack (Morrell et al. 2006). However, this impermeable conception is not completely true. For fast-growing wood, the mechanical properties, color stability, and durability are not satisfactory because of its low density and water uptake. To overcome these drawbacks, modified wood such as WPCs and wood composites are used as wood materials instead of conventional wood material (Cao et al. 2011; Chang and Chang 2001a, b). To decrease the hydrophilicity and fungal attack of wood materials, various chemical treatments have been introduced, such as acetylation, benzylation, and silane or maleic anhydride grafting. In all these cases, the chemicals used react with the –OH group of the cell lumen and cell cavity of wood (Xie et al. 2011). The use of hydrophobic thermoplastic fillers, without any reaction with wood, also improves the mechanical performance of wood by filling the cell cavities of wood. Heat treatment, hydrothermal treatment, oil heat treatment, hot pressing treatment, and in situ polymerization can also eliminate or decrease the disadvantages of wood (Islam et al. 2014; Li et al. 2013).

Polymer nanocomposites are polymers (thermoplastics, thermosets, or elastomers) that have been reinforced with small quantities (less than 5% by weight) of very high aspect ratio ($l/t > 300$) fillers in its matrix (Hegde 2009). In wood polymers, the nanocomposite nanofiller used will form an adhesive bond with wood. The adhesive bonds in wood composites hold elements of the composite together by their bond strength and transfer stress across the bond line between the bonded wood elements. For wood properties such as modulus or thermal and moisture expansion coefficients, the bond stiffness is more important than bond strength (Le and Nairn 2014). In wood polymer nanocomposites, the chemically modified wood is also photostable (Farahani and Taghizadeh 2010; Jebrane et al. 2011; Prakash and Mahadevan 2008).

Eucalyptus grandis, impregnated with styrene and methyl methacrylate monomers, does not exhibit good mechanical performance because of its low permeability for impregnation (Stolf and Lahr 2004). However, wood flour coated with the hydrophilic–hydrophobic block-copolymer based on styrene and acrylic acid shows a significant improvement in the ultimate tensile properties of the nanocomposite (Kosonen et al. 2000). The compatibility between wood and non-polar polymers such as styrene, ethylene, and vinyl acetate can be improved using glycidyl methacrylate (GMA) as a compatibilizer (Devi and Maji 2007; Dikobe and Luyt 2007). The plasticized wood nanocomposite increases the mechanical properties and decay resistance (Rahman et al. 2010) of wood. Nanoclay, by binding to hydroxyl groups of wood components, improves the interaction between wood and thermoplastic polymer.

Montmorillonite (MMT) clay has high Gibbs surface energy, so it is attracted by the monomer units, and the monomers are thus dispersed inside the galleries of clay to break the van der Waals gap until equilibrium is reached. In this process, in situ polymerization reactions occur between the layers with lower polarity, and then, the dispersed nanoclay particles reach the wood polymer (Bhattacharya and Aadhar 2014; Carroll 1970). Energy is also supplied by charge transfer between the wood and nanoclay.

With respect to a wood and copolymer macromolecular nanocomposite system, montmorillonite layers can be dispersed in wood and copolymer in three different ways. First, the particles of MMT can be dispersed homogeneously in the copolymer and the macropores of wood, leading to no change in the interlayer distance. Second, the copolymer macromolecules can enter the silicate layers and increase the interlayer distance (van der Waals forces exist here). The layers of this type of MMT will partly stay in the lumen wood and to some extent enter the wood cell wall. Third, exfoliated MMT can directly enter the wood cell wall and form chemical bonds with wood at the molecular level (Guangjie et al. 2006). The copolymer and MMT strengthen the mechanical properties of wood and make the wood resistant to water uptake and microorganisms (Baysal et al. 2007). Polymer–clay nanocomposites have emerged as a fascinating field of research because of their multiple advantages, which include an improvement in mechanical, dimensional, and thermal properties.

The present study investigates the physico-mechanical and thermal properties of clay dispersed styrene-*co*-glycidyl methacrylate (St-*co*-GMA-clay) impregnated wood polymer nanocomposites (WPNCs). The morphological properties are also reported in this study.

2 Experimental

2.1 Materials

Five wood specimens were chosen for this study. These wood specimens were impregnated by styrene (ST) (A. S. Joshi and Company, Goregaon West, Mumbai), glycidyl methacrylate (GMA) (Aldrich, USA), and nanoclay (Nanomer^{R}1.28E), modified with 25–30 wt% trimethyl stearyl ammonium (Aldrich, USA). Benzoyl peroxide (Merck Schuchardt OHG, Germany) was used in the system as a catalyst to increase the reaction rate by a free radical mechanism.

2.2 Specimen Preparation

A Kumpang tree was cut into three bolts 1.2 m long. Each bolt was quartersawn to produce planks 4 cm thick. These planks were subsequently conditioned by air drying in a room with a relative humidity of 60% and ambient temperature of approximately 25 °C for one month prior to testing. The planks were ripped and machined to 340 mm (L) × 20 mm (T) × 10 mm (R) for three-point bending and free-free vibration tests.

2.3 Preparation of Wood Polymer Nanocomposites (WPNCs)

Table 1 showed that oven-dried specimens were impregnated with 500 mL of styrene (St), a mixture of 500 mL of styrene with 10 g MMT, a mixture of 500 mL of styrene with glycidyl methacrylate (1:1 ratio) and 10 gm MMT. The four different solution mixtures were used to impregnate the wood specimens. The wood specimens were then separately placed in an impregnation vacuum chamber of 75 mm (Hg) for 30 min. In each system, approximately 10 g of benzoyl peroxide was added as the initiator. These wood nanocomposites were then removed from the chamber, and the excess chemicals were wiped off. Specimens wrapped with

Table 1 Composition of the chemicals for raw wood impregnation

SL	Specimen	ST (mL)	GMA (mL)	Clay (gm)
1	Raw wood (RW)			
2	RW	500		
3	RW	500	500	
4	RW	500		10
5	RW	500	500	10

aluminum foil were placed in an oven for 24 h at 105 °C for polymerization to take place. Weight percentage gain (WPG) was calculated from Eq. (1).

$$\mathrm{WPG\%} = \frac{W_{\mathrm{f}} - W_{\mathrm{i}}}{\mathrm{W_i}} \quad (1)$$

where W_{f} and W_{i} are the oven-dried weight of the WPNC and raw wood, respectively.

2.4 *Microstructural Characterizations*

2.4.1 Determination of Water Uptake

Percentage water uptake (WU) after two weeks at 30 °C was calculated according to the Eq. (2)

$$\mathrm{WU\%} = \frac{W_2 - W_1}{W_1} \times 100 \quad (2)$$

where

W_2 Weight of a wood sample, treated and untreated, after two weeks of water immersion and

W_1 Weight of an oven-dried sample.

2.4.2 Fourier Transform Infrared Spectroscopy (FT-IR)

The infrared spectra of all specimens were recorded on a Shimadzu FT-IR 81001 Spectrophotometer (Kyoto, Japan). The transmittance range of the scan used was 4000–600 cm^{-1}.

2.4.3 Modulus of Rupture (MOR), Modulus of Elasticity (MOE), and Dynamic Young's Modulus (E_{d}) Measurements

Three-point bending tests were conducted according to Shimadzu MSC-5/500 universal testing machine (Kyoto, Japan) operating at a crosshead speed of 5 mm/min.

MOR and MOE were calculated using Eqs. (3) and (4).

$$\mathrm{MOR} = \frac{1.5\mathrm{PL}}{bh^2} \quad (3)$$

$$\text{MOE} = \frac{L^3 m}{4bd^3} \tag{4}$$

where

W Ultimate failure load,
L Span between centers of support,
B Mean width (tangential direction) of the sample,
D Mean thickness (radial direction) of the sample, and
M Slope of the tangent to the initial line of the force displacement curve.

Determination of E_d for all specimens was carried out using a free-free flexural vibration testing system. The E_d was calculated from the resonant frequency using Eq. (5).

$$E_d = \frac{4\pi^2 f^2 l^4 A\rho}{I(m_n)^4} \tag{5}$$

where

I $bd^3/12$,
D Beam depth, b is the beam width,
L Beam length,
F Natural frequency of the specimen,
ρ Density,
A Cross-sectional area, and $n = 1$ is the first mode of vibration, where
m_1 4.730.

2.4.4 Scanning Electron Microscopy (SEM)

The interfacial bonding between wood cell and polymer with filler was examined using a Hitachi (TM 3030) Scanning Electron Microscope (SEM) supplied by JEOL (Tokyo, Japan). The SEM specimens were sputter-coated with gold prior to observation. The micrographs were acquired at a magnification of 1500×.

3 Results and Discussion

3.1 Fourier Transform Infrared Spectroscopy (FT-IR)

The FT-IR spectra of various samples were presented in Fig. 1. The peak from 4000 to 3000 cm^{-1} corresponds to stretching vibration of H-bonds in –OH groups, and C–H stretching in methyl and methylene groups range from 3000 to 2800 cm^{-1}

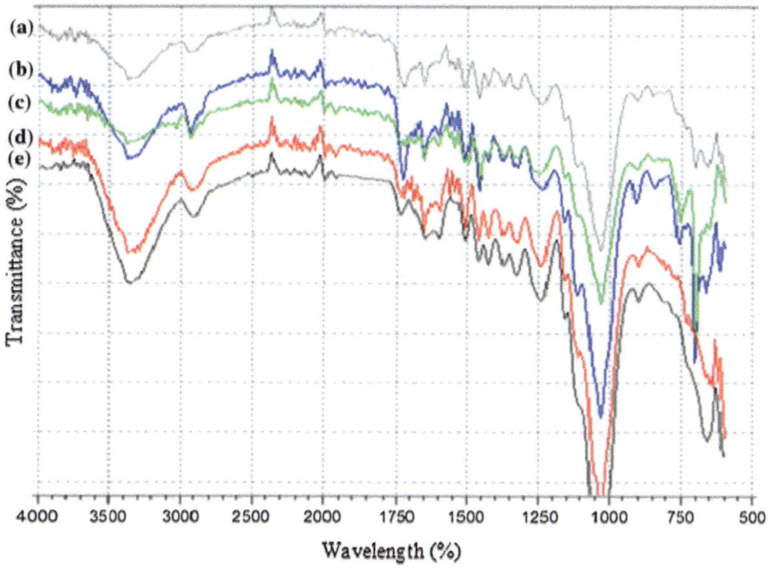

Fig. 1 FT-IR spectra of **a** ST-*co*-GMA-WPC, **b** ST-*co*-GMA-clay-WPNC, **c** ST-clay-WPNC, **d** ST-WPC and **e** RW

(Hamdan et al. 2011). The peak intensity near 1505 cm^{-1} (for aromatic double bonds) was higher for ST impregnated wood than for raw wood. This was due to the increase in the concentration of aromatic groups in the ST-WPC specimens. The IR spectra of ST-WPC samples exhibited a peak intensity above 3000 cm^{-1} for the –OH group that was lower than that of raw wood because of the removal of free-bound water from the wood lumen by ST. The peak intensity of the –OH group for ST-*co*-GMA impregnated wood was lower than that of ST impregnated wood because of the covalent bond formed between the –OH of wood and the glycidyl group of GMA. This could be attributed to the fact that ST-*co*-GMA performs better as filler than ST alone. In the case of the ST-*co*-GMA-clay-WPNC, for wave numbers above 3000 cm^{-1}, the peak intensity did not decrease compared with raw wood. The peak intensity above 3000 cm^{-1} increased with the clay ST-*co*-GMA polymer system, which removed fewer –OH groups from cellulose, hemicellulose, and water in the wood lumen. The –OH group of montmorillonite in the nanocomposite indicated absorption peak near 840 cm^{-1} for the Mg–Al–OH linkage (Sposito 1983). The absorption peak near 905 cm^{-1} was more intense in the ST-*co*-GMA-clay-WPNC system than the other ones. The peak intensity near 1730 cm^{-1} increased because of the increase in carbonyl group content in the nanocomposite compared with the raw wood. Because of the addition of montmorillonite in the ST-*co*-GMA-clay-WPNC system, the carbonyl peak intensity was higher than that in the ST-*co*-GMA system. This occurred because of the copolymerization between ST and GMA for the exfoliation of montmorillonite in the wood. The reaction scheme was shown in Fig. 2.

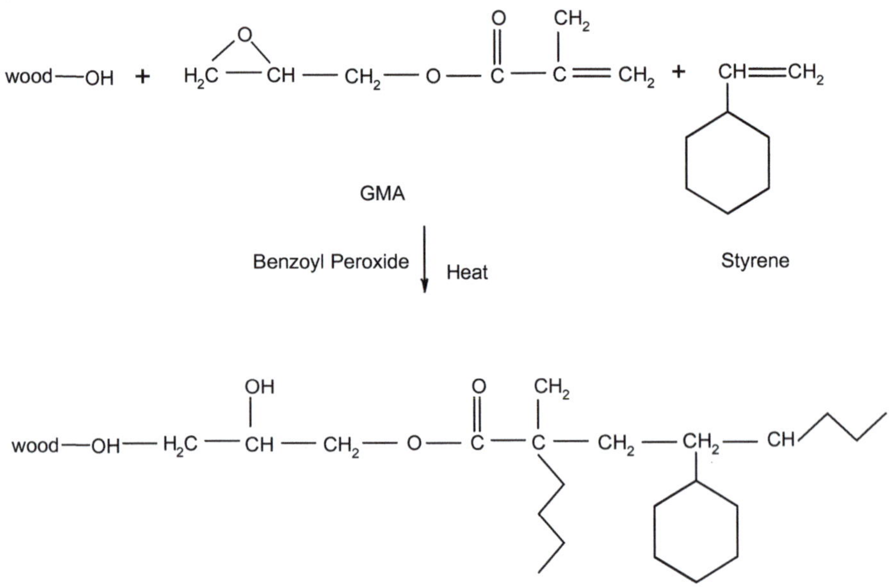

Fig. 2 Schematic reaction diagrams

3.2 *Modulus of Rupture (MOR), Modulus of Elasticity (MOE) and Dynamic Young's Modulus (E$_d$) Measurements*

The MOE, MOR, and E_d analyses of the raw wood and four different WPNC materials were given in Fig. 3.

As shown in Fig. 3, the nanocomposites exhibited higher MOE and MOR than the raw wood because of the presence of filler that exerted Van der Waals force and

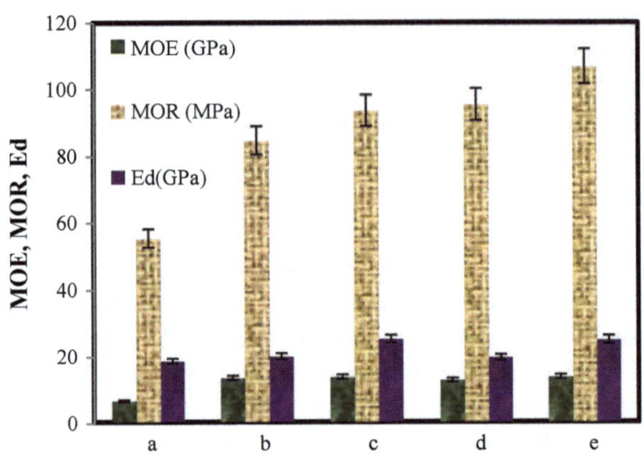

Fig. 3 MOE, MOR and E_d of **a** RW, **b** ST-WPC, **c** ST-*co*-GMA-WPC, **d** ST-clay-WPNC, and **e** ST-*co*-GMA-clay-WPNC

formed covalent bonds with the wood cell wall. In the ST-WPC, there existed only Van der Waals force between wood and ST, so it had lower MOE and MOR compared with ST-clay-WPNC and ST-clay-WPNC specimens. The clay particles had high surface area and possessed charges that form dipole bonds as well as Van der Waals forces with the wood cell wall. The MOE and MOR values were the highest for ST-*co*-GMA-clay-WPNC among these five nanocomposites. The epoxy group of GMA formed a covalent bond with the –OH group of wood and the vinyl group reacted with styrene by a free radical mechanism. Thus, the polymers and the wood cell walls achieved a sufficiently strong chemical bonding strength (Rahman et al. 2015; Hamdan et al. 2010, 2012). Also, MMT intercalated in the ST-*co*-GMA and a maximum number of MMT exfoliated in the St-*co*-GMA that had filled the wood cavities and thus increased the MOE and MOR of wood. On the other hand, in the ST-clay impregnated WPNCs showed intercalation rather than the exfoliation. Therefore, the MOE and MOR of ST-*co*-GMA-clay impregnated WPNCs were higher than that of ST-clay impregnated WPNC.

Among the nanocomposites, the highest E_d was observed for the ST-*co*-GMA-clay-WPNC, followed by ST-*co*-GMA-WPC, ST-WPC and ST-clay-WPNC. This was due to an increase in the density and shear force in the wood nanocomposites (Rahman et al. 2010). In the presence of shear force, the free energy increases and the molecules continuously oscillate from equilibrium (Toh 1979). The shear force was considered to vary sinusoidally with time. The styrene monomer could not directly react with wood cell wall components, whereas ST-*co*-GMA reacted with cell wall components to produce the covalent bond. The large clay particles entered the wood, which increased the Van der Waals forces in the nanocomposite to increase shear stress.

3.3 Weight Percentage Gain (WPG) and Water Uptake (WU)

Both WPG and WU values were shown in Fig. 4. It is clear that ST-*co*-GMA-clay-WPNC showed higher WPG% compared with the other nanocomposites. This was due to the presence of epoxy groups, which form covalent bonds with the –OH group of wood. The vinyl group present reacted with styrene by a free radical mechanism which exfoliated MMT in the ST-*co*-GMA monomer system that filled the wood cavities. ST-*co*-GMA-clay-WPNC had higher exfoliation, whereas ST-clay-WPNC showed intercalation.

The water absorption of various samples was illustrated in Fig. 4. The raw wood sample exhibited a higher percentage of water absorption compared with the modified WPNCs. This was expected because ST, ST-clay, ST-*co*-GMA, and ST-*co*-GMA-clay had reduced pore size on the surface of the wood (Rahman et al. 2010). The void area on the wood surface facilitates water penetration into the wood cell wall with hydrophilic hydroxyl groups. ST-GMA-treated samples

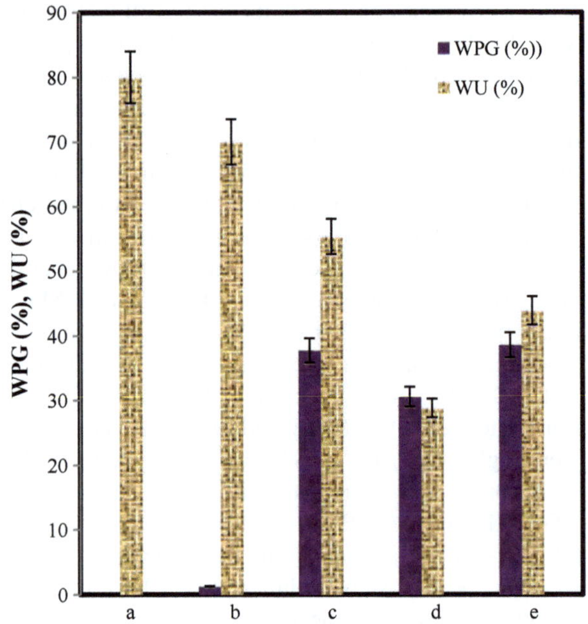

Fig. 4 WPG (%), and WU (%) of **a** RW, **b** ST-WPC, **c** ST-*co*-GMA-WPC, **d** ST-clay-WPNC, and **e** ST-*co*-GMA-clay-WPNC

absorbed less water than ST-treated samples because more void areas were present in ST-WPC compared with ST-*co*-GMA-WPC, which was also reflected in the SEM analysis. Similar results had been reported for rubber wood samples (Devi and Maji 2007). It was also apparent that ST-clay-WPNC had the lowest (28.84%) water absorption among all the samples.

3.4 Scanning Electron Microscopy (SEM)

The morphological properties were shown in Fig. 5. It was observed that the lumen and tracheids of wood were filled up by the solid polymers by in situ polymerization. In Fig. 5d, e, the wood surfaces were smoother than the ST-WPC surface because the glycidyl group from GMA formed a covalent bond with wood (–OH). However, ST could not react with wood cell wall components, whereas only Van der Waals forces were observed in ST-WPC. The poor interaction between the polymer and wood implied a more brittle nanocomposite (Li et al. 2011). The MMT was unexfoliated in the ST-clay monomer system, whereas exfoliated in ST-*co*-GMA, which shown in Fig. 5d (Xu et al. 2013). The ST-*co*-GMA polymers with MMT not only filled up the lumen of the wood cell, but also intimately contacted the wood cell walls to produce ST-*co*-GMA-WPNC. Therefore, it was concluded that there were strong interactions between the polymers and wood cell walls, as shown in Fig. 5d, e.

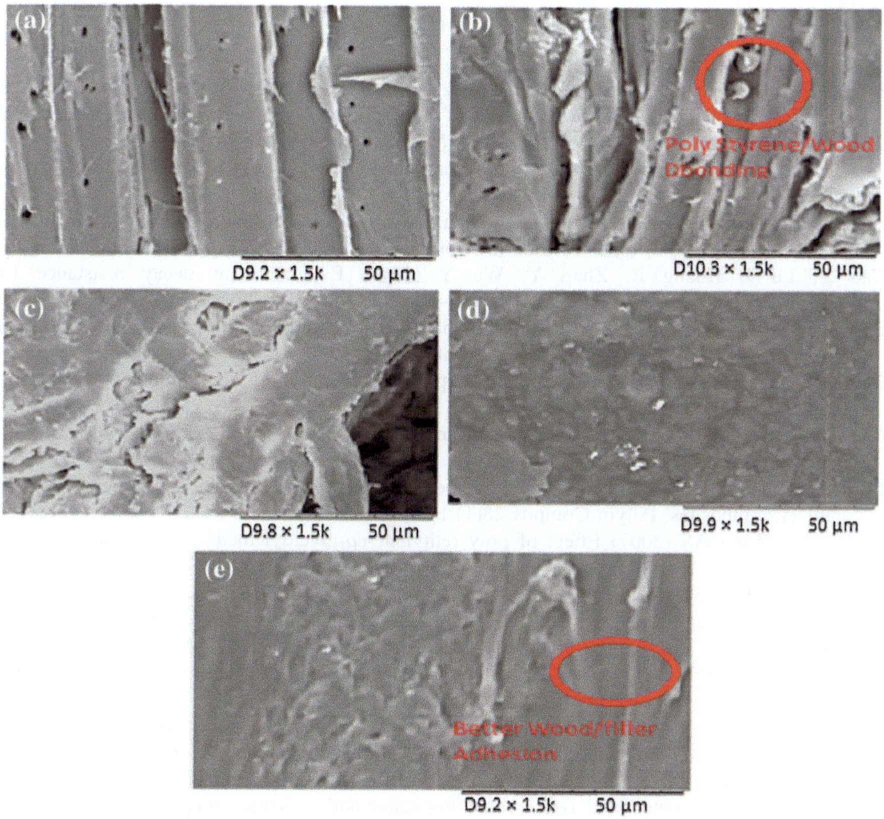

Fig. 5 SEM micrographs of **a** RW, **b** ST-WPC, **c** ST-*co*-GMA-WPC, **d** ST-clay-WPNC, and **e** ST-*co*-GMA-clay-WPNC

4 Conclusion

Kumpang wood was successfully impregnated by clay dispersed styrene-*co*-glycidyl methacrylate (St-*co*-GMA-clay) which was proved by FT-IR analysis. WPNCs had the highest MOE, MOR, and E_d values, followed by WPCs and RW. ST-*co*-GMA-clay-WPNC specimens showed the highest WPG, followed by ST-clay, ST-*co*-GMA, and ST. ST-clay-WPCs had lower water uptake than ST-*co*-GMA-clay-WPNC, ST-co GMA-WPCs, ST-WPCs, and raw wood. The SEM morphology showed that ST-*co*-GMA-clay-WPNC had a smoother surface than ST-clay-WPCs, ST-*co*-GMA-WPCs, ST-WPCs, and raw wood.

Acknowledgements The authors would like to acknowledge the financial support from Ministry of Higher Education Malaysia, for their financial support [Grant no. FRGS/SG02(01)/1085/2013 (31)] during the research.

References

Baysal E, Yalinkilic MK, Altinok M, Sonmez A, Peker H, Colak M (2007) Some physical, biological, mechanical, and fire properties of wood polymer composite (WPC) pretreated with boric acid and borax mixture. Constr Build Mater 21(9):1879–1885

Bhattacharya SS, Aadhar M (2014) Studies on preparation and analysis of organoclay nano particles. Res J Eng Sci 3(3):10–16

Bledzki AK, Mamun AA, Faruk O (2007) Abaca fibre reinforced PP composites: Comparison with jute and flax fibre composites considering fibre contents. eXPRESS Polym Lett 1(11):755–762

Cao Y, Lu J, Huang R, Zhao Y, Wu Y (2011) Evaluation of decay resistance for steam-heat-treated wood. BioRes 6(4):4696–4704

Carroll D (1970) Clay minerals: a guide to their X-ray identification, special paper 126. The Geological Society of America, Boulder, California

Chang HT, Chang ST (2001a) Inhibition of the photodiscoloration of wood by butyrylation. Holz 55(3):255–259

Chang ST, Chang HT (2001b) Comparisons of the photostability of esterified wood. Polym Degrad Stabil 71(2):261–266

Devi RR, Maji K (2007) Effect of glycidyl methacrylate on the physical properties of wood–polymer composites. Polym Compos 28(1):1–5

Dikobe DG, Luyt AS (2007) Effect of poly (ethylene-*co*-glycidyl methacrylate) compatibilizer content on the morphology and physical properties of ethylene vinyl acetate. Wood Fiber Compos 104(5):3206–3213

Farahani MRM, Taghizadeh F (2010) Roughness of esterified eastern cottonwood. BioRes 5(4):2232–2238

Feist WC, Hon DNS (eds) (1984) Chemistry of weathering and protection. American Chemical Society, Washington, DC, pp 401–451

Guang-jie Z, Wen-hua L, Zheng-hua X (2006) Preparation and characterization of wood/montmorillonite nanocomposites. Forestry Studies China 8(1):35–40

Hamdan S, Rahman MR, Ahmed AS, Islam MS, Talib ZA (2010) Dynamic young's modulus measurement of treated and post-treated tropical wood polymer composites. BioRes 5(1): 324–342

Hamdan S, Rahman MR, Ahmed AS, Islam MS, Talib ZA, Abdullah WFW, Mat MSC (2011) Thermogravimetric analysis and dynamic young's modulus measurement of *N,N*-dimethylacetamide-impregnated wood polymer composites. J Vinyl Addit Technol 17(3): 177–183

Hamdan S, Rahman RM, Ahmed A, Hasan M, Islam MS (2012) Effect of coupling reactions on the mechanical and biological properties of tropical. Int Biodeterior Biodegrad 72:108–113

Hegde RR (2009) Structure and properties of nanoclay reinforced polymer films, fibers and nonwovens. University of Tennessee, Knoxville

Islam MS, Hamdan S, Hassan A, Sobuz H, Talib ZA (2014) The chemical modification of tropical wood polymer composites. J Compos Mater 48(7):783–789

Jebrane M, Pichavant F, Sebe G (2011) A comparative study on the acetylation of wood by reaction with vinyl acetate and acetic anhydride. Carbohydr Polym 83(2):339–345

Kam TY, Lee TY (1992) Detection of cracks in structures using modal test data. J Eng Fract Mech 42(2):381–387

Kosonen ML, Wang B, Caneba GT, Gardner DJ Rials TG (2000) Polystyrene/wood composites and hydrophobic wood coatings from water-based hydrophilic-hydrophobic block. Clean Prod Proc Springer-Verlag 2:117–123

Le EA, Nairn JA (2014) Measuring interfacial stiffness of adhesively-bonded wood. Wood Sci Technol 48(6):1109–1121

Li YF, Liu YX, Wang XM, Wu QL, Yu HP, Li J (2011) Wood–polymer composites prepared by the in situ polymerization of monomers within wood. J Appl Polym Sci 119(6):3207–3216

Li Y, Liu Z, Dong X, Fu Y, Liu Y (2013) Comparison of decay resistance of wood and wood-polymer composite prepared by in situ polymerization of monomers. Int Biodeterior Biodegrad 84:401–406

Mattos BD, de Cademartori PHG, Lourencon TV, Gatto DA (2014) Colour changes of Brazilian eucalypts wood by natural weathering. Int Wood Prod J 5(1):33–38

Mohanty AK, Misra M, Drzal LT (2002) Sustainable bio-composite from renewable resources: opportunity and challenges in the green materials world. Polym Environ 10(1/2):19–26

Morrell JJ, Stark NM, Pendleton DE, McDonald AG (2006) Durability of wood-plastic composites. Wood Des Focus 16(3):7–10

Prakash GK, Mahadevan KM (2008) Enhancing the properties of wood through chemical modification with palmitoyl chloride. Appl Surf Sci 254(6):1751–1756

Rahman MR, Hamdan S, Ahmed AS, Islam MS (2010) Mechanical and biological performance of sodium metaperiodate-impregnated plasticized wood (PW). BioRes 5(2):1022–1035

Rahman MR, Lai JCH, Hamdan S, Rahman MM, Hossen MF (2015) Impact of nanoclay on physicomechanical and thermal analysis of polyvinyl alcohol/fumed silica/clay nanocomposites. J Appl Polym Sci 132(15):41843

Rowell RM (2006) Chemical modification of wood: a short review. Wood Mater Sci Eng 1(1): 29–33

Sposito G (1983) Infrared spectroscopic study of adsorbed water on reduced-charge Na/Li-montmorillonites. Clays Clay Miner 31(1):9–16

Stolf DO, Lahr FAR (2004) Wood-polymer composite: physical and mechanical properties of some wood species impregnated with styrene and methyl methacrylate. J Mater Res 7(4): 611–617

Toh HK (1979) A study of diffusion in polymers using C-14 labelled molecules. Loughborough University, Leicestershire, UK

Xie YJ, Hill CAS, Sun D, Jalaludin Z, Wang Q, Mai C (2011) Effects of dynamic aging (hydrolysis and condensation) behaviour of organofunctional silanes in the aqueous solution on their penetrability into the cell walls of wood. BioRes 6(3):2323–2339

Xu Y, Guo Z, Fang Z, Peng M, Shen L (2013) Combination of double-modified clay and polypropylene-graft-malei anhydride for the simultaneously improved thermal and mechanical properties of polypropylene. J Appl Polym Sci 128(1):283–291

Zabel RA, Morrell JJ (1992) Wood microbiology—decay and its prevention. Academic Press, San Diego, California

Physico-mechanical, Morphological, and Thermal Properties of Clay Dispersed Styrene-*co*-Maleic Acid Impregnated Wood Polymer Nanocomposites

M.R. Rahman, S. Hamdan and J.C.H. Lai

Abstract In this study, we evaluate the physical, mechanical, and morphological properties of a clay dispersed styrene-*co*-glycidyl methacrylate (ST-*co*-GMA) impregnated wood polymer nanocomposite (WPNC). The WPNC was characterized by Fourier Transform Infrared Spectroscopy (FT-IR), Scanning Electron Microscopy (SEM), 3-point bending, and free-vibration testing. The FT-IR results showed that the absorbance at 1730 cm^{-1} was increased for ST-*co*-GMA-clay-WPNC compared with other nanocomposites and the raw material. The SEM results showed that ST-*co*-GMA-clay-WPNC had a smoother surface than other nanocomposites and raw wood. The modulus of elasticity (MOE), modulus of rupture (MOR), and dynamic Young's moduli (Ed) of WPNCs were considerably increased compared to wood polymer nanocomposites (WPNCs) and raw wood. The raw wood exhibited a higher water uptake (WU) than WPNCs.

Keywords ST-*co*-GMA · Nanoclay · Mechanical properties · Morphological properties

1 Introduction

From the beginning of human history to till now, wood has been used to make shelters, boats, wheels, tools, and weapons. When the earliest humans made the first wooden wheel and it was used to carry loads, the conquering of science and civilization of human being has been started. Nowadays, the use of wood is increasing as structural materials, because wood achieves high strength-to-weight with its cellulose, hemicellulose, and lignin (Kretschmann 2010; Malkapuram et al. 2008). For sustainable development, wood is better over the non-renewable synthetic materials, because wood is renewable, available, and it contributes less

M.R. Rahman (✉) · S. Hamdan · J.C.H. Lai
Faculty of Engineering, Universiti Malaysia Sarawak,
94300 Kota Samarahan, Sarawak, Malaysia
e-mail: rmrezaur@unimas.my

M.R. Rahman, *Wood Polymer Nanocomposites*, Engineering Materials,
DOI 10.1007/978-3-319-65735-6_10

greenhouse gas emission than non-renewable materials. However, wood components can be degraded easily by microorganism attack and heat flow (Kamei et al. 2012; Rath et al. 2003). At low humidity environment, wood desorbs water to shrinkage. In outdoor applications, wood suffers from the photo degradation process. Wood in presence of UV-radiation changes its color by its lignin degradation (Pandey 2005). The high amount of hydroxyl groups may lead to poor dimensional stability. Due to the disadvantages of wood, the applications of wood are limited. Therefore, wood modification is needed to improve the weakness of wood and strengthen the properties of wood.

Impregnation of wood with polymer is one of the techniques to improve the wood properties. Generally, better compatibility between wood and polymer shows better mechanical and thermal properties of wood polymer composites (WPCs) (Fan et al. 2012; Ndiaye and Tidjani 2012). Wood impregnated with copolymers of polar and nonpolar characteristics such as styrene-*co*-glycidyl methacrylate (GMA), polyether-b-amide, glycidyl methacrylate-*co*-ethylene glycol dimethacrylate, allyl glycidyl ether-*co*-methyl methacrylate, and methacrylate-*co*-hexamethylene diisocyanate showed enhancement in dimensional stability, elevated thermostability, and biodegradable properties but the mechanical properties of wood are not showing any significant improvement (Devi and Maji 2007; Islam et al. 2012; Kartal et al. 2004; Lia et al. 2013; Sliwa et al. 2012). To increase the chemical bonding of wood, various chemical treatments have been introduced, such as acetyl, methyl methacrylate, and methacrylic anhydride grafting (Hill et al. 2001; Pawar et al. 2013; Xiao et al. 2012). Highly alkali or acid treatment of wood decreases the hydrophilicity and fungal attack but yield lower mechanical property than its raw wood due the to removal of lignin.

Recently, nanotechnological modification of wood with the mixture of copolymer and MMT has been proven to be very effective means of improving the essential properties of wood (Cai et al. 2008; Hazarika et al. 2015). Nanoclay by hydroxyl group improves the interaction between wood and thermoplastic composite polymer. Organic polymer with organically modified layered silicate (OMLS) impregnated wood demonstrates significant mechanical improvement barrier, fire resistance, thermal and environmental stability, solvent uptake, and rate of biodegradability, relative to an unmodified polymer resin impregnated wood (Alexandre and Dubois 2000; Zhao et al. 2006). The strong interfacial interactions among wood cell wall, filler polymer, and OMLS improve these properties compared to conventional composites (Chen et al. 2002). Nanoclay (≤ 5 wt%) incorporated wood shows better thermal and mechanical properties than that of conventional polymer impregnated wood due to proper dispersion in wood cell wall. For these reasons, wood polymer nanocomposites are less in weight than conventional composites and are used as a competitive with other materials for specific applications. In situ polymerization, suitable filler monomers insert into the silicate galleries and subsequent polymerization (Haraguchi et al. 2005; Jeon and Baek 2010; Usuki et al. 1993). Many studies have been carried out on physical and mechanical properties of wood and WPC (Rowell 2006). Polypropylene matrix with fiber and clay composite increases thermal and mechanical properties by decreasing the pore volume size of the system (Rahman et al. 2014).

Wood has been impregnated with polyethylene-*co*-maleic anhydride to improve the mechanical, thermal, and biodegradable properties (Bodîrlău and Teacă 2009). –COOH group of polymerized maleic acid suggests that it may react with the hydroxyl group of wood to enhance the material properties of the composite. In addition, ST-*co*-Mal plastics are recyclable from automobile parts (Simonsen et al. 1997). Little work, however, has been devoted to tropical wood species.

Although the research development in the WPCs on various wood has been increasing, there is no comprehensive work has been done on wood polymer composites of Kumpang. The importance of wood modification and the preparation of the WPCs due to the sustainable applications of wood material for the society.

In this study, Kumpang wood was impregnated with styrene or a mixture of styrene with clay or a mixture of styrene and maleic acid, or a mixture of styrene, maleic acid, and clay or to improve its physical, mechanical, and thermal properties. The prepared WPCs, WPNCs, and raw wood were characterized and compared among them.

2 Experimental

2.1 Materials

Defect-free and straight-grained Kumpang wood (RW) was obtained from Forestry Department, Sarawak, Malaysia. The chemicals used to produce WPCs and WPNs were styrene (ST), maleic acid (Mal), benzoyl peroxide, and nanoclay (Nanomer® 1.28E) which was modified with 25–30 wt% trimethyl stearyl ammonium. ST was supplied by A.S. Joshi and Company, Goregaon West, Mumbai, India while Mal and benzoyl peroxide was supplied by Merck Schuchardt OHG, Germany. Nanoclay (Nanomer® I.28E) was supplied by Aldrich, USA.

2.2 Specimen Preparation

A Kumpang tree was cut into three bolts that were 1.2 m in length. Each bolt was sawn into quarters in order to produce planks that were 4 cm thick. These planks were subsequently conditioned by air drying in a room with a relative humidity of 60% and ambient temperature of approximately 25 °C for one month prior to testing. The planks were ripped and machined to 340 mm (L) × 20 mm (T) × 10 mm (R) for three-point bending and free-free vibration tests.

2.3 Preparation of Wood Polymer Nanocomposites (WPNCs)

Oven-dried raw wood samples were impregnated with either one monomer or mixture of monomers with MMT (Table 1). To impregnate the raw wood, according to the chemical compositions of Table 1, the raw wood samples and chemicals were placed in a vacuum chamber with 75 mm (Hg) for 30 min. For each monomer(s) impregnation, approximately 10 g of benzoyl peroxide was added to initiate polymerization, and five replicate wood polymer composites were prepared. These wood composites were then removed from the chamber, and the excess chemicals were removed. Specimens were wrapped with aluminum foil and placed into a 105 °C oven for 24 h for polymerization to take place.

2.4 Microstructural Characterizations

2.4.1 Determination of Weight Percentage Gain (WPG)

Weight percentage gain (*WPG*) was calculated using the following equation:

$$\mathrm{WPG}\,(\%) = [(W_\mathrm{f} - W_\mathrm{i})/W_\mathrm{i}] \times 100 \quad (1)$$

where W_f and W_i are the oven-dried weight of the WPC (or the WPNC) and the raw wood, respectively.

2.4.2 Determination of Water Uptake (WU)

The water uptake (*WU*) was calculated using the following equation:

$$\mathrm{WU}\,(\%) = [(W_2 - W_1)/W_1] \times 100 \quad (2)$$

where W_2 is the weight of treated (or untreated control) wood samples after two weeks of water immersion, and W_1 is the weight the oven-dried sample, respectively.

Table 1 Composition of the chemicals for the raw wood (RW) impregnation

Specimen	ST (mL)	Mal (g)	Clay (g)	Initiator (BP) (g)	Name of prepared composites
Raw wood	500	–	–	10	ST-WPC
Raw wood	500	15	–	10	ST-*co*-Mal-WPC
Raw wood	500	–	10	10	ST-clay-WPNC
Raw wood	500	15	10	10	ST-*co*-Mal-clay-WPNC

2.4.3 Fourier Transform Infrared Spectroscopy (FT-IR)

The infrared spectra of all specimens were recorded on a Shimadzu FT-IR 81001 Spectrophotometer (Shimadzu; Kyoto, Japan). The wave number range of the scan used was 4000–600 cm^{-1}.

2.4.4 X-ray Diffraction (XRD) Analysis

XRD analysis for raw wood and wood polymer nanocomposites was performed with a Rigaku diffractometer (RGS Corp. Sdn. Bhd.; Selangor, Malaysia) operating with Cu$K\alpha$ radiation, the wavelength of 0.154 nm, 40 kV and 30 mA. The crystallinity index (CI_{XRD}) was calculated using the following height ratio of diffraction peaks:

$$CI_{XRD}(\%) = \frac{I_{002} - I_{am}}{I_{002}} \tag{3}$$

where I_{002} is the maximum intensity of the peak at 2θ at about 22°, and I_{am} is the minimum intensity above baseline at 2θ at about 30°, which corresponds to crystalline and amorphous parts, respectively. The d-spacing between the [002] lattice planes (d_{002}) of the sample was calculated using the Bragg equation:

$$d_{002} = \frac{n\lambda}{2\sin\theta} \tag{4}$$

where n refers to the integer wavelength number ($n = 1$), θ refers to the maximum diffraction angle, and λ is the X-ray wavelength (0.154 nm).

2.4.5 Scanning Electron Microscopy (SEM)

Fractured surfaces were examined to investigate the interfacial bonding between wood cell and polymer with filler. These surfaces were imaged using a Hitachi TM3030 Scanning Electron Microscope (SEM) supplied by JEOL (Tokyo, Japan). The SEM specimens were sputter coated with gold prior to observation. The micrographs were acquired at a magnification of 1500.

2.4.6 Modulus of Rupture (MOR), Modulus of Elasticity (MOE), and Dynamic Young's Modulus (E_d) Measurements

Three-point bending tests were conducted using a Shimadzu MSC-5/500 universal testing machine (Kyoto, Japan) operating at a crosshead speed of 5 mm/min. The modulus of rupture (MOR) and modulus of elasticity (MOE) were calculated using:

$$\mathrm{MOR} = 1.5LW/bd^2 \tag{5}$$

$$\mathrm{MOE} = L^2m/4bd^3 \tag{6}$$

where W is the ultimate failure load, L is the span between centers of support, b is the mean width (tangential direction) of the sample, d is the mean thickness (radial direction) of the sample, and m is the slope of the tangent to the initial line of the force displacement curve.

Determinations of the dynamic Young's modulus, E_d, for all specimens were carried out using a free-free flexural vibration testing system. The E_d was calculated from the resonant frequency using the following equation,

$$E_d = \frac{4\Pi^2 f^2 l^4 A\rho}{Im_n^4} \tag{7}$$

where I is the second moment of area of the beam's cross section (which is equal to $bd^3/12$), d is the beam's depth, b is the beam's width, l is the beam's length, f_n is the beam's natural frequency at the nth flexural mode, ρ is the beam's density, A is the beam's cross-sectional area, and m_n is a frequency factor which is related to the free-free beam vibration at the nth flexural mode. The value of m_n at the first mode (n = 1) is 4.730 (i.e., m_1 = 4.730).

2.4.7 Thermogravimetric Analysis (TGA)

Thermogravimetric Analysis (TGA) measurements were performed on 5–10 mg samples at a heating rate of 10 °C/min in a nitrogen atmosphere using a thermogravimetric analyzer (Model SDT Q600, TA Instruments; Selangor, Malaysia). Thermal decomposition of each sample occurred in a programmed temperature range of 30–600 °C under a constant nitrogen flow rate of 5 mL/min. The continuous weight loss and temperature were recorded and analyzed.

2.4.8 Differential Scanning Calorimetric Testing (DSC)

The differential scanning calorimetry (DSC) is used to measure and characterize the thermal properties of materials. DSC revealed the amount of material with high thermal stability. DSC measures the energy required to obtain the same temperature rise in each holder. When a transition such as melting, degradation, or crystallization i.e., an endothermic or exothermic reaction occurs, the change in power needed to keep the sample at the same temperature as the reference pan is recorded as a peak.

All differential scanning calorimetry (DSC) measurements were made on a DSC Q10 (TA instrument) thermal system, using a sealed aluminum capsule. Each test specimen was weighed to about 3–3.5 mg and was held at a single heating rate of 10 °C/min and scanning temperature from 30 to 450 °C. Each of the data reported represents an average of three runs. To assess the quantitative reproducibility, an integration of the area under the peak yields the melting endotherm.

3 Results and Discussion

3.1 FT-IR

The FT-IR spectra of raw wood and various WPCs and WPNs are presented in Fig. 1. A broad absorption peak 4000–3000 cm^{-1} and the peak 2950–2800 cm^{-1} assigned to stretch vibration of H-bonds in –OH groups and stretching vibrations of C–H of methyl and methylene groups, respectively (Popescu et al. 2013; Rahman et al. 2011). The broad peak at 2950 cm^{-1} of raw wood and St-WPC shifted to 2890 cm^{-1} with a narrower and sharper shape for St-*co*-Mal-WPC and St-*co*-Mal-clay-WPNC. This observation indicated that the wood and Mal or wood and Mal with modified nanoclay have a natural compatibility, and the molecular chains

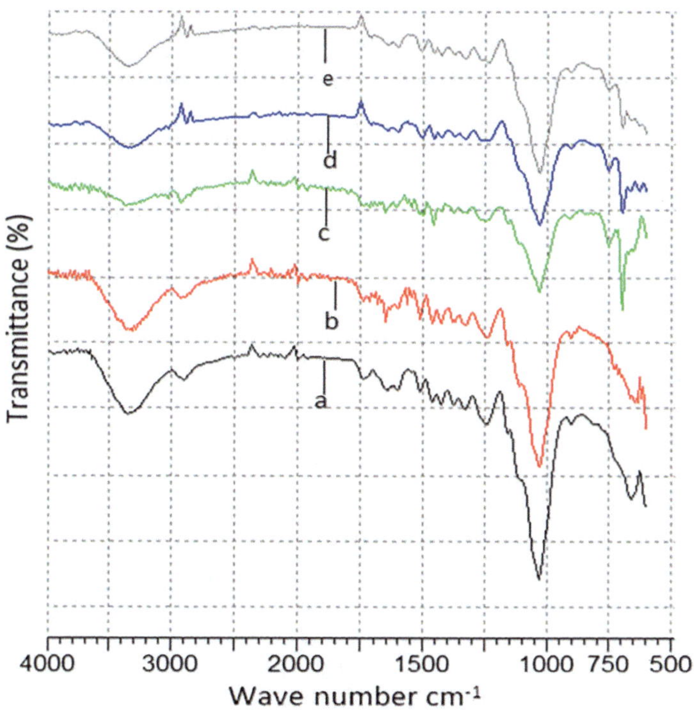

Fig. 1 FT-IR spectra of **a** Raw wood, **b** ST-WPC, **c** ST-clay-WPNC, **d** ST-*co*-Mal-clay-WPNC, and **e** ST-*co*-Mal-WPC

could interact with each other. ST impregnated WPC showed a higher peak intensity at 1505 cm^{-1} (which corresponded for aromatic double bonds) than that of raw wood. This demonstration proved that the raw wood was impregnated by ST. The intensity of the –OH group above 3000 cm^{-1} for ST–WPC was lower than that of raw wood. This was due to removal of free-bound water from the wood lumen by styrene. ST-*co*-Mal and ST-*co*-Mal-clay impregnated wood showed a lower peak intensity of –OH group above 3000 cm^{-1} and a higher peak intensity of –CO group at 1730 cm^{-1} when compared to ST-WPC. This is due to formation of covalent bond between the –OH of wood and the –OH group of maleic acid (Kim et al. 2007; Schwanninger et al. 2004). This can be attributed to the fact that ST-*co*-Mal created a network with raw wood (which was reflected in mechanical properties). The charge of Al^{3+} polarized the π-electrons of aromatic rings of wood and styrene. Consequently, C–H bending intensity at 750 cm^{-1} increased for ST-*co*-Mal-clay-WPNC and St-clay-WPNC when compared to ST-WPC (Yao et al. 2013). The peak 698 cm^{-1} is characteristic stretching vibrations of Al–O–Si and out-of-plane bending of OH of wood (Kondo and Sawatari 1996; Parker and Frost 1996). Thus, the intensity of this band was higher for WPNCs than WPCs and raw wood. These are the prove of impregnation of raw wood with clay and Maleic acid

3.1.1 X-ray Diffraction (XRD) Analysis

X-ray diffractograms of raw wood, WPCs, and WPNCs are shown in Fig. 2. The diffraction pattern of raw wood, WPCs, and WPNCs exhibited amorphous peaks.

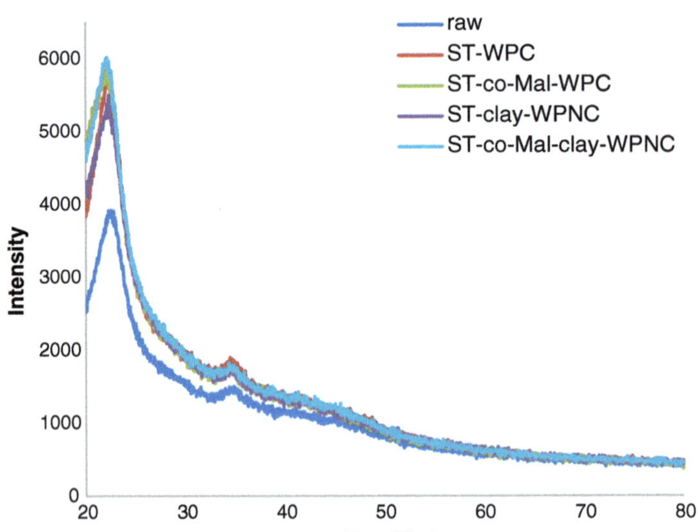

Fig. 2 X-ray diffraction (XRD) diffractograms of **a** RW, **b** ST-*co*-Mal-WPC, **c** ST-*co*-Mal-clay-WPNC, **d** ST-clay-WPNC, and **e** ST-WPC

Table 2 Crystallinity index CI_{XRD} and *D*-spacing as obtained by XRD analysis

Sample	CI_{XRD} (%)	2θ (002)	*D*-spacing (002) nm
Raw	66.59	22.6967	0.3913
ST-WPC	72.5	22.3667	0.3970
ST-*co*-Mal-WPC	73.52	22.3667	0.3970
ST-clay-WPNC	71.89	21.7067	0.4089
ST-*co*-Mal-clay-WPNC	73.03	22.2017	0.4000

The raw wood, WPCs, and WPNCs showed [002] crystalline plane and [040] crystalline plane, when the values of 2θ were around 22.4°, and 35.5°, respectively (Chen et al. 2016; Wada et al. 2004). Crystallinity index (CI_{XRD}) was determined at [002] plan which is shown in Table 2. The CI_{XRD} for ST-*co*-Mal-WPC and ST-*co*-Mal-clay-WPNC was higher than raw wood and other composite. Because ST-*co*-Mal network created bond with the –OH groups of wood and restricted the movement of amorphous chains of wood (Victor et al. 2013). In ST-WPC, styrene restricts the movement of amorphous chains of wood without bonding with wood. The CI_{XRD} of ST-clay-WPNC was lower than the other composites due to clay particles entering into crystalized region of cellulose. The *d*-spacing of [002] of WPCs and WPNCs plane increased when compared to raw wood. This was due to the 2θ values of WPCs and WPNCs were decreased compared to raw wood. The increment of *d*-spacing indicated that the wood sample was successfully impregnated by polymer (Chen et al. 2016). Among these WPCs and WPNCs, the increment of *d*-spacing was the highest for ST-clay-WPNC. Because exfoliated clay particles entered between crystalline planes (Liu et al. 2016). Though Mal reduced the –OH of raw wood, the CI_{XRD} of ST-*co*-Mal-WPC, and ST-*co*-Mal-clay-WPNC increased compared to raw woods due to the reduction of amorphous movement of wood molecules by clay and styrene. Among the composites, there was no additional peak with respect to raw wood. This was due to clay particles were exfoliated in the nanocomposites. The ST and Mal were not arranged in the composites.

3.2 Scanning Electron Microscopy (SEM)

The morphological properties of RW, ST-WPC, ST-*co*-Mal-WPC, ST-clay-WPNC and ST-*co*-Mal-clay-WPNC were shown in Fig. 3. It was observed that the lumen and tracheids of wood were filled up by the polymers and nanoclay. In Fig. 6, the surfaces of ST-clay-WPNC, and ST-*co*-Mal-clay-WPNC are smoother than that of raw wood, ST-WPC, and ST-*co*-MAL-WPC. This was due to nanoclay was exfoliated by ST, Mal, and wood cell wall to fill the cavities of raw wood (Xu et al. 2015). ST-*co*-MAL-WPC had smother surface than ST-WPC surface because the first one not only extracts the Van der Waals forces among them but also there was a covalent bond between wood maleic acid. ST could not react with wood cell wall components, only Van der Waals forces were created in ST-WPC. The poor

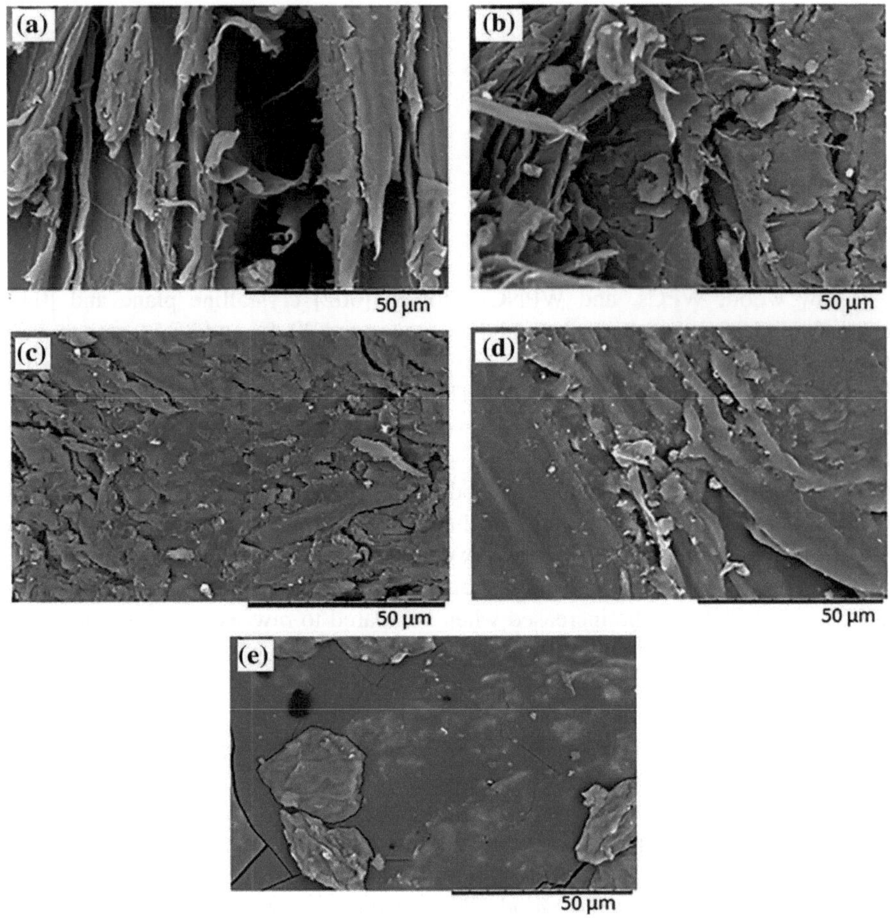

Fig. 3 SEM micrographs of **a** RW, **b** ST-WPC, **c** ST-*co*-Mal-WPC, **d** ST-clay-WPNC; and **e** ST-*co*-Mal-clay-WPNC

interaction between the polymer and wood implied a more brittle composite (Li et al. 2011). Therefore, it was concluded that there were strong interactions among Mal, ST, nanoclay, and wood cell walls, as shown in Fig. 3e.

3.3 MOR, MOE, and E_d

The MOE, MOR, and E_d analyzes of (the average values of five samples of each kind) raw wood and four different types of composite are given in Fig. 4. As shown in Fig. 4, the composites exhibited higher MOE and MOR than the raw wood because wood components sheared the stress with fillers. ST-WPC showed better

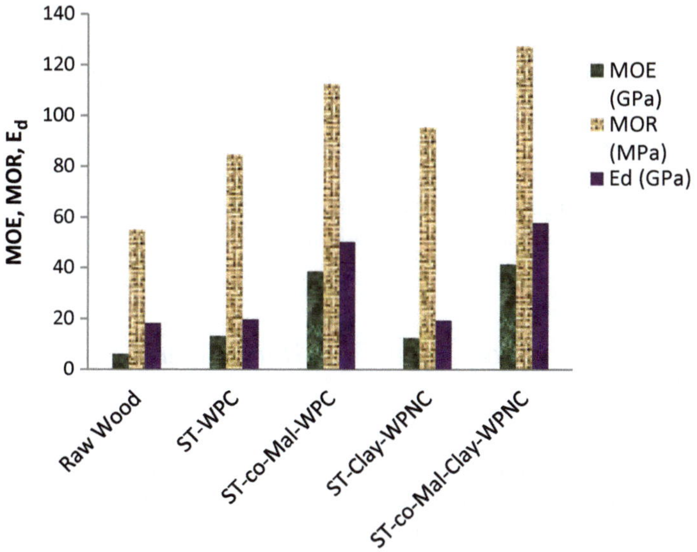

Fig. 4 MOE, MOR, and E_d of **a** RW, **b** ST-WPC, **c** ST-*co*-GMA-WPC, **d** ST-clay-WPNC, and **e** ST-*co*-GMA-clay-WPNC

mechanical properties than that of raw wood due to the presence of moisture in the cell wall of raw wood served as plasticizer and thus lowered the strength and modulus of wood (Hazarika et al. 2015). ST-clay-WPNC showed better MOE and MOR when compared to ST–WPC. This was due to in ST–WPC, polystyrene without any charge extracted only a weak Van der Waals force between wood and ST. Nanoclay particles exfoliated in ST-clay-WPNC, had high surface area, possessed charges that formed dipole bonds with the wood cell; organic part of modified clay extracted Van der Waals forces with wood cell wall and styrene (Chowdary and Niranjan 2015).

The MOE and MOR values were the highest for ST-*co*-Mal-clay-WPNC among these four composites. This was due to –COOH groups and alkenes group of maleic acid created covalent bonds with the –OH group of wood and with vinyl group of styrene, respectively. The π-electrons of aromatic groups of styrene and lignin were induced by Al^{3+} of nanoclay, and the high surface area of the nanoclay in the composite resulted in a higher degree of interaction of the polymer chains with the nanoclay. Thus, the polymers and the wood cell walls achieved a sufficiently strong chemical bonding strength (Rahman et al. 2014). The MOE and MOR values of ST-*co*-Mal-WPC were lower than that of ST-*co*-Mal-clay-WPNC; without organically modified clay the compatibility between wood molecules and ST-*co*-Mal was not good.

Prepared composites showed higher amount of E_d than raw wood. This was due to an increase in the density and shear force in the wood composite (Greil et al. 1998; Lopes et al. 2000). The shear force is considered to vary sinusoidally with time. In the presence of shear force, the free energy increases, and the molecules continuously oscillate from equilibrium and thus increased E_d (Liu et al. 2016; Toh 1979). Among the composites, the highest was observed for the ST-*co*-Mal-clay-WPNC, followed by ST-*co*-Mal-WPC, ST-WPC, and ST-clay-WPNC. ST-*co*-Mal reacts with cell wall components to produce the covalent bond, whereas styrene monomer cannot react with wood cell wall components (Wolfenden and Kinzy 1997). The large clay particles enter the wood, which increase the Van der Waals forces in the composite to increase shear stress.

3.4 Weight Percentage Gain (WPG) and Water Uptake (WU)

Both WPG and WU values are shown in Fig. 5. It was clear that ST-*co*-clay-WPNC showed higher WPG when compared to the ST-*co*-Mal-clay-WPNC. This was due to the high polar maleic acid was first priority in the mixture of ST, Mal, and modified nanoclay to fill the polar raw wood cavity when compared to nonpolar styrene and organically modified clay. WPG of raw wood by the impregnation with ST-*co*-Mal-clay is higher than by the impregnation with ST-*co*-Mal. This was due to nonpolar styrene and polar maleic acid. Besides, the molecular weight of nanoclay is higher than that of Mal.

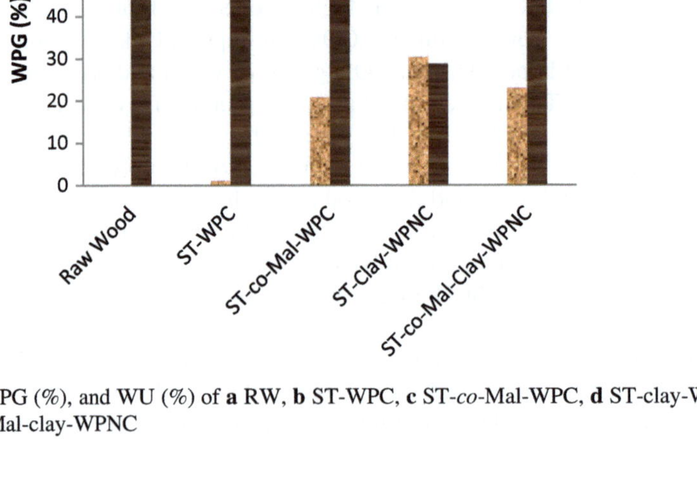

Fig. 5 WPG (%), and WU (%) of **a** RW, **b** ST-WPC, **c** ST-*co*-Mal-WPC, **d** ST-clay-WPNC, and **e** ST-*co*-Mal-clay-WPNC

The water absorption of various samples is illustrated in Fig. 5. The raw wood sample exhibited a higher percentage of water absorption compared with the WPCs. This was expected because the pore volume on raw wood surface was reduced by ST, ST-clay, ST-*co*-Mal, and ST-*co*-Mal-clay (Rahman et al. 2010). The wood cell wall with hydrophilic hydroxyl groups facilitates water penetration into the void areas of wood. Styrene-Mal-treated wood absorbed higher water than styrene-treated wood because ST-*co*-Mal has polar carbonyl group. ST-clay-WPNC absorbed less water than ST-*co*-Mal-WPC. This was due to styrene and clay modified with trimethyl stearyl ammonium are both hydrophobic.

3.5 Thermogravimetric Analysis (TGA)

Thermogravimetric analyzes of raw wood, WPC, and WPNC are shown in Fig. 6, and the corresponding thermal characteristic are given in Table 3. Figure 6 shows that raw wood and ST-WPC exhibited two stages of thermal degradation and decomposition. However, ST-*co*-Mal-WPC, ST-clay-WPNC, and ST-*co*-Mal-clay-WPNC exhibited three stages of thermal degradation and decomposition.

The first stage of weight loss occurred at 70–110 °C and was attributed to absorbed moisture evaporation from all the samples (Liu et al. 2013). The initial weight loss of the samples below 110 °C was the lowest for ST-clay-WPNC followed by, ST-*co*-Mal-WPC, ST-*co*-Mal-clay-WPNC, ST-WPC, and RW. This was attributed to the reaction between –COOH of Mal and the –OH groups of wood cell

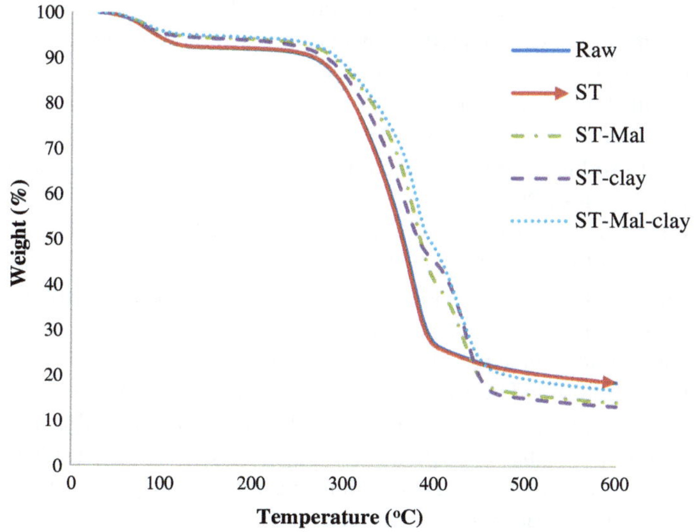

Fig. 6 TGA of **a** RW, **b** ST-WPC, **c** ST-*co*-Mal-WPC, **d** ST-clay-WPNC, and **e** ST-*co*-Mal-clay-WPNC

Table 3 Results for T_g analysis for raw wood and prepared composites

Sample name	Number of transition	Transition temperature			Weight loss at corresponding transition (%)	Residual weight (%) at last stage transition
		T_i	T_m	T_f		
RW	1st 2nd	37.80 127.58	80.36 354.51	121.70 425.59	7.30 67.65	24.81
ST-WPC	1st 2nd	43.04 205.27	80.04 353.07	134.33 425.21	7.30 67.46	24.59
ST-*co*-Mal-WPC	1st 2nd 3rd	40.00 189.00 407.00	79 362 433	127.00 404.00 545.00	5.20 54.00 24.00	15.00
ST-clay-WPNC	1st 2nd 3rd	43.70 129.65 396.34	74.15 343.83 434.27	112.22 387.42 494.34	4.83 45.92 31.03	15.22
ST-*co*-Mal-clay-WPNC	1st 2nd 3rd	37.00 139.00 414.00	79 359 437	131.00 408.00 559.00	6.30 54.00 22.70	14.00

T_i Onset temperature
T_m Temperature corresponding to the maximum rate of mass loss; and T_f End temperature

wall reduced the hydrophilicity of wood polymer composite; the surface modified MMT filled the fiber walls and wood cavities.

The second stage of thermal decomposition of raw wood and ST-WPC occurred at 127–425 °C. Styrene did not covalently bonded with wood, so there were no appreciable differences in thermal properties. On the other hand, the second-step decomposition rate (at 129–390 °C) of ST-*co*-Mal-clay-WPNC was lower than that of ST-*co*-Mal-WPC, ST-clay-WPNC, and raw wood, respectively. This was due to the –COOH group of Mal created a covalent bond with –OH group of wood; the compatibility of clay with lignin of wood and ST-*co*-Mal was higher than that of ST-*co*-Mal with lignin of wood. The covalent bond is thermally more stable than dipole bond. Well-distributed MMT layers could prevent the passage of volatile decomposition product throughout the composite during TGA (García et al. 2009; Sanchez-Jimenez et al. 2012).

In the third stage of decomposition of ST-*co*-Mal-WPC, ST-clay-WPNC, and ST-*co*-Mal-clay-WPNC started at 392–500 °C. This was due to the thermal decomposition of the wood-polymer structure, which produced combustible gases like carbon monoxide, hydrogen, and methane, along with carbon dioxide. In this third thermal degradation stage, the cellulose and lignin were degraded (Kim et al. 2006; Yang et al. 2007). The weight loss rate of ST-*co*-Mal-WPC was higher than ST-*co*-Mal-clay-WPC and ST-clay-WPNC. This was attributed that nanoclay did not decompose in this temperature range. The final residual weights ST-clay-WPNC were significantly less than the raw wood, which was due to the dehydroxylation of the aluminosilicate (Xie et al. 2001).

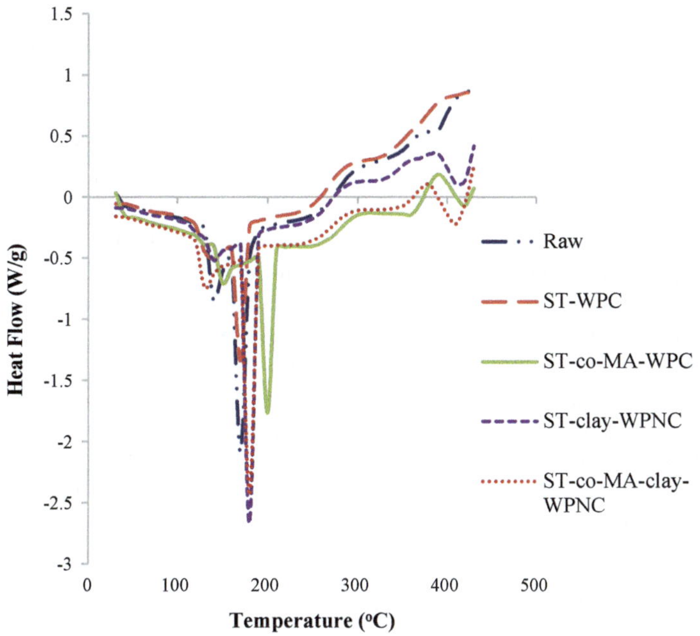

Fig. 7 Differential scanning calorimetry (DSC) thermogram of **a** Raw wood, **b** ST-WPC, **c** ST-clay-WPNC, **d** ST-*co*-MA-WPC, and **e** ST-*co*-MA-clay-WPNC

3.6 Differential Scanning Calorimetry (DSC)

Figure 7 showed the DSC curves of the raw wood, WPCs, and WPNCs. In the DSC graph, the raw wood exhibit three distinctly separated endothermic peaks at 120–160, 160–200, and 370–410 °C, corresponding to amorphous, paracrystalline, and crystalline phases respectively. The amorphous and paracrystalline components started early to decompose by heat whereas the crystalline components were attacked at the last stage. The molecular movement of raw wood, WPCs, and WPNCs is more in its non-crystalline in comparison to that in the crystalline part. The motion of the macromolecules chain in the composite is rigid. This change in movement has been attributed to the intramolecular and intermolecular hydrogen bonding.

The firs endothermic peak at 120–160 °C showed for amorphous movement and expulsion of absorbed water from the raw wood and wood composites. The peak area of wood composites is lower than that of raw wood. The filler restricted the amorphous movement and reduced the void area of raw wood. Thus, the WPCs and WPNCs contained less water than the raw wood. The first endothermic peak of ST-*co*-MA-WPC shifted to higher temperature than the raw wood, because unreacted –COOH group held the water molecule more tightly than that of raw wood.

The paracrystalline endothermic stability of raw wood and ST-WPC is samilar, whereas the paracrystalline enthlphy of raw wood is higher than ST-WPC. This was due to ST did not created any bond with wood molecule but restricted the paracrystalline movement to increase the crystallinity of ST-WPC. ST-clay-WPNC and ST-*co*-MA-clay-WPNC showed higher paracrystalline enthalphy among all the composites; cations of nanoclay created an induced dipole bond with the aromatic group of wood and styrene. The paracrystalline dissociation enthalpy of ST-*co*-MA is lower than raw wood due to the disruption of paracrystalline lattice by ST-*co*-MA. The paracrystalline endothermic temperature of ST-*co*-MA shifted to higher temperature compared to raw wood. This observation proved that ST-*co*-MA created a bond with wood, which reflected in FT-IR result.

In the third endothermic peak at 370–410 °C, ST-*co*-MA-WPC, ST-clay-WPNC, and ST-*co*-MA-clay-WPNC showed the higher enthalpy to dissociate their crystallinity when compared to raw wood because some amorphous region of raw wood converted to crystalline region by ST-*co*-MA, ST-clay, and ST-*co*-MA-clay. Amorphous property of lignin and styrene decreased to increase their crystallinity because the cations of nanoclay induced π electron of lignin and styrene. Between 330 and 400 °C, ST-WPC showed an exothermic peak due to dissociation polystyrene.

4 Conclusion

By studying FT-IR, XRD, and SEM it is concluded that wood was impregnated by a monomer or a mixture of monomer with nanoclay. The modulus of elasticity (MOE), modulus of rupture (MOR), and dynamic Young's moduli (E_d) of WPCs and WPCs were higher than that of raw wood. The raw wood exhibited a higher water uptake (WU) than WPNCs and WPCs. Between 129 and 390 °C, the thermal decomposition rate of ST-*co*-Mal-clay-WPNC was lower than that of other wood polymer composites and raw wood.

Acknowledgements The authors would like to acknowledge the financial support from Research and Innovation Management Centre, Universiti Malaysia Sarawak under Fund with Grant no. (FRGS/SG02(01)/1085/2013(31)).

References

Alexandre M, Dubois P (2000) Polymer-layered silicate nanocomposites: preparation, properties and uses of a new class of materials. Mater Sci Eng, R 28(1):1–63

Bodirlau R, Teaca CA (2009) Fourier transform infrared spectroscopy and thermal analysis of lignocellulosic fillers treated with organic anhydrides. Romanian Rep Phy 54(1–2):93–104

Cai X, Riedl B, Zhang SY, Wan H (2008) The impact of interphase between wood, melamine-urea-formaldehyde and layered silicate on the performance of wood polymer nanocomposites. Compos Part A 39:727–737

Chen JS, Poliks MD, Ober CK, Zhang Y, Wiesner U, Giannelis E (2002) Study of the interlayer expansion mechanism and thermal-mechanical properties of surface-initiated epoxy nanocomposites. Polym 43(18):4895–4904

Chen T, Wu Z, Niu M, Xie Y, Wang X (2016) Effect of Si-Al molar ratio on microstructure and mechanical properties of ultra-low density fiberboard. Eur J Wood Wood Prod 74(2):151–160

Chowdary MS, Niranjan MSR (2015) Effect of nanoclay on the mechanical properties of polyester and S-glass fiber (Al). Int J Adv Sci Technol 47:35–42

Devi RR, Maji TK (2007) Effect of glycidyl methacrylate on the physical properties of wood polymer composites. Polym Compos 28(1):1–5

Fan M, Ndikontar MK, Zhou X, Ngamveng JN (2012) Cement-bonded composites made from tropical wood: compatibility of wood and cement. Const Build Mater 36:135–140

Garcia M, Garmendia HI, Garcia-Jaca J (2009) Wood-plastics composites with better fire retardancy and durability performance. Compos Part A 40(11):1772–1776

Greil P, Lifka T, Kaindl A (1998) Biomorphic cellular silicon carbide ceramics from wood: II. Mechanical properties. J Eur Ceram Soc 18(14):1975–1983

Haraguchi K, Li HJ, Matsuda K, Takeisa T, Elliott E (2005) Mechanism of forming organic/inorganic network structures during in-situ free-radical polymerization in PNIPA-clay nanocomposite hydrogels. Macromol 38(8):3482–3490

Hazarika A, Deka BK, Maji TK (2015) Melamine-formaldehyde acrylamide and gum polymer impregnated wood polymer nanocomposite. J Bio Eng 12(2):304–315

Hill CAS, Cetin NS, Quinney H, Ewen RJ (2001) An investigation of the potential for chemical modification and subsequent polymeric grafting as a means of protecting wood against photodegradation. Polym Degrad Stabil 72(1):133–139

Islam MS, Hamdan S, Hasan M, Ahmed AS, Rahman MR (2012) Effect of coupling reactions on the mechanical and biological properties of tropical wood polymer composites (WPC). Int Biodeter Biodegrad 72:108–113

Jeon IY, Baek JB (2010) Nanocomposites derived from polymers and inorganic nanoparticles. Mater 3(6):3654–3674

Kamei I, Hirota Y, Meguro S (2012) Integrated delignification and simultaneous saccharification and fermentation of hardwood by a white-rot fungus, Phlebia sp. MG-60. Biores Technol 126:137–141

Kartal SN, Yoshimura T, Inamura Y (2004) Decat and termite resistance of boron-treated and chemically modified wood by in situ co-polymerization of allyl glycidyl ether (AGE) with methyl methacrylate (MMA). Int Biodeter Biodegrad 53(2):111–117

Kim HS, Kim S, Kim HJ, Yang HS (2006) Thermal properties of bio-flour-filled polyolefin composites with different compatibilizing agent type and content. Thermochim Acta 451(1–2): 181–188

Kim HS, Lee BH, Choi SW, Kim S, Kim HJ (2007) The effect of types of maleic anhydride-grafted polypropylene (MAPP) on the interfacial adhesion properties of bio-flour-filled polypropylene composites. Compos Part A 38(6):1473–1482

Kondo T, Sawatari C (1996) A Fourier transform infra-red spectroscopic analysis of the character of hydrogen bonds in amorphous cellulose. Polym 37(3):393–399

Krestchmann DE (2010) Mechanical properties of wood. Wood Handbook-Wood as an engineering material. Forest Products Laboratory, United States, pp 1–46

Li YF, Liu YX, Wang XM, Wu QL, Yu HP, Li J (2011) Wood polymer composites prepared by the in situ polymerization of monomers within wood. J Appl Polym Sci 119(6):3207–3216

Li Y, Liu Z, Dong X, Fu Y, Liu Y (2013) Comparison of decay resistance of wood and wood polymer composite prepared by in-situ polymerization of monomers. Int Biodeter Biodegrad 84:401–406

Liu Z, Jiang Z, Fei B, Liu X (2013) Thermal decomposition characteristic of chinese fir. BioRes 8(4):5014–5024

Liu R, Sun W, Cao J, Wang J (2016) Surface properties of in situ organo-montmorillonite modified wood flour and the influence on mechanical properties of composites with polypropylene. Appl Surf Sci 361:234–241
Lopes MA, Silva RF, Monteiro FJ, Santos JD (2000) Microstructural dependence of Young's and shear moduli of P_2O_5 glass reinforced hydroxyapatite for biomedical applications. Biomater 21(7):749–754
Malkapuran R, Kumar V, Negi YS (2008) Recent development in natural fiber reinforced polypropylene composites. J Reinf Plast Compos 28(10):1169–1189
Ndiaye D, Tidjani A (2012) Effects of coupling agents on thermal behavior and mechanical properties of wood flour/polypropylene composites. J Compos Mater 46(24):3067–3075
Pandey KK (2005) Study of the effect of photo-irritation on the surface chemistry of wood. Polym Degrad Stabil 90(1):9–20
Parker RW, Frost RI (1996) The application of drift spectroscopy to the multicomponent analysis of organic chemicals adsorbed on montmorillonite. Clay Miner 44(1):32–40
Pawar PM, Koutaniemi S, Tenkanen M, Mellerowicz EJ (2013) Acetylation of woody lignocellulosic: significance and regulation. Front Plant Sci 21(4):118
Popescu MC, Froidevaux J, Navi P, Popescu CM (2013) Structural modifications of Tilia cordata wood during heat treatment investigated by FT-IR and 2D IR correlation spectroscopy. J Mol Struct 1033:176–186
Rahman MR, Hamdan S, Islam MS (2010) Mechanical and biological performance of sodium metaperiodate-impregnated plasticized wood (pw). BioRes 5(2):1022–1035
Rahman RM, Hamdan S, Ahmed AS, Islam MS, Talib ZA, Abdullah WFW, Che Mat MS (2011) Thermogravimetric analysis and dynamic Young's modulus measurement of *N*,*N*-dimethylacetamide-impregnated wood polymer composites. J Vinyl Addit Technol 17(3):177–183
Rahman MR, Hamdan S, Hashim DMA, Islam MS, Takagi H (2014) Bamboo fiber polypropylene composites: Effect of fiber treatment and nanoclay on mechanical and thermal properties. J Vinyl Addit Technol 21(4):253–258
Rath J, Wolfinger MG, Steiner G, Krammer G, Barontini F, Cozzani V (2003) Heat of wood pyrolysis. Fuel 82(1):81–91
Rowell RM (2006) Chemical modification of wood: a short review. Wood Mater Sci Eng 1:29–33
Sanchez-Jimenez PE, Perez-Maqueda LA, Crespo-Amoros JE, Lopez J, Perejon A, Criado JM (2012) Nanoclay nucleation effect in the thermal stabilization of a polymer nanocomposite: a kinetic mechanism change. J Phy Chem C 116(21):11797–11807
Schwanninger M, Rodrigues JC, Pereira H, Hinterstoisser B (2004) Effects of short-time vibratory ball milling on the shape of FT-IR spectra of wood and cellulose. Vib Spectrosc 36(1):23–40
Silwa F, Bounia NE, Charrier F, Marin G, Malet F (2012) Mechanical and interfacial properties of wood and bio-based thermoplastic composite. Compos Sci Technol 72(14):1733–1740
Simonsen J, Jacobson R, Rowell R (1997) Wood fiber reinforcement of styrene-maleic anhydride copolymers. J Appl Polym Sci 68(10):1567–1573
Toh HK (1979) Dissertation Loughborough University, 1979
Usuki A, Kojima Y, Kawasumi M, Okada A, Fukus Y (1993) Swelling behaviour of montmorillonite cation exchanged for ώ-amino acids by caprolactam. J Mater Res 8(5): 1174–1178
Victor A, Agubra VA, Owuor PS, Hosur MV (2013) Influence of nanoclay dispersion methods on the mechanical behaviour of E-glass/epoxy nanocomposites. Nanomater 3(3):550–563
Wada M, Heux L, Sugiyama J (2004) Polymorphism of cellulose I family: reinvestigation of cellulose IVI. Biomacromol 5(4):1385–1391
Wolfenden JE, Kinzy MS (1997) Measurements of dynamic Young's modulus and damping of a chemical-vapor infiltrated SiC/SiC composites. J Mater Sci Lett 16(9):708–711
Xiao Z, Xie Y, Adamopoulos S, Mai C (2012) Effects of chemical modification with glutaraldehyde on the weathering performance of Scots pine sapwood. Wood Sci Technol 46(4):749–767

Xie W, Gao Z, Pan WP, Hunter D, Singh A, Vaia R (2001) Thermal degradation chemistry of alkyl quaternary ammonium montmorillonite. Chem Mater 13(9):2979–2990

Xu B, Ding J, Feng L, Ding Y, Ge F, Cai Z (2015) Self-cleaning cotton fabrics via combination of photocatalytic TiO_2 and superhydrophobic SiO_2. Surf Coating Technol 262:70–76

Yang H, Yan R, Chen H, Lee DH, Zheng C (2007) Characteristic of hemicellulose, cellulose and lignin pyrolysis. Fuel 86(12–13):1781–1788

Yao H, You Z, Li L, Goh SW, Lee CH, Yap YK, Shi X (2013) Rheological properties and chemical analysis of nanoclay and carbon microfiber modified asphalt with Fourier transform infrared spectroscopy. Const Build Mater 38:327–337

Zhao Y, Wang K, Zhu F, Xue P, Jia M (2006) Properties of poly(vinyl chloride)/wood flour/montmorillonite composites: effects of coupling agents and layered silicate. Polym Degrad Stabil 91(12):2874–2883

Preparation and Characterizations of Clay-Dispersed Styrene-co-Ethylene Glycol Dimethacrylate-Impregnated Wood Polymer Nanocomposites

M.R. Rahman, S. Hamdan and J.C.H. Lai

Abstract In this study, physico-mechanical, thermal, and morphological properties of wood polymer nanocomposites (WPNCs) of styrene-co-ethylene glycol dimethacrylate with clay (ST-co-EGDMA-clay) and styrene with clay (ST-clay) wood polymer composite (WPC) of styrene-co-ethylene glycol dimethacrylate (ST-co-EGDMA) and styrene (ST) were investigated. The WPNC was characterized by Fourier Transform Infrared Spectroscopy (FT-IR), X-ray diffraction (XRD), Scanning Electron Microscopy (SEM), 3-point bending and free-free vibration testing, and thermogravimatric analysis (TGA). The FT-IR results showed that the absorbance at 1730 cm^{-1} increased for ST-co-EGDMA-clay-WPNC and ST-co-EGDMA-WPC compared to other composites and the raw wood. The XRD result revealed that the *d*-spacing of ST-clay-WPNC, ST-co-EGDMA-WPC, ST-WPC, and ST-co-EGDMA-clay-WPNC was higher than the raw wood. The SEM results showed that ST-co-EGDMA-clay-WPNC and ST-clay-WPNC had a smoother surface compared to the other composites and raw wood. The modulus of elasticity (MOE), modulus of rupture (MOR), and dynamic Young's moduli (E_d) of ST-co-EGDMA-clay-WPNC were higher than those of ST-co-EGDMA-WPC, ST-clay-WPNC, ST-WPCs, and raw wood, respectively. The raw wood exhibited a higher water uptake (WU) compared to WPNCs and WPCs. ST-co-EGDMA-WPC was more thermally stable, compared to other composites and raw wood.

Keywords ST-co-EGDMA · Nanoclay · Mechanical properties · Morphological properties

M.R. Rahman (✉) · S. Hamdan · J.C.H. Lai
Faculty of Engineering, Universiti Malaysia Sarawak,
94300 Kota Samarahan, Sarawak, Malaysia
e-mail: rmrezaur@unimas.my

M.R. Rahman, *Wood Polymer Nanocomposites*, Engineering Materials,
DOI 10.1007/978-3-319-65735-6_11

1 Introduction

For sustainable development, wood is more favorable than synthetic polymer because wood is renewable whereas synthetic polymer will be run out. Growing environmental awareness and new rules and regulations are forcing industries to seek more eco-friendly materials for their products (Nuthong et al. 2013). Regarding the problem from the greenhouse gas effect, tree acts as sinks to diminish CO_2 which produces by burning of synthetic polymer and the natural polymer. Wood is used for the construction of the building, transportation, and sports instrument, music instrument due to its excellent specific strength and high specific modulus, biodegradability, and low density (Xie et al. 2013). It also reduces the price of the material and energy consumption when compared to glass material (Malkapuram et al. 2008). The main drawbacks of wood are dimensional unstable, and susceptible to microorganism attack for taking water from the environment by its hydrophilic –OH and void cavities (Hill and Papadopoulos 2001; Rautkari et al. 2013). At low-humidity environment, wood desorbs water to shrinkage. Oxidation of wood lignin disclosures wood materials when uses in outdoor application (Srinivas and Pandey 2012). However, the consumption of wood has been increasing every year. Wood scientists have been emphasizing to convert the low-quality fast growing raw wood to a value-added wood composite by polymer impregnation (Thybring 2013). Wood composites are prepared by occurring reaction between –OH group of wood and reagent functional groups which have the ability for responding of the fiber structures and changing their composition.

As a result, modified wood reduces the moisture absorption and decreases the attacking of microorganism. Both, wood and synthetic polymer, have polar and nonpolar properties. These properties increase incompatibility between them (Ray and Bousmina 2005). It is also reported that the thermoplastic polymers in the presence of crosslinker emphasize the enhancement in dimensional stability, superior mechanical strength, and elevated thermostability (Nachtigall et al. 2007; Nourbakhsh et al. 2008). Vinyl monomer in the presence of a free radical initiator turns into the polymer (Murinov et al. 2013). In the presence of crosslinker, the polymerization rate of the vinyl monomer is very high (Jiang et al. 2012). Polymer from this process is superior to condensation reaction to impregnate the wood because to occur this reaction no acid or base, which degrades the wood polymer, is used (Dong et al. 2015). Physical properties of vinyl-type polymer change in a range from soft rubber to hard, brittle solid, which depends on the types of monomer that form the polymer backbone such as styrene, ethylene, and acrylates (Trey et al. 2012; Ding et al. 2008). Wood flour coated with the hydrophilic–hydrophobic block copolymer based on styrene and acrylic acid shows significant improvement in the ultimate tensile properties of composite (Kosonen et al. 2000). The compatibility between wood and nonpolar polymer, like styrene, ethylene, vinyl acetate, can be improved using ethyl glycol dimethyl methacrylate (EGDMA) (Dong et al. 2015).

In situ polymerization thermoplastic monomer with nanoclay entering into raw wood cavities enhanced wood's mechanical properties and decay resistance (Cai et al. 2008; Rahman et al. 2014; Saba et al. 2014; Wenhua Lu 2008). An organic polymer with organically modified layered silicate (OMLS)-impregnated wood demonstrates significant mechanical improvement, barrier, fire resistance, thermal and environmental stability, solvent uptake, and rate of biodegradability, relative to an unmodified polymer resin-impregnated wood (Alamri et al. 2012; Hazarika et al. 2014; Leszczyńska et al. 2007). Recent researchers have found that commercial organoclay could be used to make wood nanocomposites, which possess excellent mechanical strength and low coefficient of thermal expansion with relatively low cost and ease of fabrication. The thermal stability of wood composite depends on its fire retardant chemicals and chemical modification of wood cell wall (Lin et al. 2008). The fire retardant chemicals such as montmorillonite, phosphoric acid, boric acid $Al(OH)_3$ and $Mg(OH)_2$, TiO_2, Sb_2O_3, alumino-silicates, vermiculite, perlite produce char and water which reduce the effective heat of combustion to increase the thermal stability (Ahmed Hared et al. 2007; Griffin 2011; Liang et al. 2015; Chang et al. 2012; Qu et al. 2011). Silicates play an important role in wood composite to produce silicon carbide compound which creates a barrier, during the heat flow (Xu et al. 2015). Chemically modified and nanofiller-impregnated wood enhanced mechanical and thermal properties (Liodakis et al. 2003).

In this study, ST-co-EGDMA was synthesized in the presence of nanoclay or without nanoclay by in situ polymerization into raw wood. The prepared composites were characterized by physico-mechanical and thermal analysis. The morphological properties were also reported.

2 Experimental

2.1 *Materials*

Defect-free and straight-grained Kumpang raw wood (RW) was obtained from Forestry Department, Sarawak. The chemicals used to produce WPCs were styrene (ST), ethylene glycol dimethacrylate (EGDMA), benzoyl peroxide, and nanoclay (Nanomer® 1.28E). Styrene was supplied by A.S. Joshi and Company, Goregaon West, Mumbai, India. Ethylene glycol dimethacrylate (EGDMA) and benzoyl peroxide were supplied by Merck Schuchardt OHG, Germany. Nanoclay (Nanomer® 1.28E) that had been modified with 25–30 wt% trimethyl stearyl ammonium was supplied by Sigma-Aldrich (USA).

2.2 Specimen Preparation

The tree was cut into three bolts with 1.2 m long. Each bolt was quartersawn to produce planks of 4 cm thick. These planks were subsequently conditioned by air-drying in a room with a relative humidity of 60% and ambient temperature of approximately 25 °C for one month prior to testing. The planks were ripped and machined to 340 mm (L) × 20 mm (T) × 10 mm (R) for three-point bending and free-free vibration tests.

2.3 Preparation of Wood Polymer Nanocomposites (WPNCs)

Oven-dried specimens were impregnated with either one monomer with nanoclay or mixture of monomers with nanoclay, which is shown in Table 1. The wood specimens were placed separately into an impregnation vacuum chamber (75 mm Hg) for 30 min. For each monomer(s) impregnation, approximately 10 g of benzoyl peroxide was added to initiate polymerization. These wood composites were then removed from the chamber, and the excess chemicals were removed. Specimens were wrapped with aluminum foil and placed into a 105 °C oven for 24 h for polymerization to take place.

2.4 Microstructural Characterizations

2.4.1 Determination of Weight Percentage Gain (WPG)

Weight percentage gain (WPG) was calculated using the following equation:

$$\mathrm{WPG}\,(\%) = \left[\left(W_f - W_i\right)/W_i\right] \times 100 \quad (1)$$

where W_f and W_i are the oven-dried weight of the WPC (or the WPNC) and the raw wood, respectively.

Table 1 Composition of the chemicals for the raw wood (RW) impregnation

Specimen	ST (mL)	EDGMA (g)	Clay (g)
Raw wood	500	–	–
Raw wood	500	500	–
Raw wood	500	–	10
Raw wood	500	500	10

2.4.2 Determination of Water Uptake (WU)

The water uptake (WU) was calculated using the following equation:

$$\mathrm{WU}\,(\%) = [(W_2 - W_1)/W_1] \times 100 \quad (2)$$

where W_2 is the weight of the treated (or untreated control) wood samples after two weeks of water immersion and W_1 is the weight of the oven-dried sample, respectively.

2.4.3 Fourier Transform Infrared Spectroscopy (FT-IR)

The infrared spectra of all specimens were recorded on a Shimadzu FT-IR 81001 Spectrophotometer (Shimadzu; Kyoto, Japan). The wave number range of the scan used was 4000–600 cm^{-1}.

2.4.4 X-Ray Diffraction (XRD) Analysis

XRD analysis for raw wood and wood polymer nanocomposites was performed with a Rigaku diffractometer (RGS Corp. Sdn. Bhd.; Selangor, Malaysia) operating with Cu$K\alpha$ radiation, wavelength of 0.154 nm, 40 kV, and 30 mA. The crystallinity index ($\mathrm{CI_{XRD}}$) was calculated using the following height ratio of diffraction peaks:

$$\mathrm{CI_{XRD}}\,(\%) = \frac{I_{002} - I_{\mathrm{am}}}{I_{002}} \quad (3)$$

where I_{002} is the maximum intensity of the peak at 2θ at about 22° and I_{am} is the minimum intensity above baseline at 2θ at about 30°, which corresponds to crystalline and amorphous parts, respectively. The *d*-spacing between the [002] lattice planes (d_{002}) of the sample was calculated using the Bragg equation:

$$d_{002} = \frac{n\lambda}{2\sin\theta} \quad (4)$$

where n refers to the integer wavelength number ($n = 1$), θ refers to the maximum diffraction angle, and λ is the X-ray wavelength (0.154 nm).

2.4.5 Scanning Electron Microscopy (SEM)

Fractured surfaces were examined to investigate the interfacial bonding between wood cell and polymer with filler. These surfaces were imaged using a Hitachi

TM3030 Scanning Electron Microscope (SEM) supplied by JEOL (Tokyo, Japan). The SEM specimens were sputter-coated with gold prior to observation. The micrographs were acquired at a magnification of 1500.

2.4.6 Modulus of Rupture (MOR), Modulus of Elasticity (MOE), and Dynamic Young's Modulus (E_d) Measurements

Three-point bending tests were conducted using a Shimadzu MSC-5/500 universal testing machine (Kyoto, Japan) operating at a crosshead speed of 5 mm/min. The modulus of rupture (MOR) and modulus of elasticity (MOE) were calculated using the following formulae:

$$\text{MOR} = 1.5\,LW/bd^2 \tag{5}$$

$$\text{MOE} = L^2m/4bd^3 \tag{6}$$

where W is the ultimate failure load, L is the span between centers of support, b is the mean width (tangential direction) of the sample, d is the mean thickness (radial direction) of the sample, and m is the slope of the tangent to the initial line of the force displacement curve.

Determinations of the dynamic Young's modulus, E_d, for all specimens were carried out using a free-free flexural vibration testing system. The E_d was calculated from the resonant frequency using the following equation,

$$E_d = \frac{4\Pi^2 f^2 l^4 A\rho}{Im_n^4} \tag{7}$$

where I is the second moment of area of the beam's cross section (which is equal to $bd^3/12$), d is the beam's depth, b is the beam's width, l is the beam's length, f_n is the beam's natural frequency at the nth flexural mode, ρ is the beam's density, A is the beam's cross-sectional area, and m_n is a frequency factor which is related to the free-free beam vibration at the nth flexural mode. The value of m_n at the first mode (n = 1) is 4.730 (i.e., m_1 = 4.730).

2.4.7 Thermogravimetric Analysis (TGA)

Thermogravimetric Analysis (TGA) measurements were performed on 5–10 mg samples at a heating rate of 10 °C/min in a nitrogen atmosphere using a thermogravimetric analyzer (Model SDT Q600, TA Instruments; Selangor, Malaysia). Thermal decomposition of each sample occurred in a programmed temperature range of 30 to 600 °C under a constant nitrogen flow rate of 5 mL/min. The continuous weight loss and temperature were recorded and analyzed.

2.4.8 Differential Scanning Calorimetric Testing (DSC)

The differential scanning calorimetry (DSC) is used to measure and characterize the thermal properties of materials. DSC revealed the amount of material with high thermal stability. DSC measures the energy required to obtain the same temperature rise in each holder. When a transition such as melting, degradation, or crystallization, i.e., an endothermic or exothermic reaction occurs, the change in power needed to keep the sample at the same temperature as the reference pan is recorded as a peak.

All differential scanning calorimetry (DSC) measurements were made on a DSC Q10 (TA instrument) thermal system, using a sealed aluminum capsule. Each test specimen was weighed to about 3–3.5 mg and was held at a single heating rate of 10 °C/min and scanning temperature from 30 to 450 °C. Each of the data reported represents an average of three runs. To assess the quantitative reproducibility, an integration of the area under the peak yields the melting endotherm.

3 Results and Discussion

3.1 *FT-IR*

The FT-IR spectra of RW and raw wood impregnated with ST, or ST-co-EGDMA, or ST-clay, or ST-co-EGDMA-clay are shown in Fig. 1. The peaks from 4000 to 3000 cm^{-1} and from 2900 to 2800 cm^{-1} correspond to stretching vibration of H-bonds in –OH groups and stretching vibration of C–H in methyl and methylene groups, respectively (Rahman et al. 2015). The peak intensity near 1505 cm^{-1} (for aromatic double bonds) was higher for ST-impregnated WPC compared to raw wood. This was due to the increasing double bond. The IR spectra of ST-WPC exhibited a peak intensity above 3000 cm^{-1} for the –OH group that was lower than that of raw wood. The ST plasticized within the wood lumens to displace the free-bound water contained in the wood matrix. The peak intensity of the –OH group for ST-co-EGDMA and ST-co-EGDMA with clay-impregnated wood was lower than that of ST-impregnated wood. This was due to the creation of covalent between –OH group of wood and –COOH group of hydrolyzed EGDMA, which is shown in Fig. 2. This can be attributed that ST-co-EGDMA performs better polymer than ST alone. The characteristic peak at 698 cm^{-1} was bending vibrations of Mg_3OH, and out-of-plane vibration for Fe–O. However, the vibration of Al–O–Si was out-of-plane whereas the hydrolyzed EGDMA showed the bending of –OH at 2900–3500 cm^{-1} (Kondo and Sawatari 1996; Madejova 2003). Thus, the intensity of this band is higher for WPNCs and ST-co-EGDMA-WPC than ST-WPC and raw wood. C-H bending vibration at 750 cm^{-1} increased for ST-clay-WPNC compared to other composites due to polarization capacity of Al^{3+} to π electron of styrene (Yao et al. 2013). The peak intensity near 1730 cm^{-1} of ST-co-EGDMA-clay-WPC and ST-co-EGDMA-clay-WPNC increased compared to raw wood, ST-WPC, and ST-clay-WPNC.

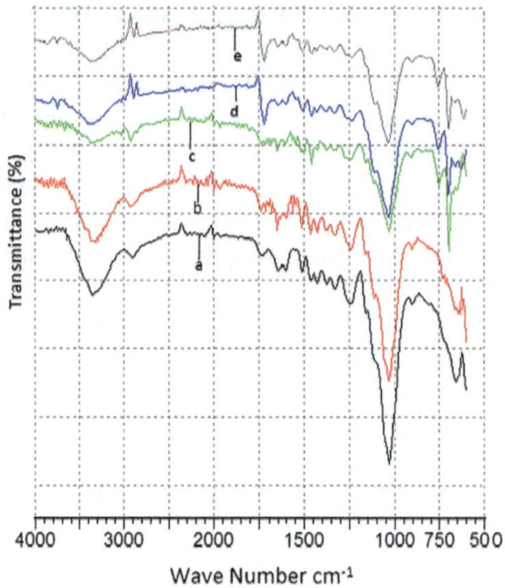

Fig. 1 FT-IR spectra of **a** raw wood, **b** ST-WPC, **c** ST-clay-WPNC, **d** ST-co-EGDMA-clay-WPNC, and **e** ST-co-EGDMA-WPC

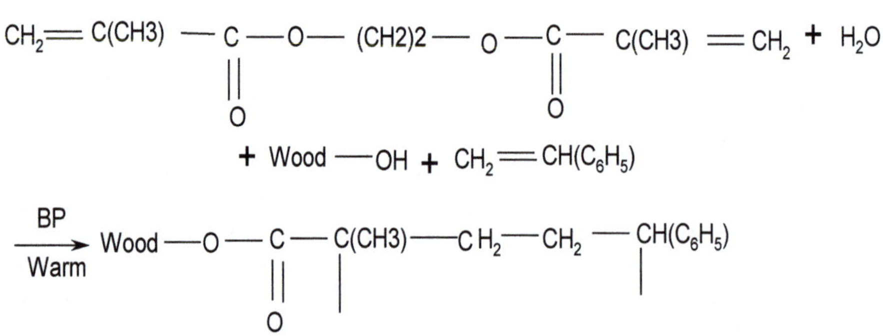

Fig. 2 Schematic reaction diagrams

3.2 *X-Ray Diffraction (XRD) Analysis*

X-ray diffractograms of raw wood, WPCs, and WPNCs are shown in Fig. 3. The diffraction pattern of WPCs, WPNCs, and raw wood exhibited a broad amorphous region. The values of 2θ were around 22.4° for raw wood, WPCs, and WPNCs assigned to the [002] crystalline plane. However, the values of $2\theta = 35°$ shown as [040] plane and crystallinity index CI_{XRD} ware determined at [002] plane which is shown in Table 2 (Chen et al. 2016; Wada et al. 2004). The crystallinity of ST-WPC and ST-clay-WPNC increased compared to ST-co-EGDMA-clay-WPNC,

ST-co-EGDMA-WPC, and raw wood. This is due to ST-co-EGDMA was amorphous and reduced the –OH group from cellulose. This explains the crystallinity index decreasing for crosslinked composites (Mattos et al. 2015). Among them, the CI_{XRD} was the highest for ST-WPC because styrene does not react with the OH group. It restricts the movement of amorphous chains (Agubra et al. 2013). The values of 2θ for WPCs and WPNCs decreased compared to raw wood, whereas the *d*-spacing of [002] plane is increased. The increment of *d*-spacing indicated that the wood sample was successfully impregnated by the polymer (Ramos-Filho et al. 2005). Although some –OH groups were reduced by EGDMA, the CI_{XRD} of all composites increased compared to the raw wood. This is due to the reduction of amorphous movement. Among the composites, there was no additional peak with respect to raw wood.

3.3 Scanning Electron Microscopy (SEM)

The scanning electron micrographs of raw wood, WPCs, and WPNCs are shown in Fig. 4a–d. It was observed that the lumen and tracheids of wood had been filled by the solid polymers by in situ polymerization. Figures 4d and 6e show smoother surfaces compared to surfaces as shown in Fig. 6c–d. This is due to the void spaces of raw wood filled by the intercalated or exfoliated MMT and formed the composite systems, which reflected in the XRD analysis (Chen and Baird 2012). However, Fig. 4a shows that there was unfilled cell wall and pore volume, whereas ST-WPC partially filled cell wall and pore volume (Fig. 4b). On the other hand,

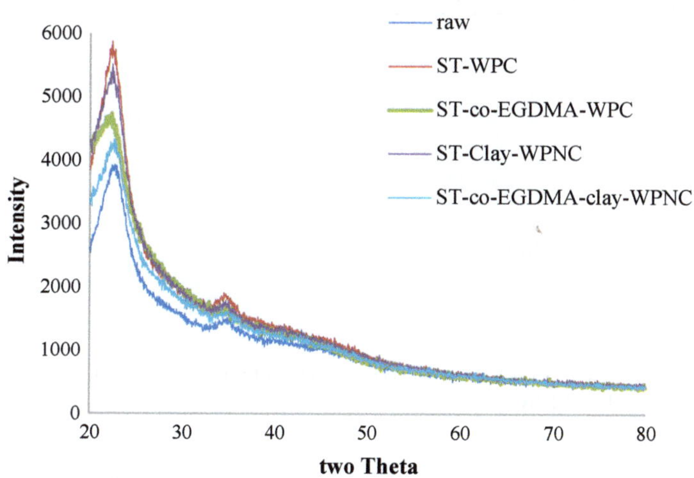

Fig. 3 X-ray diffraction (XRD) diffractograms of **a** RW, **b** ST-co-GMA-WPC, **c** ST-co-GMA-clay-WPNC **d** ST-clay-WPNC, and **e** ST-WPC

Table 2 Crystallinity index CI_{XRD} and *d*-spacing as obtained by XRD analysis

Sample	CI_{XRD} (%)	*d*-spacing (002) nm	2θ (002)
Raw	66.59	0.3913	22.6967
ST-WPC	72.5	0.3970	22.3667
ST-co-EGDMA-WPC	66.89	0.398	22.3007
ST-clay-WPNC	71.89	0.4089	21.7067
ST-co-EGDMA-clay-WPNC	66.73	0.394	22.5647

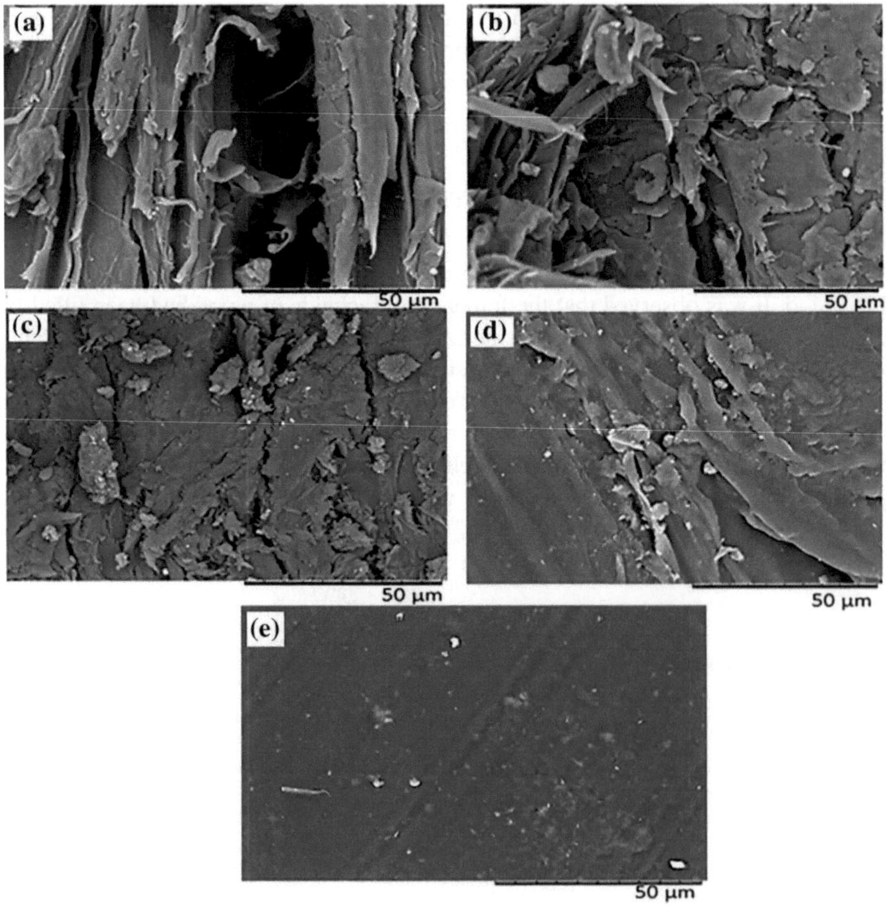

Fig. 4 SEM micrographs of **a** RW; **b** ST-WPC; **c** ST-co-EGDMA-WPC; **d** ST-clay-WPNC; and **e** ST-co-EGDMA-clay-WPNC

ST-co-EGDMA moderately filled the cell walls and pores, as shown in Fig. 4c. Styrene and ST-co-EGDMA could not fill the maximum number of void spaces of raw wood compared to ST-clay and ST-co-EGDMA-clay. The poor interaction between the polymer and wood implied a more brittle composite. Therefore, it was concluded that there were strong interactions among the polymers, nanoclay, and wood cell walls, as shown in Fig. 4d and e.

3.4 Modulus of Rupture (MOR), Modulus of Elasticity (MOE), and Dynamic Young's Modulus (E_d) Measurements

The results of MOE, MOR, and E_d analyses of the raw wood, WPNCs, and WPCs are shown in Fig. 5. In Fig. 5, the composites exhibited higher MOE and MOR than the raw wood. This was due to the fillers exert adhesive forces with the wood cell wall. The MOE and MOR values were the highest for ST-co-EGDMA-clay-WPNC followed by ST-co-EGDMA-WPC, ST-clay-WPNC, ST-WPC, and RW, respectively. This is due to the –COOH group of hydrolyzed EGDMA formed a covalent bond with the –OH group of wood, and the vinyl group of EGDMA reacted with styrene by a free radical mechanism. Also, nanoclay exfoliated into wood cavities and ST-co-EGDMA. Thus, the polymers and the wood cell walls achieved a sufficiently strong chemical bonding strength (Lai et al. 2015; Tabari et al. 2011). ST-WPCs showed lower MOE and MOR compared to

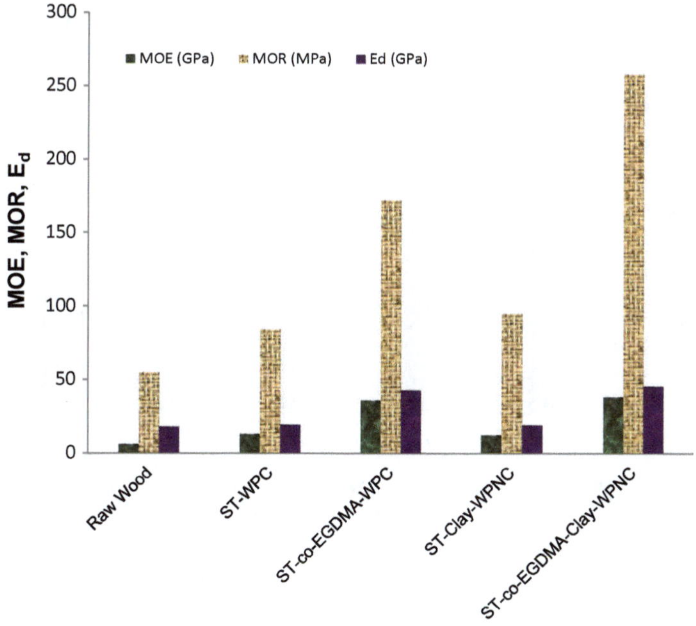

Fig. 5 MOE, MOR, and Ed of **a** RW, **b** ST-WPC, **c** ST-co-EGDMA-WPC, **d** ST-clay-WPNC, and **e** ST-co-EGDMA-clay-WPNC

ST-clay-WPNCs. This was expected because the compatibility of styrene with wood cell wall was less compared to the compatibility of modified nanoclay with wood cell wall. Modified nanoclay with high surface area had polar and nonpolar properties whereas styrene had only nonpolar property.

Among the composites, the highest Ed was observed for the ST-co-EGDMA-clay-WPNC, followed by ST-co-EGDMA-WPC, ST-WPC, and ST-clay-WPNC. This is due to an increase in the density and shear force in the wood composite (Rahman et al. 2010). In the presence of shear force, the free energy increases and the molecules continuously oscillate from equilibrium (Toh 1979). The shear force is considered to vary sinusoidally with time. The styrene monomer cannot directly react with wood cell wall components, whereas ST-co-EGDMA reacts with cell wall components to produce the covalent bond. The clay particles enter into the wood, which increases the Van der Waals forces in the composite to increase shear stress.

3.5 *Weight Percentage Gain (WPG) and Water Uptake (WU)*

Both WPG and WU values are shown in Fig. 6. It is clear that ST-co-EGDMA-clay-WPNC showed higher WPG% compared to the other composites. This is due to EGDMA created covalent bonds with the –OH group of wood and vinyl group of styrene. This formation energy might help to break the

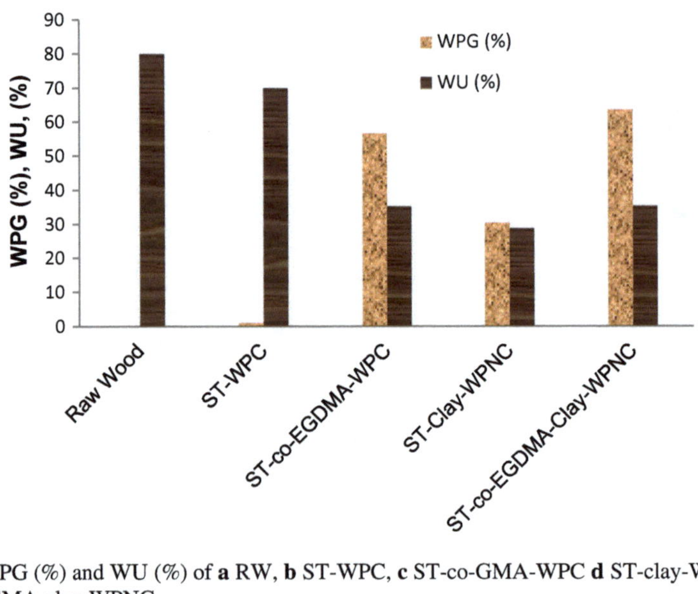

Fig. 6 WPG (%) and WU (%) of **a** RW, **b** ST-WPC, **c** ST-co-GMA-WPC **d** ST-clay-WPNC, and **e** ST-co-GMA-clay-WPNC

gallery of clay which entered into wood cell wall and wood cavities. ST-WPC and ST-clay-WPNC showed lower WPG compared to ST-co-EGDMA-WPC and ST-co-EGDMA-clay-WPNC. This is due to the attraction of wood with the mixture of styrene and EDGMA had stronger than that of wood with the mixture of ST and clay.

The water absorption of prepared samples is illustrated in Fig. 6. The raw wood exhibited a higher percentage of water absorption compared to the WPCs and WPNCs. This was expected because ST, ST-clay, ST-co-EGDMA, and ST-co-EGDMA-clay had reduced pore size on the surface of the wood (Jinshu et al. 2007). The void area on the wood surface facilitates water penetration into the wood cell wall with hydrophilic hydroxyl groups. Styrene with EGDMA-treated samples absorbed less water compared to styrene-treated samples. Because ST-WPCs had more void areas compared to ST-co-EGDMA-WPCs, which is also reflected in the SEM analysis. Similar results have been reported for rubber wood samples (Devi and Maji 2007). It is also apparent that ST-clay-WPNC had the lowest (28.84%) water absorption among all the composites and raw wood. This is due to styrene and MMT modified with alkyl group are hydrophobic. ST-co-EGDMA-clay-WPNC had the higher hydrophilicity than the ST-clay-WPNC due to the containing of carbonyl group in ST-co-EGDMA-clay.

3.6 Thermogravimetric Analysis (TGA)

Thermogravimetric Analysis of raw wood, WPC, and WPNC is shown in Fig. 7 and the corresponding thermal characteristics are given in Table 3. The thermal stability of RW, ST-WPC, ST-co-EGDMA-WPC, ST-clay-WPNC, and ST-co-EGDMA-clay-WPNC was carried out using TGA. Figure 7 shows that raw wood and ST-WPC showed two-step thermal degradation and decomposition. ST-co-EGDMA-WPC, ST-clay-WPNC, and ST-co-EGDMA-clay-WPNC showed three-step thermal degradation and decomposition.

The first-step weight loss occurred at 70–110 °C due to evaporation of absorbed moisture from RW, WPCs, and WPNCs (Zhao et al. 2014). The initial weight loss below 110 °C was the lowest for ST-clay-WPNC, compared to ST-co-EGDMA-clay-WPNC, ST-co-EGDMA-WPC, ST-WPC and RW, respectively. This is due to hydrophobic styrene and modified clay filled the cavity of wood cell wall.

The second-stage thermal decomposition of raw wood and ST-WPC occurred at 127–425 °C. In this region, there was no appreciable difference because styrene did not create any chemical bond with wood. Generally, the existence of inorganic materials in polymer matrix enhances the thermal stability of the nanocomposite (Srinivasulu et al. 2013). On the contrary, at 129–404 °C the thermal stability of ST-co-EGDMA-WPC was higher than that of ST-co-EGDMA-clay-WPNC, ST-clay-WPNC. This is due to the formation energy of ST-co-EGDMA-WPC was higher than the formation energy ST-co-EGDMA-clay-WPNC and

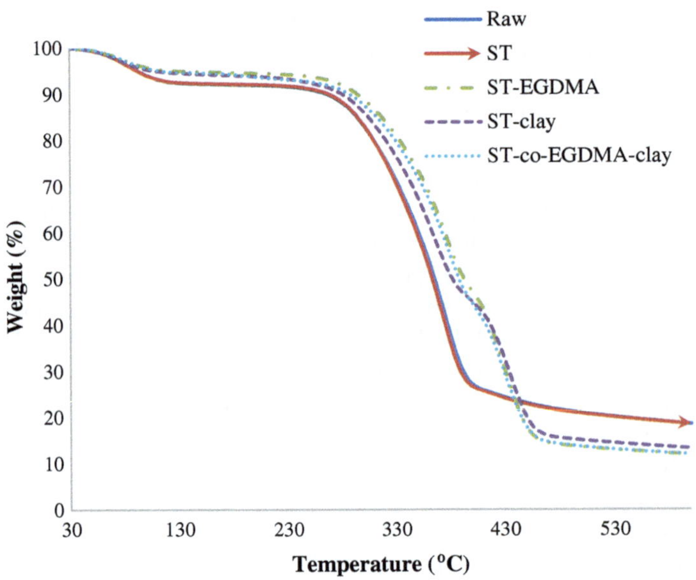

Fig. 7 TGA of **a** RW; **b** ST-WPC; **c** ST-co-EGDMA-WPC; **d** ST-clay-WPNC; and **e** ST-co-EGDMA-clay-WPNC

Table 3 Results for TG analysis for raw wood and prepared composites

Sample name	Number of transition	Transition temperature			Weight loss at corresponding transition (%)	Residual weight (%) at last stage transition
		T_i	T_m	T_f		
RW	1st	37	80	122	7.30	24.81
	2nd	127	354	426	67.65	
ST-WPC	1st	43	80	134	7.30	24.59
	2nd	205	353	425	67.46	
ST-co-EDGMA-WPC	1st	40	77	120	4.74	1308
	2nd	240	378	392	43.41	
	3rd	395	430	526	36.13	
ST-clay-WPNC	1st	43	74	112	4.83	15.22
	2nd	129	344	387	45.92	
	3rd	396	434	494	31.03	
ST-co-EDGMA-clay-WPNC	1st	40	78	144	5.25	13.05
	2nd	168	357	405	51.14	
	3rd	407	431	529	29.43	

T_i onset temperature
T_m temperature corresponding to the maximum rate of mass loss; and
T_f end temperature

ST-clay-WPNC. In the third-step decomposition of ST-co-EGDMA-WPC, ST-clay-WPNC, and ST-co-EGDMA-clay-WPNC started at 395–529 °C. This is due to breaking of the wood polymer structure which produces carbon dioxide and tars. In the third stage, cellulose and lignin were degraded (Kim et al. 2006; Yang et al. 2007). However, the temperature is in the range of 395 to 529 °C and the weight loss rate of ST-co-EGDMA-WPC and ST-co-EGDMA-clay-WPNC was higher than that of ST-clay-WPNC. This was due to the breaking of bond between wood and EGDMA to produce volatile CO_2 gas. At 525–600 °C, the final weights of ST-co-EGDMA-clay-WPNC and ST-clay-WPNC were significantly less than the raw wood due to the dehydroxylation of the aluminosilicate (Assaedi et al. 2016; Xie et al. 2001). It was also cleared that in this region the dehydroxylation of the aluminosilicate of ST-co-EGDMA-clay-WPNC was lower than that of ST-clay-WPNC because ST-clay-WPNC had more clay than that of ST-co-EGDMA-clay-WPNC.

3.7 Differential Scanning Calorimetry (DSC)

Figure 8 shows the DSC curves of the raw wood, WPCs, and WPNCs. The raw wood exhibited three separated endothermic peak ranges 120–160 °C, 160–200 °C, and 370–410 °C, corresponding to amorphous, paracrystalline, and crystalline phases, respectively (Mahato et al. 2013; Mattos et al. 2015). The amorphous and paracrystalline components are first affected by heat, whereas the crystalline components are attacked at the last stage. The molecular movement of wood is more in its non-crystalline in comparison with that in the crystalline part (Bao et al. 2003). The motion of the polymer chain is rigid. So it can be attributed that the intramolecular hydrogen bonding and intermolecular hydrogen bonding change the internal movement of wood.

The first endothermic peak which corresponds to amorphous movement and expulsion of absorbed water from the raw wood and wood composites. The peak area of raw wood in this region is higher than that of wood composites. This is due to the fillers restricted the amorphous movement and reduced the amount of water absorption.

The thermal stability of second endothermic region of paracrystalline phase of raw wood and ST-WPC is similar, whereas the paracrystalline enthalpy of raw wood is higher than ST-WPC. This is due to ST did not create any bond with wood molecule but only filled the wood cavities to restrict the paracrystalline movement. ST-clay-WPNC showed higher paracrystalline enthalpy among all the composites because some amorphous region of raw wood and ST converted to paracrystalline and crystalline regions. This conversion was created by cations of nanoclay, which induced a dipole bond with the aromatic group of wood and styrene. The paracrystalline endothermic peak of ST-co-EGDMA-WPC completely disappeared; however, that peak for ST-co-EGDMA-clay-WPN partially disappeared, respectively. This is due to the disruption of paracrystalline lattice by ST-co-EGDMA.

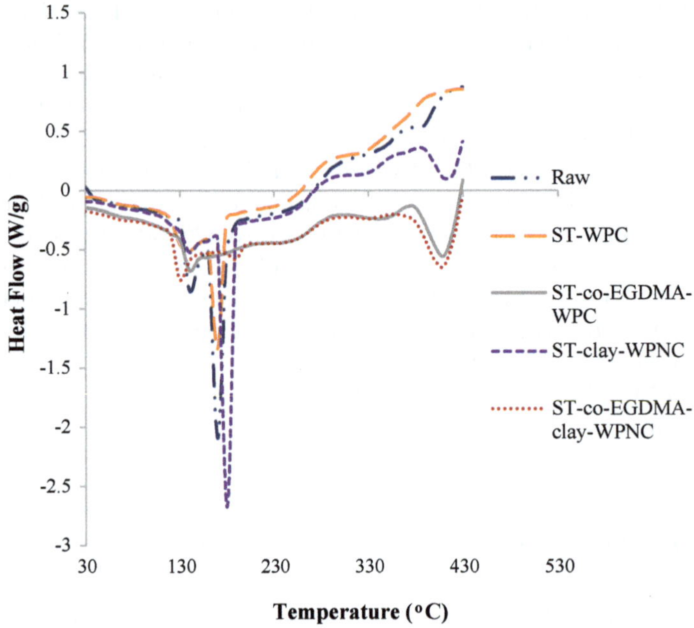

Fig. 8 Differential scanning calorimetry (DSC) thermogram of **a** RW, **b** ST-WPC, **c** ST-clay-WPNC **d** ST-co-EGDMA-WPC, and **e** ST-co-EGDMA-clay-WPNC

The third endothermic peak of ST-co-EGDMA-WPC and ST-co-GMA-clay-WPNC showed the higher enthalpy for breaking of crystallinity region when compared to raw wood and ST-clay-WPNC. This is due to ST-co-EGDMA and ST-co-GMA-clay entered into amorphous, and paracrystalline region of raw wood to convert the crystalline region (Mehrotra et al. 2010; Phetkaew et al. 2009). Between 360 and 400 °C, ST-WPC showed an exothermic peak due to dissociation polystyrene. ST-clay-WPNC showed a higher bond breaking enthalpy of crystalline region than that of raw wood. The cations of nanoclay induced π electron of lignin and styrene, thus increased the crystallinity of this region (Chand et al. 1987; Zhu et al. 2004). The temperature between 330 and 400 °C, ST-WPC showed an exothermic due to dissociation polystyrene.

4 Conclusion

Kumpang wood was successfully impregnated by ST, ST-co-EGDMA, ST-clay, and ST-co-EGDMA-clay, which was proved by FT-IR and XRD analysis. The SEM results showed that ST-co-EGDMA-clay-WPNC and ST-clay-WPNC had a smoother surface compared to the other composites and raw wood. The modulus of elasticity (MOE), modulus of rupture (MOR), and dynamic Young's moduli (Ed) were the highest for ST-co-EGDMA-clay-WPNC followed by ST-co-EGDMA-WPC, ST-clay-WPNC, ST-WPC, and raw wood. The raw wood exhibited a higher water uptake (WU) compared to WPNCs and WPCs. Between 120 and 400 °C, ST-co-EGDMA-WPC was more thermally compared to the other composites and raw wood.

Acknowledgements The authors would like to acknowledge the financial support from Research and Innovation Management Centre, Universiti Malaysia Sarawak, under fund with Grant No (FRGS/SG02(01)/1085/2013(31)) during the research.

References

Agubra VA, Owuor PS, Hosur MV (2013) Influence of nanoclay dispersion methods on the mechanical behavior of E-glass/epoxy nanocomposites. Nanomater 3(3):550–563

Ahmed Hared I, Dirion JL, Salvador S, Lacroix M, Rio S (2007) Pyrolysis of wood impregnated with phosphoric acid for the production of activated carbon: kinetics and porosity development studies. J Anal Appl Pyrol 79(1–2):101–105

Alamri H, Low IM, Alothman BZ (2012) Mechanical, thermal and microstructural characteristics of cellulose fibre reinforced epoxy/organoclay nanocomposites. Compos Part B 43(7): 2762–2771

Assaedi H, Shaikh FUA, Low IM (2016) Effect of nanoclay on mechanical and thermal properties of geopolymer. J Asian Cer Soc 4(1):19–28

Bao S, Daunch WA, Sun Y, Rinaldi PL, Marcinko JJPC (2003) Solid state two-dimensional NMR studies of polymeric diphenylmethane diisocyanate (PMDI) reaction in wood. Forest Prod J 53(6):63–71

Cai X, Riedl B, Zhang SY, Wan H (2008) The impact of interphase between wood, melamine-urea-formaldehyde and layered silicate on the performance of wood polymer nanocomposites. Compos Part A 39(5):727–737

Chand N, Sood S, Singh DK, Rohatgi PK (1987) Structural and thermal studies on sisal fibre. J Ther Anal 32(2):595–599

Chang MK, Wei HL, Wu KS (2012) The strength and thermal stability of low-density polyethylene grafted maleic anhydride/montmorillonite nanocomposites. Adv Sci Lett 13(1):240–244

Chen C, Baird DG (2012) Dispersion of nanoclay at higher levels into polypropylene with carbon dioxide in the presence of maleated polypropylene. Polym 53(19):4178–4185

Chen T, Wu Z, Niu M, Xie Y, Wang X (2016) Effect of Si-Al molar ratio on microstructure and mechanical properties of ultra-low density fiberboard. Eur J Wood Wood Prod 74(2):151–160

Devi RR, Maji TK (2007) Effect of glycidyl methacrylate on the physical properties of wood-polymer composites. Polym Compos 28(1):1–5

Ding WD, Koubaa A, Chaala A, Belem T, Krause C (2008) Relationship between wood porosity, wood density and methyl methacrylate impregnation rate. Wood Mater Sci Eng 3(1/2):62–70

Dong X, Sun T, Liu Y, Li C, Li Y (2015) Structure and properties of polymer-impregnated wood prepared by in-situ polymerization of reactive monomers. BioRes 10(4):7854–7864

Griffin GJ (2011) The effect of fire retardants on combustion and pyrolysis of sugar-cane bagasse. Bioresour Technol 102(17):8199–8204

Hazarika A, Mandal M, Maji TK (2014) Dynamic mechanical analysis, biodegradability and thermal stability of wood polymer nanocomposites. Compos Part B 60:568–576

Hill CAS, Papadopoulos AN (2001) A review of methods used to determine the size of the cell wall microvoids of wood. J Ins Wood Sci 15(6):337–345

Jiang L, Huang W, Xue X, Yang H, Jiang B, Zhang D, Fang J, Chen J, Yang Y, Zhai G, Kong L, Wang S (2012) Radical polymerization in the presence of chain transfer monomer: an approach to branched vinyl polymers. Macromol 45(10):4092–4100

Jinshu S, Jianzhang L, Wenrui Z, Derong Z (2007) Improvement of wood properties by urea-formaldehyde resin and nano-SiO_2. Front Forest China 2:104

Kim HS, Kim S, Kim HJ, Yang HS (2006) Thermal properties of bio-flour-filled polyolefin composites with different compatibilizing agent type and content. Thermochim Acta 451(1–2): 181–188

Kondo T, Sawatari C (1996) A fourier transform infrared spectroscopic analysis of the character of hydrogen bonds in amorphous cellulose. Polym 37(3):393–399

Kosonen ML, Wang B, Caneba GT, Gardner DJ, Rials TG (2000) Polystyrene/wood composites and hydrophobic wood coatings from water-based hydrophilic-hydrophobic block. Clan Prod Proc 2(2):117–123

Lai JCH, Rahman MR, Hamdan S, Rahman MM, Hossen MF (2015) Impact of nanoclay on physicomechanical and thermal analysis of polyvinyl alcohol/fumed silica/clay nanocomposites. J Appl Polym Sci 132(15)

Leszczyńska A, Njuguna J, Pielichowski K, Banerjee JR (2007) Polymer/montmorillonite nanocomposites with improved thermal properties: part I. Factors influencing thermal stability and mechanisms of thermal stability improvement. Thermochim Acta 453(2):75–96

Liang JZ, Feng JQ, Tsui CP, Yang CY, Liu DF, Zhang SD, Huang WF (2015) Mechanical properties and flame-retardant of PP/MRP/$Mg(OH)_2$/$Al(OH)_3$ composites. Compos Part B 71:74–81

Lin Z, Renneckar S, Hindman DP (2008) Nanocomposite-based lignocellulosic fibers. 1. Thermal stability of modified fibers with clay-polyelectrolyte multilayers. Cell 15(2):333–346

Liodakis S, Bakirtzis D, Dimitrakopoulos AP (2003) Autoignition and thermogravimetric analysis of forest species treated with fire retardants. Thermochim Acta 399(1–2):31–42

Lu WGZ (2008) Structure and characterization of Chinese fir (Cunninghamia Ianceolata) wood/MMT intercalation nanocomposite (WMNC). Front Forest China 3(1):121–126

Madejova J (2003) FTIR techniques in clay mineral studies. Vibrat Spec 31(1):1–10

Mahato DN, Mathur BK, Bhattacherjee S (2013) DSC and IR methods for determination of accessibility of cellulosic coir fibre and thermal degradation under mercerization. Ind J Fibre Texttile Res 38:96–106

Malkapuram R, Kumar V, Negi YS (2008) Recent development in natural fiber reinforced polypropylene composites. J Reinf Plast Compos 28(10):1169–1189

Mattos BD, Cademartori PHGAL, Missio AL, Gatto DAMWLE (2015) Wood-polymer composites prepared by free radical in situ polymerization of methacrylate monomers into fast-growing pine wood. Wood Sci Technol 49(6):1281–1294

Mehrotra R, Singh P, Kandpal H (2010) Near infrared spectroscopic investigation of the thermal degradation of wood. Thermochim Acta 507–508:60–65

Murinov YI, Grabovskiy SA, Islamova RM, Kabalnova NN (2013) Mechanism of methyl methacrylate polymerization initiated by benzoyl peroxide and ferrocene in the presence of oxygen. Mendelev Commun 23(1):53–55

Nachtigall SMB, Cerveira GS, Rosa SML (2007) New polymeric-coupling agent for polypropylene/wood-flour composites. Polym Test 26(5):619–628

Nourbakhsh A, Kokta BV, Ashori A, Jahan-Latibari A (2008) Effect of a novel coupling agent, polybutadiene isocyanate on mechanical properties of wood-fiber polypropylene composites. J Reinf Plast Compos 27(16–17):1679–1687

Nuthong W, Uawongsuwan P, Pivsa-Art W, Hamada H (2013) Impact property of flexible epoxy treated natural fiber reinforced PLA composites. Energ Proc 34:839–847

Phetkaew W, Kyokong B, Khongtong S, Mekanawakul M (2009) Effect of pre-treatment and heat treatment on tensile and thermal behaviour of Parawood strands. Songklanakarin J Sci Technol 31(3):323–330

Qu H, Wu W, Zheng Y, Xie J, Xu J (2011) Synergistic effects of inorganic tin compounds and Sb_2O_3 on thermal properties and flame retardancy of flexible poly(vinyl chloride). Fire Saf J 46(7):462–467

Rahman MR, Hamdan S, Islam MS (2010) Mechanical and biological performance of sodium metaperiodate-impregnated plasticized wood (pw). BioRes 5(2):1022–1035

Rahman MR, Hamdan S, Hashim DMA, Islam MS, Takagi H (2014) Bamboo fiber polypropylene composites: effect of fiber treatment and nanoclay on mechanical and thermal properties. J Vinyl Addit Technol 21(4):253–258

Rahman MM, Rahman MR, Hamdan S, Hossen MF, Lai JCH, Liew FK (2015) Synthesis of cotton from tossa jute fiber and comparison with original cotton. Int J Polym Sci 2015:1–4

Ramos-Filho FG, Melo TJA, Rabello MS, Silva SML (2005) Thermal stability of nanocomposites based on polypropylene and bentonite. Polym Degrad Stabil 89(3):383–392

Rautkari L, Hill CAS, Curling S, Jalaludin Z, Ormondroyd G (2013) What is the role of accessibility of wood hydroxyl groups in controlling moisture content? J Mater Sci 48(18):6352–6356

Ray SS, Bousmina M (2005) Biodegradable polymers and their layered silicate nanocomposites: in greening the 21st century materials world. Prog Mater Sci 50(8):962–1079

Saba N, Tahir PM, Jawaid M (2014) A review on potentially of nanofiller/natural fiber filled polymer hybrid composites. Polym 6:2247–2273

Srinivas K, Pandey KK (2012) Photodegradation of thermally modified wood. J Photochem Photobio 117:140–145

Srinivasulu R, Madhusudan S, Ashok Kumar M, Nikil Murthy V, Kartikeyan N (2013) Synthesis and characterization of polymer nanocomposites filled with nanoclay/aerosol on mechanical and thermal properties. Int J Nanomater Biostruct 3(4):51–56

Tabari HZ, Nourbakhsh A, Asho A (2011) Effects of nanoclay and coupling agent on the physico-mechanical, morphological and thermal properties of wood flour/polypropylene composites. Polym Eng Sci 51(2):272–277

Thybring EE (2013) The decay resistance of modified wood influenced by moisture exclusion and swelling reduction. Int Biodeter Biodegrad 82:87–95

Toh HK (1979) Dissertation Loughborough University

Trey S, Jafarzadeh S, Johansson M (2012) In situ polymerization of polyaniline in wood veneers. ACS Appl Mater Interfaces 4(3):1760–1769

Wada M, Heux L, Sugiyama J (2004) Polymorphism of cellulose I family: reinvestigation of cellulose IVI. Biomacromol 5(4):1385–1391

Xie W, Gao Z, Pan WP, Hunter D, Singh A, Vaia R (2001) Thermal degradation chemistry of alkyl quaternary ammonium montmorillonite. Chem Mater 13(9):2979–2990

Xie Y, Fu Q, Wang Q, Xiao Z, Militz H (2013) Effects of chemical modification on the mechanical properties of wood. Eur J Wood Wood Prod 71(4):401–416

Xu B, Ding J, Feng L, Ding Y, Ge F, Cai Z (2015) Self-cleaning cotton fabrics via combination of photocatalytic TiO_2 and superhydrophobic SiO_2. Surf Coating Technol 262:70–76

Yang H, Yan R, Chen H, Lee DH, Zheng C (2007) Characteristic of hemicellulose, cellulose and lignin pyrolysis. Fuel 86(12–13):1781–1788

Yao H, You Z, Li L, Goh SW, Lee CH, Yap YK, Shi X (2013) Rheological properties and chemical analysis of nanoclay and carbon microfiber modified asphalt with Fourier transform infrared spectroscopy. Const Build Mater 38:327–337

Zhao J, Wang X, Hu J, Liu Q, Shen D, Xiao R (2014) Thermal degradation of softwood lignin and hardwood lignin by TG-FTIR and Py-GC/MS. Polym Degrad Stabil 108:133–138

Zhu P, Sui S, Wang B, Sun K, Sun G (2004) A study of pyrolysis and pyrolysis products of flame-retardant cotton fabrics by DSC, TGA and PY-GC-MS. J Anal Pyrol 71:645–655

Physico-Mechanical, Thermal, and Morphological Properties of Styrene-co-3-(Trimethoxysilyl)Propyl Methacrylate with Clay Impregnated Wood Polymer Nanocomposites

M.R. Rahman, S. Hamdan and J.C.H. Lai

Abstract In this study, the physico-mechanical, thermal, and morphological properties of styrene-co-3-(trimethoxysilyl)propyl methacrylate (ST-co-SPMA) with clay impregnated wood polymer nanocomposites (WPNCs) were investigated. The WPNCs were characterized by Fourier Transform Infrared Spectroscopy (FT-IR), X-ray diffraction (XRD), Scanning Electron Microscopy (SEM), 3-point bending and free-vibration testing, and Thermogravimetric Analysis (TGA). The FT-IR results showed that the absorbance intensity at 698 cm^{-1} was higher for ST-co-MSPM-clay-WPNC and ST-clay-WPNC when compared to styrene wood polymer composite (ST-WPC), ST-co-MSPA-WPC, and raw wood. The XRD result revealed that the *d*-spacing of WPNCs and WPCs was higher than that of raw wood. The SEM results showed that ST-co-MSPM-clay-WPNC had smoother surfaces when compared to other nanocomposites and raw wood. The modulus of elasticity (MOE), modulus of rupture (MOR), dynamic Young's moduli (Ed), and thermal stability of ST-co-MSPM-clay-WPNC were considerably higher when compared to wood polymer nanocomposites (WPNCs) and raw wood.

Keywords Wood fibers · Nanocomposites · Mechanical properties · Thermal properties

1 Introduction

For sustainable development, wood is more favorable than synthetic polymer because of its renewability. Wood is used worldwide due to its advantages such as its availability, excellent specific strength, high specific modulus, ability to be recycled, and aesthetics properties. Wood has lower costs compared to other

M.R. Rahman (✉) · S. Hamdan · J.C.H. Lai
Faculty of Engineering, Universiti Malaysia Sarawak,
94300 Kota Samarahan, Sarawak, Malaysia
e-mail: rmrezaur@unimas.my

M.R. Rahman, *Wood Polymer Nanocomposites*, Engineering Materials,
DOI 10.1007/978-3-319-65735-6_12

materials, such as glass, which have higher energy consumption to produce them (Donnell et al. 2004; Ku et al. 2011). Increased carbon dioxide gas emissions have been proposed to be attributing to global climate change by burning of fossil fuel. Natural forests, which contain various types of wood species, can reduce the carbon dioxide in the atmosphere. The main drawbacks of wood as a building material are its dimensional instability and its susceptibility to attack by microorganisms (Fengel and Wegener 1989; Mattos et al. 2014; Rowell 2006). At low humidity, wood desorbs water, which results in its shrinkage. In outdoor applications, wood suffers from photo degradation. Wood also becomes discolored when it is used indoor, which is attributed to lignin oxidation (Feist and Hon 1984).

In the past decades, in order to overcome the drawbacks of wood, wood composites were prepared with different thermoplastic polymers in the presence of a crosslinker. These wood composites have enhanced in dimensional stability, superior mechanical strength, and higher thermal stability (Banks and Lawther 1994; Kumar 1994; Rowell 2014). Wood composites are prepared by various chemical reactions, which involve the reactions between wood hydroxyl groups (–OH) and functional groups of filler, which have the ability to change their structures and compositions. As a result, modified wood absorbs less moisture and is less prone to attack by microorganisms. Both wood and synthetic polymers have polar and nonpolar properties, which lead to an excellent enhancement incompatibility between the wood and synthetic polymer (Ray and Bousmina 2005).

Eucalyptus grandis when impregnated with styrene and methyl methacrylate does not exhibit good mechanical performance because permeability of these chemical into that wood is low (Stolf and Lahr 2004). Wood flour coated with hydrophilic–hydrophobic block copolymers of styrene and acrylic acid appreciably enhanced the ultimate tensile properties of the composite (Kosonen et al. 2000). The compatibility between wood and 3-(trimethoxysilyl)propyl methacrylate can be improved (Lu et al. 2013; Rangel-Vázquez and Leal-García 2010). Plasticized wood composites have higher mechanical properties and decay resistance (Rahman et al. 2015).

Clay can act as a plasticizer to reduce the pore volume and improve the thermal properties of wood composites. The polar parts of nanoclay form dipole bonds with the walls of the wood fibers, whereas the nonpolar parts of nanoclay interact with thermoplastic resin of the composite. Wood impregnated with a mixture of organic monomer and organically modified layered silicate (OML) demonstrates significant mechanical improvement, fire resistance, thermal stability, solvent uptake, and rate of biodegradability. Monomers can easily enter into wood cell wall and clay galleries for subsequent polymerization (Gaboune et al. 2006; Kojima et al. 2011; Usuki et al. 1993). A few weight percent of OML (i.e., ≤ 5 wt%) when properly disperse throughout the low-density polymer can create a strong interfacial interaction with the resin through the high surface area of the OMLS. For these reasons, wood polymer nanocomposites are lightweight compared to conventional composites, which make them competitive with other materials for specific applications. The unmodified polymer resin does not exhibit strong interfacial interactions with the walls of the wood fibers (Chen et al. 2002; Ray and Okamoto 2003).

In the Thermogravimetric Analysis, heat flow can reduce or increase heat production. The thermal stability of wood composites depends upon the fire retardant chemicals used and the chemical modifications to the fiber walls of the wood (Islam et al. 2013). Fire retardant chemicals in wood composites produce char and water when subjected to Thermogravimetric Analysis, which reduces the effective heat of combustion and increases the thermal stability of the composites (Levan and Tran 1990; Yunchu et al. 2000). Phosphoric acid is one of the most commonly used fire retardants used to reduce the amount of volatiles generated and to increase the amount of residual char produced during heating (Baysal 2011). $Al(OH)_3$ and $Mg(OH)_2$ are used as heat retardants for their endothermic and dehydration properties (García et al. 2009). Silicates play an important role in wood composite to produce silicate carbide compounds, which create a barrier to heat flow (Lu and Hamerton 2002; Xu et al. 2015). Montmorillonite clays, TiO_2, Sb_2O_3, alumino-silicates, vermiculite, perlite, and organoclays are used as heat retardants to increase the thermal stability of wood (Zhang and Horrocks 2003; Chang and Li 2011). Chemically modified wood along with nanofiller enhances the mechanical and thermal properties (Liodakis et al. 2003).

The present study evaluates the effect of clay dispersed styrene-co-3-(trimethoxysilyl)propyl methacrylate on the physico-mechanical and thermal properties of wood. The morphological properties of the wood nanocomposites are also reported.

2 Experimental

2.1 Materials

Five replicate wood specimens were chosen for each series. To prepare the wood polymer nanocomposites, styrene (ST) was supplied by A.S. Joshi and Company (Goregaon West, Mumbai, India). 3-(trimethoxysilyl)propyl methacrylate (MSPM) and montmorillonite (MMT) nanoclay (Nanomer® I.28E)modified with 25–30 wt% trimethylstearylammonium were supplied by Aldrich (Sigma-Aldrich, St. Louis, MO, USA). Benzoyl peroxide (BP) was supplied by Merck (Merck Schuchardt OHG, Hohenbrunn, Germany) which was used as a free radical catalyst to induce polymerization of monomers.

2.2 Specimen Preparation

A Kumpang tree was cut into three bolts that were 1.2 m in length. Each bolt was sawn into quarters in order to produce planks that were 4 cm thick. These planks were subsequently conditioned by air drying in a room with a relative humidity of 60% and ambient temperature of approximately 25 °C for one month prior to testing. The planks were ripped and machined to 340 mm (L) × 20 mm (T) × 10 mm (R) for three-point bending and free-free-vibration tests.

2.3 Preparation of Wood Polymer Nanocomposites

Oven-dried specimens were impregnated with either one monomer or mixture of monomers with MMT (Table 1). The wood specimens were placed separately into an impregnation vacuum chamber (75 mm Hg) for 30 min. For each monomer(s) impregnation, approximately, 10 g of benzoyl peroxide was added to initiate polymerization. These wood nanocomposites were then removed from the chamber, and the excess chemicals were removed. Specimens were wrapped with aluminum foil and placed into a 105 °C oven for 24 h for polymerization to take place.

2.4 Characterizations

2.4.1 Determination of Weight Percentage Gain (WPG)

Weight percentage gain (WPG) was calculated using the following equation:

$$\mathrm{WPG}\,(\%) = \left[\left(W_f - W_i\right)/W_i\right] \times 100 \tag{1}$$

where W_f and W_i are the oven-dried weight of the WPC (or the WPNC) and the raw wood, respectively.

2.4.2 Determination of Water Uptake (WU)

The water uptake (WU) was calculated using the following equation:

$$\mathrm{WU}\,(\%) = \left[(W_2 - W_1)/W_1\right] \times 100 \tag{2}$$

where W_2 is the weight of treated (or untreated control) wood samples after two weeks of water immersion, and W_1 is the weight the oven-dried sample, respectively.

Table 1 Composition of the chemicals for the raw wood (RW) impregnation

Specimen	Volume of ST (mL)	Volume of MSPM (mL)	Amount of clay (g)	Amount of initiator (BP) (g)	Name of prepared nanocomposites
Raw wood	500	–	–	10	ST-WPC
Raw wood	500	200	–	10	ST-co-MSPA-WPC
Raw wood	500	–	10	10	ST-clay-WPNC
Raw wood	500	200	10	10	ST-co-MSPM-clay-WPNC

2.4.3 Fourier Transform Infrared Spectroscopy (FT-IR)

The infrared spectra of all specimens were recorded on a Shimadzu FT-IR 81,001 Spectrophotometer (Shimadzu, Kyoto, Japan). The wave number range of the scan used was 4000–600 cm^{-1}.

2.4.4 X-ray Diffraction (XRD) Analysis

XRD analysis for raw wood and wood polymer nanocomposites was performed with a Rigaku diffractometer (RGS Corp. Sdn. Bhd., Selangor, Malaysia) operating with Cu*K* radiation, wavelength of 0.154 nm, 40 kV, and 30 mA. The crystallinity index (CI_{XRD}) was calculated using the following height ratio of diffraction peaks:

$$CI_{XRD}\,(\%) = \frac{I_{002} - I_{am}}{I_{002}} \tag{3}$$

where I_{002} is the maximum intensity of the peak at 2θ at about 22°, and I_{am} is the minimum intensity above baseline at 2θ at about 30°, which corresponds to crystalline and amorphous parts, respectively. The *d*-spacing between the [002] lattice planes (d_{002}) of the sample was calculated using the Bragg equation:

$$d_{002} = \frac{n\lambda}{2\sin\theta} \tag{4}$$

where λ is the X-ray wavelength (0.154 nm).

2.4.5 Scanning Electron Microscopy (SEM)

Fractured surfaces were examined to investigate the interfacial bonding between wood cell and polymer with filler. These surfaces were imaged using a Hitachi TM3030 Scanning Electron Microscope (SEM) supplied by JEOL (Tokyo, Japan). The SEM specimens were sputter coated with gold prior to observation. The micrographs were acquired at a magnification of 1500.

2.4.6 Modulus of Rupture (MOR), Modulus of Elasticity (MOE), and Dynamic Young's Modulus (E_d) Measurements

Three-point bending tests were conducted using a Shimadzu MSC-5/500 universal testing machine (Kyoto, Japan) operating at a crosshead speed of 5 mm/min. The modulus of rupture (MOR) and modulus of elasticity (MOE) were calculated using Eqs. (5) and (6).

$$\text{MOR} = \frac{1.5LW}{bd^2} \quad (5)$$

$$\text{MOE} = \frac{L^s m}{4bd^s} \quad (6)$$

where

W ultimate failure load,
L span between centers of support,
b mean width (tangential direction) of the sample,
d mean thickness (radial direction) of the sample, and
m slope of the tangent to the initial line of the force displacement curve.

Determinations of the dynamic Young's modulus, E_d, for all specimens were carried out using a free-free flexural vibration testing system. The E_d was calculated from the resonant frequency using Eq. (7).

$$E_d = \frac{4\pi^2 f^2 l^4 A\rho}{I(m_n)^4} \quad (7)$$

where

I second moment of area of the beam's cross section (which is equal to $bd^3/12$),
d beam's depth, *b* is the beam's width,
l beam's length, f_n is the beam's natural frequency at the *n*th flexural mode,
ρ beam's density,
A beam's cross-sectional area, and
m_n constant related to the free-free beam vibration at the nth flexural mode. The value of m_n at the first mode (n = 1) is 4.730 (i.e., m_1 = 4.730).

2.4.7 Thermogravimetric Analysis (TGA)

Thermogravimetric Analysis (TGA) measurements were performed on 5–10 mg samples at a heating rate of 10 °C/min in a nitrogen atmosphere using a Thermogravimetric Analyzer (Model SDT Q600, TA Instruments, Selangor, Malaysia). Thermal decomposition of each sample occurred in a programmed temperature range of 30–600 °C under a constant nitrogen flow rate of 5 mL/min. The continuous weight loss and temperature were recorded and analyzed.

3 Results and Discussion

3.1 Fourier Transform Infrared Spectroscopy (FT-IR) Analysis

The FT-IR spectra of various nanocomposites and raw wood are presented in Fig. 1. The peak at 4000–3000 cm^{-1} corresponded to stretching vibration of H-bonds in –OH groups, whereas C–H stretching in methyl and methylene groups occurred at 3000–2800 cm^{-1} (Rahman et al. 2011). The peak near 1505 cm^{-1} (which corresponded to aromatic double bonds) had higher intensity for ST-impregnated WPC when compared to raw wood. This was due to the increased amount of aromatic double bonds from the styrene. The IR spectra of the ST-impregnated WPC exhibited lower peak intensity above 3000 cm^{-1} than that of raw wood (which corresponded to –OH groups). The ST plasticized within the wood lumens to displace the free-bound water contained in the wood matrix. The intensity of the –OH group peak for the ST-co-MSPM impregnated wood was lower than that of the ST-impregnated wood. This observation is ascribed to the formation of covalent bonds between the –OH groups of the wood and the partially hydrolyzed ether groups of the MSPM (i.e., silylation with MSPM). This can be attributed that ST-co-MSPM-WPC will show better mechanical performance than that of ST-WPC. ST-co-MSPM-clay-WPNC showed lower peak intensity at 3500–3000 cm^{-1} when compared to ST-co-MSPM-WPC. This was attributed to the removal of free water. The peak intensity at 698 cm^{-1} was assigned to the stretching vibrations of Al–O–Si of the clay and to out-of-plane bending of –OH groups of the wood (Catauroa et al. 2015; Kondo 1996; Parker and Frost 1996). Thus, the intensity of this band was higher for WPNCs when compared to WPCs and raw wood. The C-H bending vibration peak at 750 cm^{-1} had higher intensity for ST-co-MSPM-clay-WPNC and ST-clay-WPNC when compared to ST-WPC and ST-co-MSPM-WPC. This was attributed to the polarization capacity of Al^{3+} cations to π-electrons of styrene (Yao et al. 2013). Among the nanocomposites, the peak at 819 cm^{-1} was visible in ST-co-MSPM-clay-WPNC and ST-co-MSPM-WPC. This peak is assigned to the vibration of Si–O–C and Mg–Al–OH groups (Karesoja et al. 2009; Lu et al. 2013). The peak intensity of the carbonyl group at 1730 cm^{-1} indicated that MSPM was impregnated into the wood. The reaction scheme is shown in Fig. 2.

3.2 X-ray Diffraction (XRD) Analysis

X-ray diffractograms of raw wood, WPCs, and WPNCs are shown in Fig. 3. The diffraction patterns of raw wood, WPCs, and WPNCs indicated that these materials

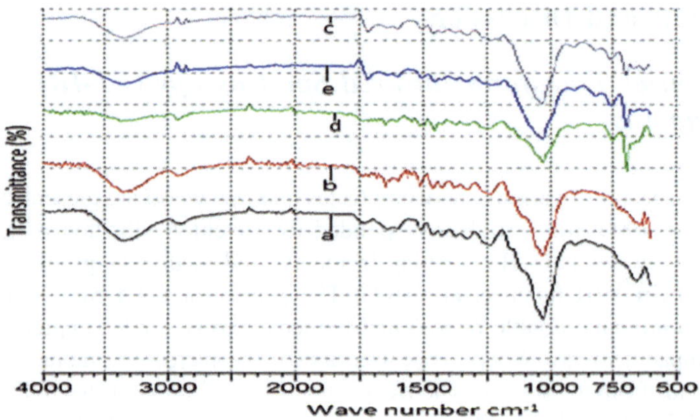

Fig. 1 FT-IR spectra of **a** raw wood, **b** ST-WPC, **c** ST-co-MSPA-WPC, **d** ST-clay-WPNC, and **e** ST-co-MSPM-clay-WPNC

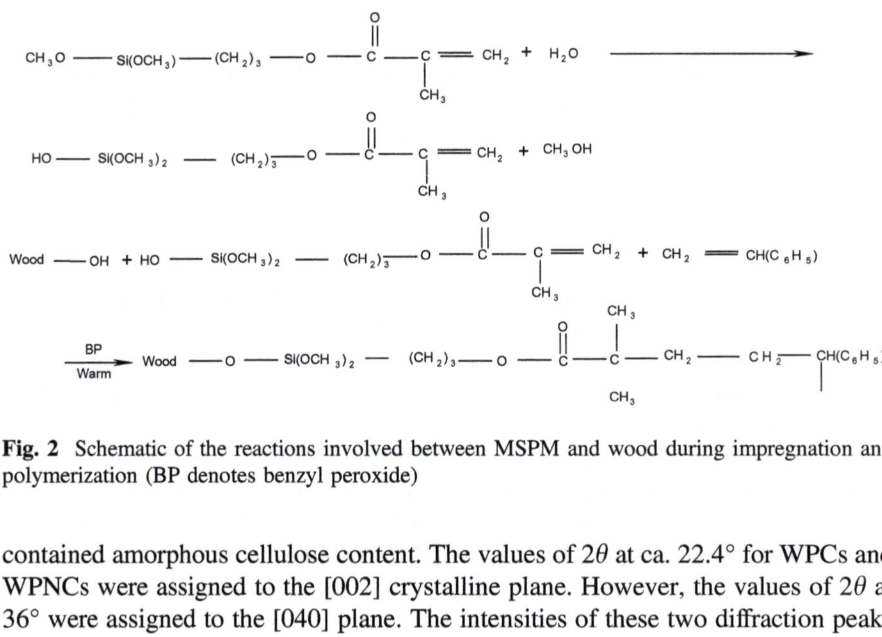

Fig. 2 Schematic of the reactions involved between MSPM and wood during impregnation and polymerization (BP denotes benzyl peroxide)

contained amorphous cellulose content. The values of 2θ at ca. 22.4° for WPCs and WPNCs were assigned to the [002] crystalline plane. However, the values of 2θ at 36° were assigned to the [040] plane. The intensities of these two diffraction peaks were used to calculate the crystallinity index (CIXRD) in Table 2 according to Eq. (3) (Chen et al. 2016; Wada et al. 2004). The CIXRD was the highest for ST-WPC because ST entered into amorphous region of wood and reduced the amorphous movement of wood molecule (Agubra et al. 2013). The CIXRD of ST-clay-WPNC was higher than ST-co-MSPM-clay-WPNC due to the removal of –OH groups from cellulose when it reacted with hydrolyzed MSPM. The organically

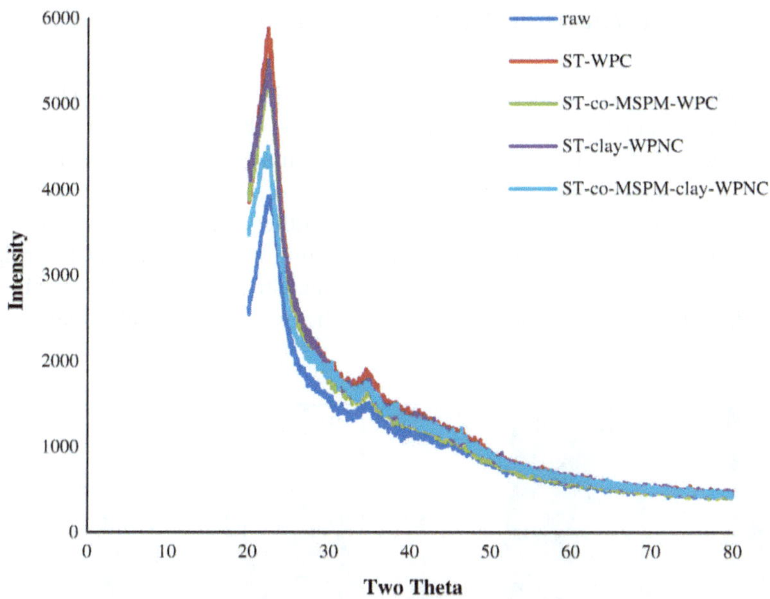

Fig. 3 X-ray diffraction (XRD) diffractograms of **a** RW, **b** ST-WPC, **c** ST-co-MSPM-WPC, **d** ST-clay-WPNC, and **e** ST-co-MSPM-clay-WPNC

Table 2 Crystallinity index (CI_{xrd}) and d_{002} as obtained by XRD analysis

Sample	CI_{XRD} (%)	2θ [002] (°)	d_{002} (nm)
RW	66.59	22.6967	0.3913
ST-WPC	72.50	22.3667	0.3970
ST-co-MSPM-WPC	71.52	22.4657	0.3952
ST-clay-WPNC	71.89	21.7067	0.4089
ST-co-MSPM-clay-WPNC	66.81	22.3667	0.3970

modified clay particle entered into amorphous region of the wood component. X-ray diffractograms of nanocomposites did not show extra peak when compared with raw wood. This can be concluded that the interaction among wood, styrene, ST-co-MSPM, and clay broke the clay gallery.

The values of 2θ for WPCs and WPNCs were lower when compared to raw wood, whereas the *d*-spacing of [002] plane was higher (Table 2). The increment of *d*-spacing indicated that the wood sample was successfully impregnated by the polymer (Chen et al. 2016). Although the –OH groups of wood had reacted with MSPM, the CIXRD of all the nanocomposites was higher than that of the raw wood. This was due to the reduction of amorphous cellulose region.

3.3 Scanning Electron Microscopy (SEM)

The morphological properties raw wood, ST-WPC, ST-co-MSPA-WPC, ST-clay-WPC, and ST-co-MSPM-clay-WPNC are shown in Fig. 4. The lumens and tracheids of the wood fibers were filled by the polymer by in situ polymerization of the monomers. In Fig. 4d, e, the ST-clay-WPC and ST-co-MSPM-clay-WPNC surfaces were smoother than the ST-WPC surface (Fig. 4b). This was due to the clay particles acting as filler material to bridge the void volumes between the wood fiber pores and the polymer (Xu et al. 2013). ST-co-MSPM formed covalent bonds with the wood fiber walls that filled the

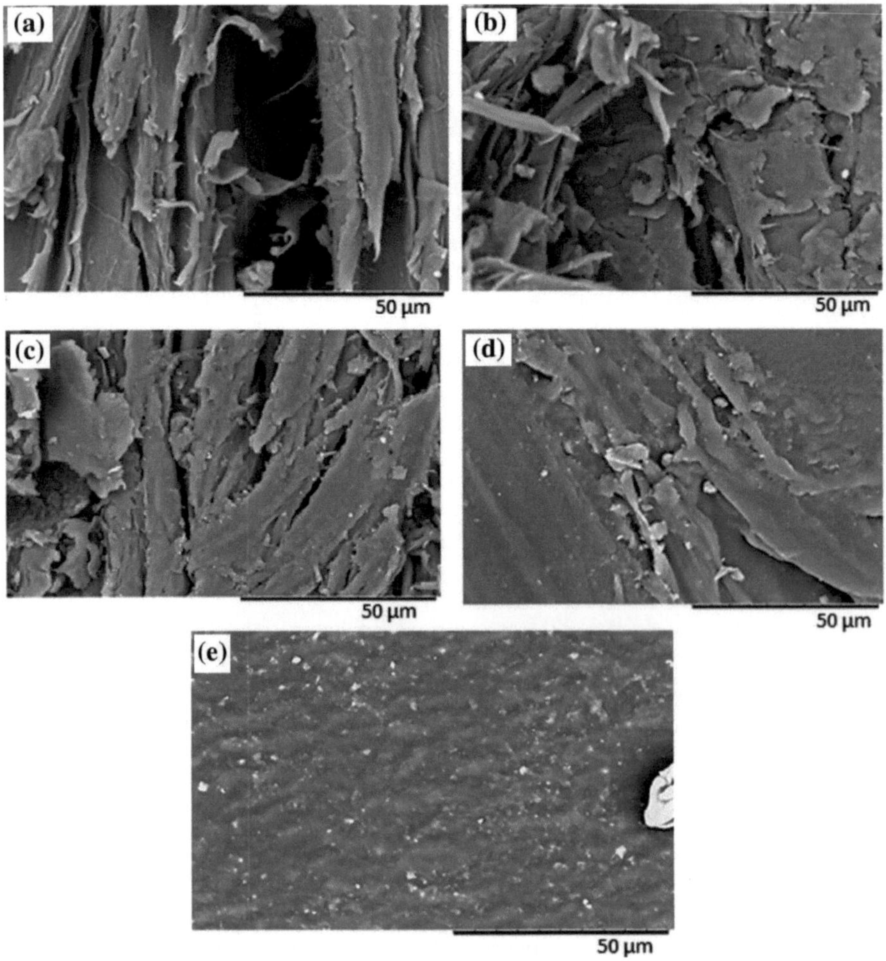

Fig. 4 SEM micrographs of **a** RW, **b** ST-WPC, **c** ST-co-MSPM-WPC, **d** ST-clay-WPNC, and **e** ST-co-MSPM-clay-WPNC

wood's void spaces when compared to the raw wood. The better interaction between filler and wood implied a more rigid composite (Li et al. 2011), which is reflected in composite's mechanical properties.

3.4 MOR, MOE, and Ed

The MOE, MOR, and Ed analyses of the raw wood, ST-WPC, ST-co-MSPA-WPC, ST-clay-WPC, and ST-co-MSPM-clay-WPNC are shown in Fig. 5. The wood polymer nanocomposites exhibited higher MOE and MOR when compared to the raw wood. The filler exerted adhesive force as well as formed covalent bonds with the wood's fiber walls. In the ST–WPC composite, there were only adhesive forces between the wood and the ST, which had lower MOE and MOR when compared to the ST-co-MSPM-WPC, ST-clay-WPNC, and ST-co-MSPM-clay-WPNCs. The WPNCs showed higher MOE and MOR when compared to raw wood due to the clay particles, which had high surface area and possessed ionic charges that create dipole bonds. However, when the clay particles enter the pores and voids of the wood fiber walls, they improve the shearing resistance of the nanocomposites. The MOE and MOR values were the highest for ST-co-MSPM-clay-WPNC among the nanocomposites. The hydroxysilyl group of the hydrolyzed MSPM can form a covalent bond with the –OH group of wood (via silylation reaction), and the vinyl group of MSPM can react with styrene by a free radical mechanism during polymerization. Thus, the enthalpy of these reactions broke the clay galleries for

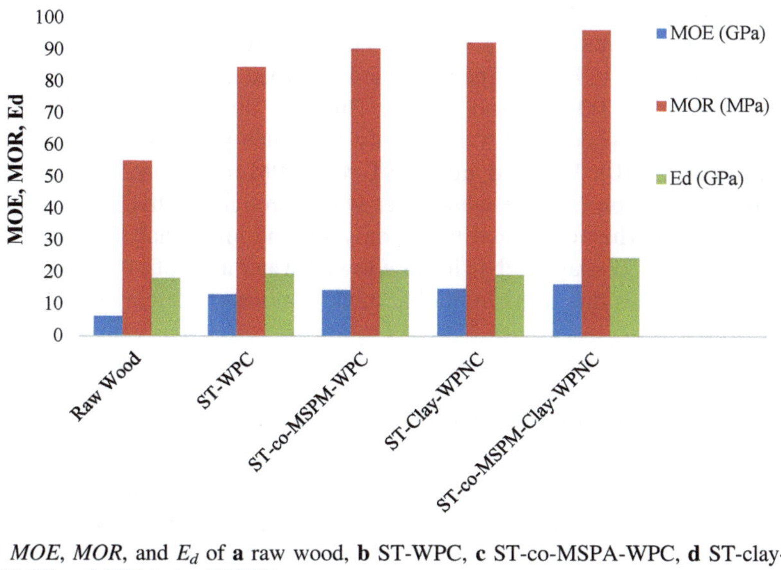

Fig. 5 *MOE*, *MOR*, and E_d of **a** raw wood, **b** ST-WPC, **c** ST-co-MSPA-WPC, **d** ST-clay-, and **e** WPNC ST-co-MSPM-clay-WPNC

exfoliation (Yao et al. 2013). The manufactured nanocomposites had sufficiently strong covalent bonding strength, which enhanced the MOR and MOE for WPNC (Rahman et al. 2015; Rashmi et al. 2011).

Ed of all the WPNCs was higher than the raw wood. This was due to the higher density and shear resistance for all the wood polymer nanocomposites (Rahman et al. 2010a, b). In the presence of shear force, the free energy increases and the molecules continuously oscillate from equilibrium (Toh 1979). The shear force is considered to vary sinusoidally with time. ST-co-MSPM-WPC showed higher Ed when compared to ST-WPC, which was due to covalent crosslinking between wood, styrene, and MSPM (Ismail et al. 2002). The interaction between raw wood and clay extracts the adhesive forces to increase shear stress. Among these nanocomposites, the highest Ed was observed for ST-co-MSPM-clay-WPNC. This could be attributed to the increased interaction between wood, ST-co-MSPM, and clay, which reduced the mobility of the intercalated polymer chains (Rashmi et al. 2011).

3.5 Weight Percentage Gain (WPG) and Water Uptake (WU)

Both WPG and WU values are shown in Fig. 6. It was clear that ST-co-MSPM-clay-WPNC showed higher WPG when compared to the other nanocomposites. This was due to the formation energy of ST-co-MSPM broke the clay particles to enter into wood cavities and its cell wall. Thus, ST-co-MSPM-clay-WPNC had higher WPG, followed by ST-co-MSPM-WPC, ST-clay-WPNC, and ST-WPC.

The water absorption of the various wood polymer nanocomposites and raw wood are illustrated in Fig. 6. The raw wood absorbed a higher percentage of water when compared to WPNCs and WPCs. This was due to ST, MSPM, and clay interacts with the wood fiber walls greatly to reduce pore size and pore volume (Hamdan et al. 2010). The void area of ST-co-MSPM impregnated WPC increased the water penetration into the wood fiber walls through the hydrophilic hydroxyl groups of wood when compared to the only styrene impregnated WPC. It is also observed that the ST-clay-WPNC had the lowest water absorption (28.84%) among the nanocomposites due to hydrophobic modifications of the clay filler.

3.6 Thermogravimetric Analysis (TGA)

Thermogravimetric Analyses of raw wood, WPC, and WPNC are shown in Fig. 7, and the corresponding thermal characteristic is given in Table 3. Figure 7 shows

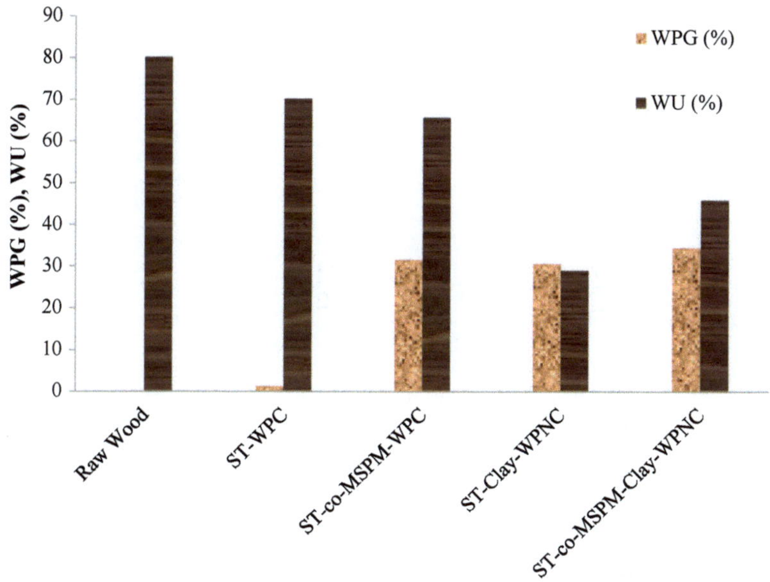

Fig. 6 *WPG* (%), and *WU* (%) of: (a) RW, (b) ST-WPC, (c) ST-co-GMA-WPC, (d) ST-clay-WPNC, and (e) ST-co-GMA-clay-WPNC

that raw wood and ST-WPC exhibited two stages of thermal degradation and decomposition.

The first stage of weight loss occurred at 70–110 °C and was attributed to absorbed moisture evaporation from all the samples (Liu et al. 2013). The initial weight loss of the samples below 110 °C was the lowest for ST-co-MSPM-clay-WPNC, followed by ST-clay-WPNC, ST-co-MSPM-WPC, ST-WPC, and RW. This was attributed to the reaction of hydrolyzed MSPM with the –OH groups of wood fiber wall and with the surface modified MMT filler within the fiber walls.

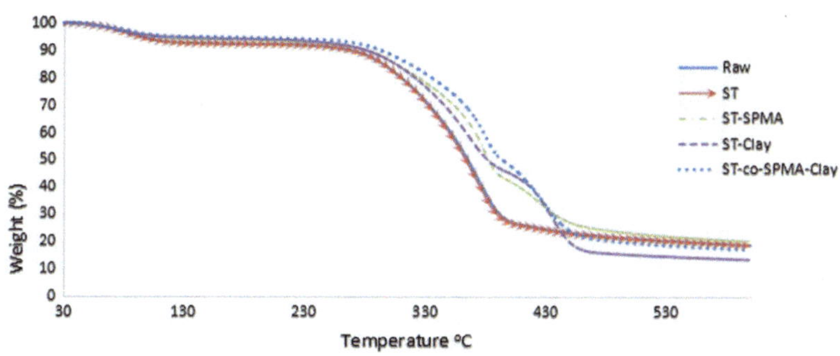

Fig. 7 TGA of (a) RW, (b) ST-WPC, (c) ST-co-MSPM-WPC, (d) ST-clay-WPNC, and (e) ST-co-MSPM-clay-WPNC

Table 3 Results for TG analysis for raw wood and prepared nanocomposites

Sample name	Number of transition	Transition temperature			Weight loss at corresponding transition (%)	Residual weight (%) at last stage transition
		T_i	T_m	T_f		
RW	1	37.80	80.36	121.70	7.30	24.81
	2	127.58	354.51	425.59	67.65	
ST-WPC	1	43.04	80.04	134.33	7.30	24.59
	2	205.27	353.07	425.21	67.46	
ST-co-MSPM-WPC	1	42.35	79.31	125.90	5.81	20.62
	2	277.99	357.57	386.48	43.50	
	3	392.56	430.92	573.12	23.48	
ST-clay-WPNC	1	43.70	74.15	112.22	4.83	15.22
	2	129.65	343.83	387.42	45.92	
	3	396.34	434.27	494.34	31.03	
ST-co-MSPM-clay-WPNC	1	43.69	77.95	128.14	4.80	17.89
	2	233.43	357.38	391.18	43.06	
	3	390.72	428.05	550.25	33.26	

T_i: onset temperature; T_m: temperature corresponding to the maximum rate of mass loss; and T_f: end temperature

The second stage of thermal decomposition of raw wood and ST-WPC occurred at 127–425 °C. In this region, there were no appreciable differences in thermal properties, which were because the styrene did not covalently bond with wood. On the other hand, the second-step decomposition at 129–282 °C of ST-co-MSPM-clay-WPNC was higher than ST-clay-WPNC, ST-co-MSPM-WPC, ST-WPC, and raw wood. This was due to both exfoliated MMT and ST-co-MSPM had polar and nonpolar properties, and this polar–polar and nonpolar–nonpolar interaction reduced the surface energy of that elements. Well-distributed MMT layers could prevent the passage of volatile decomposition product throughout the composite during TGA. Thus, the addition of MMT improved the thermal stability for the second decomposition stage of the nanocomposites (García et al. 2009; Sanchez-Jimenez et al. 2012).

In the third stage of decomposition of ST-co-MSPM-WPC, ST-clay-WPNC and ST-co-MSPM-clay-WPNC started at 392–573 °C. This was due to the thermal decomposition of the wood polymer structure, which produced combustible gases like carbon monoxide, hydrogen, and methane, along with carbon dioxide. In this third thermal degradation stage, the cellulose and lignin were degraded (Kim et al. 2006; Yang et al. 2007). The weight loss rate of ST-co-MSPM-WPC and ST-co-MSPM-clay-WPC was higher than ST-clay-WPNC. This was attributed to the thermal decomposition of the Si–O–CH_3 group to produce volatile carbon dioxide gas. The final residual weights of ST-co-MSPM-clay-WPNC and ST-clay-WPNC were significantly less than the raw wood and ST-co-MSPM-WPC, which was due to the dehydroxylation of the aluminosilicate (Xie et al. 2001). However, the final residual weight of the ST-co-MSPM-WPC was higher than the ST-co-clay-WPNC because of Si–O–C bond thermally decomposes above 800 °C (Torrey and Bordia 2008).

4 Conclusion

Kumpang wood was successfully impregnated with clay dispersed styrene-co-3-(trimethoxysilyl)propyl methacrylate (ST-co-MSPM), which was confirmed by FT-IR and XRD analyses. The SEM morphology showed that ST-co-MSPM-clay-WPNCs and ST-clay-WPCs had smoother surfaces when compared to ST-co-MSPM-WPCs, ST-WPCs, and raw wood. WPNCs had the highest MOE, MOR, and Ed values, followed by WPCs and RW. ST-co-MSPM-clay-WPNC showed the highest WPG, followed by ST-co-MSPM-WPC, ST-clay-WPNC ,and ST-WPC. ST-clay-WPNC had lower water uptake than ST-co-MSPM-clay-WPNC, ST-co-MSPM-WPC, ST-WPC, and raw wood. WPNCs showed better thermal stability when compared to WPCs and raw wood.

Acknowledgments The authors would like to acknowledge the financial support from Ministry of Higher Education Malaysia, for their financial support (Grant no. FRGS/SG02(01)/1085/2013 (31)).

References

Agubra VA, Owuor PS, Hosur MV (2013) Influence of nanoclay dispersion methods on the mechanical behavior of E-glass/epoxy nanocomposites. Nanomater 3:550–563

Banks WB, Lawther ML (eds) (1994) Cellulosic polymers blends and composites. Hanser/Garder Publications Inc., Cincinnati

Baysal E (2011) Combustion properties of wood impregnated with commercial fertilizers. African J Biotechnol 10:19255–19560

Catauroa M, Papalea F, Lamannaa G, Bollinoa F (2015) Geopolymer/PEG hybrid materials synthesis and investigation of the polymer influence on microstructure and mechanical behavior. Mater Res 18:698–705

Chang MK, Li SJ (2011) A study of strength and thermal stability of low-density polyethylene grafted maleic anhydride/montmorillonite nanocomposites. In: IEEE international conference on industrial engineering and operational management. Kuala Lumpur, Malaysia, 22–24 Jan 2011

Chen JS, Poliks MD, Ober CK, Zhang Y, Wiesner U, Giannelis E (2002) Study of the interlayer expansion mechanism and thermal-mechanical properties of surface-initiated epoxy nanocomposites. Polym 43:4895–4904

Chen T, Wu Z, Niu M, Xie Y, Wang X (2016) Effect of Si–Al molar ratio on microstructure and mechanical properties of ultra-low density fiberboard. Eur J Wood Wood Prod 74:151–160

Donnell A, Dweib MA, Wool RP (2004) Natural fiber composites with plant oil-based resin. Compos Sci Technol 64:1135–1145

Feist WC, Hon DNS (1984) Chemistry of weathering and protection. Chem Solid Wood 11: 401–451

Fengel D, Wegener G (1989) Wood: chemistry, ultrastructure, reactions. Walter de Gruyter, Berlin

Gaboune A, Ray SS, Ait-Kadi A, Riedl B, Bousmina M (2006) Polyethylene/clay nanocomposites prepared by polymerization compounding method. J Nanosci Nanotechnol 6:530–535

García M, Hidalgo J, Garmendia I, García-Jaca J (2009) Wood-plastics composites with better fire retardancy and durability performance. Compos Part A Appl Sci Manuf 40:1772–1776

Hamdan S, Rahman R, Ahmed AS, Talib ZA, Islam S (2010) Influence of *N,N*-dimethylacetamid on the thermal and mechanical properties of polymer-filled wood. BioRes 5:2611–2624

Islam MS, Hamdan S, Rusop M, Rahman MR (2013) Thermal stability and decay resistance properties of tropical wood polymer nanocomposites (WPNC). Adv Mater Res 667:482–489

Ismail H, Shuhelmy S, Edyham MR (2002) The effects of a silane coupling agent on curing characteristics and mechanical properties of bamboo fibre filled natural rubber composites. Eur Polym J 38:39–47

Karesoja M, Jokinen H, Karjalainen E, Pulkkinen P, Torkkeli M, Soininen A, Ruokolainen J, Tenhu H (2009) Grafting of montmorillonite nano-clay with butyl acrylate (BuA) and methacrylate (MMA) by ATRP. Blends with poly(BuA-co-MMA. J Polym Sci Part A 47(12): 3086–3097

Kim HS, Kim S, Kim HJ, Yang HS (2006) Thermal properties of bio-flour-filled polyolefin composites with different compatibilizing agent type and content. Thermochim Acta 451: 181–188

Kojima Y, Usuki A, Kawasumi M, Okada A, Fukushima Y, Kurauchi T (2011) Mechanical properties of nylon 6-clay hybrid. J Mater Res 8:1185–1189

Kondo T (1996) A Fourier transform infra-red spectroscopic analysis of the character of hydrogen bonds in amorphous cellulose. Polym 37:393–399

Kosonen ML, Wang B, Caneba GT, Gardner DJ (2000) Polystyrene/wood composites and hydrophobic wood coatings from water-based hydrophilic-hydrophobic block. Clan Prod Process 2:117–123

Ku H, Wang H, Pattarachaiyakoop N, Trada M (2011) A review on the tensile properties of natural fiber reinforced polymer composites. Compos Part B Eng 42:856–873

Kumar S (1994) Chemical modification of wood. Wood Fiber Sci 26:270–280

Levan SL, Tran HC (1990) The role of boron in flame-retardant treatments. Forest Prod Res Soc 1990:39–41

Li YF, Liu YX, Wang XM, Wu QL, Yu HP, Li J (2011) Wood–polymer composites prepared by the in situ polymerization of monomers within wood. Appl Polym Sci 119:3207–3216

Liodakis S, Bakirtzis D, Dimitrakopoulos AP (2003) Autoignition and thermogravimetric analysis of forest species treated with fire retardants. Thermochim Acta 399:31–42

Liu Z, Jiang Z, Fei B, Liu X (2013) Thermal decomposition characteristics of chinese fir. BioRes 8:5014–5024

Lu SY, Hamerton I (2002) Recent developments in the chemistry of halogen-free flame retardant polymers. Prog Polym Sci 27:1661–1712

Lu T, Jiang M, Jiang Z, Hui D, Wang Z, Zhou Z (2013) Effect of surface modification of bamboo cellulose fibers on mechanical properties of cellulose/epoxy composites. Compos Part B 51(51):28–34

Mattos BD, Cademartori PHG, Louren TV, Gatto DA, Magalhaes WLE (2014) Biodeterioration of wood from two fast-growing eucalypts exposed to field test. Int Biodeterior Biodegrad 93: 210–215

Parker RW, Frost RL (1996) The application of drift spectroscopy to the multicomponent analysis of organic chemicals adsorbed on montmorillonite. Clay Miner 44:32–40

Rahman MR, Hamdan S, Ahmed AS, Islam MS (2010a) Mechanical and biological performance of sodium metaperiodate-impregnated plasticized wood (PW). BioRes 5:1022–1035

Rahman MR, Hamdan S, Ahmed AS, Islam MS (2010b) Mechanical and biological performance of sodium metaperiodate-impregnated plasticized wood (PW). BioRes 5:1022–1035

Rahman MR, Hamdan S, Ahmed AS, Islam MS, Talib ZA, Islam MS, Abdullah WFW, Mat MSC (2011) Thermogravimetric analysis and dynamic Young's modulus measurement of *N,N*-dimethylacetamide-impregnated wood polymer composites. J Vinyl Addit Technol 17: 177–183

Rahman MM, Rahman MR, Hamdan S, Hossen MF, Lai JCH, Liew FK (2015) Synthesis of cotton from tossa jute fiber and comparison with original cotton. Int J Polym Sci 2015:1–4

Rangel-Vázquez NA, Leal-García T (2010) Spectroscopy analysis of chemical modification of cellulose fibers. J Mex Chem Soc 54:192–197

Rashmi Renukappa NM, Suresha B, Devarajaiah RM, Shivakumar KN (2011) Dry sliding wear behaviour of organo-modified montmorillonite filled epoxy nanocomposites using Taguchi's techniques. Mater Des 32:4528–4536

Ray SS, Bousmina M (2005) Biodegradable polymers and their layered silicate nanocomposites: in greening the 21st century materials world. Prog Mater Sci 50:962–1079

Ray SS, Okamoto M (2003) Polymer/layered silicate nanocomposites: a review from preparation to processing. Prog Polym Sci 28:1539–1641

Rowell RM (2006) Chemical modification of wood: a short review. Wood Mater Sci Eng 1:29–33

Rowell RM (2014) Acetylation of wood—a review. Intl J Lignocellul Prod 1:1–27

Sanchez-Jimenez PE, Perez-Maqueda LA, Crespo-Amoros JE, Lopez J, Perejon A, Criado JM (2012) Nanoclay nucleation effect in the thermal stabilization of a polymer nanocomposite: a kinetic mechanism change. J Phys Chem 116(21):11797–11807

Stolf DO, Lahr FAR (2004) Wood-polymer composite: physical and mechanical properties of some wood species impregnated with styrene and methyl methacrylate. J Mater Res 7:611–617

Toh HK (1979) A study of diffusion in polymers using C-14 labelled molecules. Diss Loughbrgh Univ Leicestershire, United Kingdom, pp 49–50

Torrey JD, Bordia RK (2008) Mechanical properties of polymer-derived ceramic composite coatings on steel. J Eur Ceram Soc 28:253–257

Usuki A, Kawasumi M, Kojima Y, Okada A, Kurauchi T, Kamigaito O (1993) Swelling behavior of montmorillonite cation exchanged for ω-amino acids by -caprolactam. J Mater Res 8: 1174–1178

Wada M, Heux L, Sugiyama J (2004) Polymorphism of cellulose I family: reinvestigation of cellulose IVl. Biomacromol 5:1385–1391

Xie W, Gao Z, Pan WP, Hunter D, Singh A, Vaia R (2001) Thermal degradation chemistry of alkyl quaternary ammonium Montmorillonite. Chem Mater 13:2979–2990

Xu Y, Guo Z, Fang Z, Peng M, Shen L (2013) Combination of double-modified clay and polypropylene-graft-maleic anhydride for the simultaneously improved thermal and mechanical properties of polypropylene. J Appl Polym Sci 128:283–291

Xu B, Ding J, Feng L, Ding Y, Ge F, Cai Z (2015) Self-cleaning cotton fabrics via combination of photocatalytic TiO_2 and superhydrophobic SiO_2. Surf Coatings Technol 262:70–76

Yang H, Yan R, Chen H, Lee DH, Zheng C (2007) Characteristics of hemicellulose, cellulose and lignin pyrolysis. Fuel 86:1781–1788

Yao H, You Z, Li L, Goh SW, Lee CH, Yap YK (2013) Rheological properties and chemical analysis of nanoclay and carbon microfiber modified asphalt with Fourier transform infrared spectroscopy. Constr Build Mater 38:327–337

Yunchu H, Peijang Z, Songsheng Q (2000) TG-DTA studies on wood treated with flame-retardants. Holz Als Roh- Und Werkst 58:35–38

Zhang S, Horrocks AR (2003) A review of flame retardant polypropylene fibres. Prog Polym Sci 28:1517–1538

Acrylonitrile/Butyl Methacrylate/Halloysite Nanoclay Impregnated Sindora Wood Polymer Nanocomposites (WPNCs): Physico-mechanical, Morphological and Thermal Properties

M.R. Rahman, J.C.H. Lai and S. Hamdan

Abstract In this study, physical, morphological, mechanical, and thermal properties of acrylonitrile/butyl methacrylate/halloysite nanoclay wood polymer nanocomposites (AN-*co*-BMA-HNC WPNCs) were investigated. AN-*co*-BMA-HNC WPNCs were prepared via impregnation method, and the effect of different ratio between the polymers was subsequently investigated. The properties of nanocomposites were characterized through weight percent gain, Fourier Transform Infrared Spectroscopy (FT-IR), Scanning Electron Microscopy (SEM), three-point flexural test, dynamic mechanical thermal analysis (DMTA), Thermogravimetric Analysis (TGA), differential scanning calorimetry (DSC) analysis, and moisture absorption test. The weight percent gain in 50:50 AN-*co*-BMA-HNC WPNCs was the highest compared to raw wood (RW) and other WPNCs. FT-IR results confirmed the polymerization took place in the nanocomposites, especially 50:50 AN-*co*-BMA-HNC WPNCs with reducing hydroxyl groups. SEM result revealed that the 50:50 AN-*co*-BMA-HNC WPNCs showed the best surface morphology among all the samples. Through three-point flexural test, 50:50 AN-*co*-BMA-HNC WPNCs showed the highest flexural strength and modulus of elasticity. The results revealed that the storage modulus and loss modulus of AN-*co*-BMA-HNC WPNCs were higher while the tan δ of AN-*co*-BMA-HNC WPNCs was lower compared to RW. AN-*co*-BMA-HNC WPNCs exhibited the higher thermal stability through TGA and DSC analysis. 50:50 AN-*co*-BMA-HNC WPNCs exhibited significantly lower moisture absorption compared to RW. Overall, this study proved that the ratio 50:50 AN-*co*-BMA was the most suitable to be introduced in the RW.

Keywords Morphology · Strength · Thermal · Clay

M.R. Rahman (✉) · J.C.H. Lai · S. Hamdan
Faculty of Engineering, Universiti Malaysia Sarawak,
94300 Kota Samarahan, Sarawak, Malaysia
e-mail: rmrezaur@unimas.my

M.R. Rahman, *Wood Polymer Nanocomposites*, Engineering Materials,
DOI 10.1007/978-3-319-65735-6_13

1 Introduction

Wood has been used as a raw material in many technical and artistic applications (Vorreiter 1949). Wood fiber is widely used as fillers due to their low cost and abundant in production (Bodirlau et al. 2009).

In recent years, wood-based renewable composites especially wood polymer nanocomposites (WPNCs) have gained much attention from both scientific and commercial application. This is because the wood-based composites are light in weight, eco-friendly, and low cost. Besides, wood fibers are considered as one of the alternatives to glass or carbon fibers due to their environmental friendliness, biodegradability as well as lower in cost (Ashori 2008). WPNCs are materials which are composed of lignocellulosic fibers and thermoplastic polymers in varying percentages (Stark and Matuana 2007). WPNCs offer interesting improvement in various properties. Besides, the introduction of wood fiber which acts as reinforcing fillers in WPNCs showed great advantages such as high specific strength and better toughness (Pakeyangkoon and Ploydee 2013).

However, the application of wood fiber in composite materials is limited due to the hydroxyl groups inside the wood that leads to poor compatibility with the hydrophobic polymer matrix (Kallakas et al. 2015). With poor compatibility, the properties of the WPNCs could not be enhanced. In order to ensure the WPNCs can be well fabricated, impregnation of wood can be applied to alter the structure and at the same time improve the interfacial adhesion between polymer matrix and wood fibers. This can lead to the improvement in physical and mechanical properties of the WPNCs (Tesinova 2011).

Sindora wood is one of the species that are abundantly available in Malaysia. It is widely applied in the production of furniture, agriculture implements, and heavy construction (Smitinand and Larsen 1984). However, the research of the WPNCs on Sindora wood is still limited in both research and industry areas.

One of the polymer matrices that is often used in WPNCs includes acrylonitrile or polyamide. Acrylonitrile usually works with different polymer matrix to enhance the mechanical and thermal properties of WPNCs (Prochon et al. 2007). Butyl methacrylate is usually applied in WPNCs for better thermal properties. It can be well intercalated with different fillers such as silica or clay in small quantity (Kobayashi et al. 2002). Clay usually acted as filler for WPNCs. It helps to produce a smooth surface finishing, lower water absorption as well as improving mechanical properties (Mekhemer et al. 2006).

The addition of silica into in situ emulsion polymerization acrylonitrile-methyl methacrylate copolymer improved thermal properties of the WPNCs (Bao et al. 2013). Poly(acrylonitrile-*co*-methyl methacrylate) copolymer undergo precipitation polymerization technique which had significantly improved the thermal stability as well as better surface morphology of the nanoparticles (Mohy Eldin et al. 2014). The presence of nanoclay had improved the modulus of elasticity of acrylonitrile-butadiene-styrene (ABS) nanocomposites through melt compounding (Singh and Ghosh 2014).

The presence of different clay loadings through in situ polymerization into poly (styrene-*co*-butyl methacrylate) copolymer matrix had greatly improved the thermal properties (Siddiqui et al. 2013). The addition of surface modified natural montmorillonite improved the mixing between the clay and poly(*n*-butylacrylate) matrix through in situ polymerization and thermal properties of the nanocomposites (Herrera-Alonso et al. 2010).

However, there is no detailed work on the combination of special polymer matrix, such as acrylonitrile-*co*-butyl methacrylate-halloysite nanoclay (AN-*co*-BMA-HNC) and wood fiber produce a WPNC. Therefore, the aim of this present study is to fabricate AN-*co*-BMA-HNC WPNCs via impregnation method and to investigate the effect of different ratios of the polymer matrix on physical, mechanical, morphological, and thermal properties of WPNCs.

TGA analysis of Pour et al. (2015) showed that the incorporation of graphene particles enhanced the thermal stability of melt extruded polycarbonate/acrylonitrile butadiene styrene (ABS) nanocomposites in the ratio of 70/30 wt%. With the addition of silica into in situ acrylonitrile/methyl methacrylate copolymer, the pore diameter and total pore volume increased during nitrogen adsorption–desorption. Besides, the SEM analysis of this nanocomposite showed disordered mesopores due to the templating effect of silica nanoparticles (Bao et al. 2013).

2 Experimental

2.1 Materials

Defect-free and straight-grained Sindora glabra wood (RW) was obtained from Lundu, Sarawak, Malaysia. Both control and modified wood samples were obtained from the same timber and machined at the same time with a dimension of 30 cm × 2 cm × 1 cm. The chemicals used to produce WPNCs were acrylonitrile (AN), butyl methacrylate (BMA), benzoyl peroxide, and halloysite nanoclay (HN). Acrylonitrile and halloysite nanoclay were supplied by Sigma Aldrich (USA) while butyl methacrylate and benzoyl peroxide were supplied by Merck Millipore (USA). Halloysite nanoclay was nanopowder with the diameter of 30–70 nm, and length was in between 1 and 3 μm.

*2.2 Preparation of Acrylonitrile/Butyl Methacrylate/Halloysite Nanoclay Wood Polymer Nanocomposites (AN-*co*-BMA-HNC WPNCs)*

The polymer system was prepared using AN-*co*-BMA-HNC in the presence of benzoyl peroxide. Benzoyl peroxide acted as an initiator to influence the reaction

Table 1 Preparation of polymer system at different ratio

Volume of acrylonitrile (AN) (mL)	Volume of butyl methacrylate (BMA) (mL)	Amount of halloysite nanoclay (HNC) (g)	Amount of benzoyl peroxide (g)
0	200	2	5
100	100	2	5
140	60	2	5
200	0	2	5

between acrylonitrile and butyl methacrylate. Acrylonitrile, butyl methacrylate, and halloysite nanoclay were mixed with different ratios as shown in Table 1. The mixtures were covered with aluminum foil and placed in an autoclave for 15 min to complete the reaction.

2.3 *Impregnation of AN-*co*-BMA-HNC WPNCs*

Four wood samples prepared according to ASTM D1037 were weighed using an electronic balance. The samples were then placed in the vacuum chamber (box). The mixtures were poured into the vacuum box with pressure of 0.1 bar for 60 min. After impregnation, the excessive solution was removed from the wood samples with tissue paper. Wood samples were wrapped with aluminum foil and kept in an oven at 100 °C for 24 h for polymerization and formed wood polymer nanocomposites (WPNCs). After polymerization, the aluminum foil was removed and WPNCs were dried at 105 °C until constant weight was obtained. The weight percent gain (WPG) of all the WPNCs were then measured using Eq. (1),

$$\mathrm{WPG} = \frac{W_f - W_o}{W_o} \times 100\% \quad (1)$$

where

W_f over-dried weight after modification of the WPNCs,
W_o over-dried weight before modification of the WPNCs.

2.4 *Microstructural Characterizations*

2.4.1 Fourier Transform Infrared Spectroscopy (FT-IR)

The infrared spectra of the WPNCs were recorded on a Shimadzu IRAffinity-1. The technique used was attenuated total reflection (ATR). The spectral resolution of this equipment is 12000 spectra. The transmittance range of the scan was 4000–600 cm^{-1}.

2.4.2 Scanning Electron Microscopy (SEM)

The interfacial bonding between the acrylonitrile, butyl methacrylate, and halloysite nanoclay were examined using a Scanning Electron Microscope (SEM) (JSM-6710F) supplied by JEOL Company Limited, Japan. The AN-*co*-BMA-HNC WPNCs were cut in the dimension of 5 mm × 5 mm × 5 mm (length × width × height). The accelerating voltage of this instrument was 15 kV. The specimens were first fixed with Karnovsky's fixative and then taken through a graded alcohol dehydration series. Once dehydrated, the specimen was coated with a thin layer of gold before being viewed microscopically. The micrographs were taken at magnifications ranging from 500–1000.

2.4.3 Three-Point Flexural Test

The static flexural tests of the AN-*co*-BMA-HNC WPNCs were carried out using AG-X plus series precision universal testers (300 kN Floor Model) supplied by Shimadzu Corporation, Japan. Dimensions of flexural test specimen were 80 mm (length) × 30 mm (width) × 10 mm (thickness). Flexural tests were conducted following ASTM D 7900-00 at a crosshead speed of 10 mm/min. Four specimens of each composition were tested and the average values were reported. The flexural strength (Q_{fs}) and modulus of elasticity (E_{m}) were calculated using Eqs. (2) and (3).

$$Q_{\mathrm{fs}} = \frac{3PL}{2bd^2} \tag{2}$$

where

- P maximum load on the load-deflection curve,
- L support span, 63.7 mm,
- b width of beam tested, 10 mm,
- d thickness of beam, 4 mm.

$$E_{\mathrm{m}} = \frac{L^3 m}{4bd^3} \tag{3}$$

where

- L support span, 63.7 mm,
- m slope of the tangent to the initial straight-line portion of the load-deflection curve, N/mm of deflection,
- b width of beam tested, 10 mm,
- d thickness of beam, 4 mm.

2.4.4 Dynamic Mechanical Thermal Analysis (DMTA)

Dynamic mechanical thermal analysis was applied to study the effect of the temperature on the storage modulus ($\log E'$) and loss tangent ($\tan \delta$) of AN-*co*-BMA-HNC WPNCs. It is a mechanical test that allowed molecules in woods to interact with mechanical stress. This test was carried out using Perkin Elmer dynamic mechanical thermal analyzer (PE-DMTA) supplied by Perkin Elmer, with frequency of 10 Hz, strain of ×4, and temperature rise of 2 °C min^{-1}. The rectangular specimens with moisture content around 15% were tested using a dual-cantilever bending mode on a standard bending head. The chamber surrounding the specimen at 65% RH was cooled by liquid nitrogen. The system provided a simple thermal scan at 2 °C min^{-1} with various temperatures ranging from 0 to 200 °C. T_g was determined from the graph of tan δ versus T.

2.4.5 Thermogravimetric Analysis (TGA)

Thermogravimetric Analysis (TGA) measurements were carried out on 5–10 mg of AN-*co*-BMA-HNC WPNCs at a heating rate of 10 °C/min in a nitrogen atmosphere using a thermogravimetric analyzer (TA Instrument SDT Q600) supplied by TA Instruments. AN-*co*-BMA-HNC WPNCs were subjected to TGA in high-purity nitrogen under a constant flow rate of 5 mL/min. Thermal decomposition of each sample occurred in a programmed temperature range of 30–700 °C. The continuous weight loss and temperature were recorded and analyzed.

2.4.6 Differential Scanning Calorimetry (DSC) Analysis

AN-*co*-BMA-HNC WPNCs were analyzed using a Perkin Elmer thermal analyzer. All the measurements were made under a N_2 flow (30 mL/min), keeping a constant heating rate of 10 °C/min.

2.4.7 Moisture Absorption Test

Moisture absorption was carried out using electronic moisture balance (MOC-120H) supplied by Shimadzu Corporation, Kyoto, Japan. Dry wood samples (dried at 25 °C) were immersed in distilled water. AN-*co*-BMA-HNC WPNCs were removed and were placed on the pan and weighted. The moisture water absorbed, W_{ab} was calculated by using Eq. (4).

Moisture absorbed percentage, W_{ab} (%)

$$W_{ab} = \frac{W_w - W_d}{W_d} \times 100 \tag{4}$$

where

W_{ab} absorbed weight,
W_w weight of wet WPNCs,
W_d weight of dry WPNCs.

3 Results and Discussion

3.1 Weight Percent Gain (WPG %)

The values of WPG for RW and AN-*co*-BMA-HNC WPNCs were measured before and after modification, as given in Table 2. Results indicated that WPG significantly increased due to the ratios of AN-*co*-BMA-HNC that reacted well with the hydroxyl groups in wood fiber (Rahman et al. 2013). Besides, AN-*co*-BMA-HNC polymer matrix aids in removing the impurities from the wood fiber surfaces, which enhanced the adhesion between wood fibers and polymer matrix, resulting in a higher WPG. According to the findings, the 50:50 AN-*co*-BMA-HNC WPNCs had the highest WPG compared with the other monomer systems.

3.2 Fourier Transform Infrared Spectroscopy (FT-IR)

Figure 1 showed the infrared spectra of RW and AN-*co*-BMA-HNC WPNCs. From Fig. 1, RW showed the highest peak intensity at 3300 cm^{-1} compared to WPNCs. This peak was related to the hydroxyl groups stretching vibration in wood fiber

Table 2 Average weight percent gain, WPG (%) of RW and different ratio of AN-*co*-BMA-HNC WPNCs

Samples	Average weight before impregnation (g)	Average weight after impregnation (g)	Weight percent gain, WPG (%)
RW	70.756	71.023	0.376
AN-HNC WPNCs	57.880	61.310	5.595
50:50 AN-*co*-BMA-HNC WPNCs	46.654	56.502	17.429
70:30 AN-*co*-BMA-HNC WPNCs	62.170	70.924	12.343
BMA-HNC WPNCs	66.488	69.789	4.730

(Han et al. 2010). The appearance of a specific peak at 2240 cm^{-1} was observed in all WPNCs, which were associated with the –CN groups stretching vibration of acrylonitrile. The peak at about 2360 cm^{-1} was clearly observed which was attributed to the carbon dioxide absorption band. The high-intensity peak at 1745 cm^{-1} was detected in both RW and WPNCs due to the carboxyl groups in the C=O stretching vibration (Han et al. 2010). The peak at 1200 cm^{-1} was clearly shown in Fig. 1 that was associated with C–O stretching vibration in all the samples. Overall, it could be concluded that the AN-*co*-BMA-HNC was well introduced into the RW by reducing the hydroxyl groups as well as forming stronger chemical bonds between the polymer matrix and wood fiber. Besides, 50:50 AN-*co*-BMA-HNC was the optimum ratio to be impregnated to enhance the RW. The reaction scheme of AN-*co*-BMA-HNC with wood fiber was shown in Fig. 2.

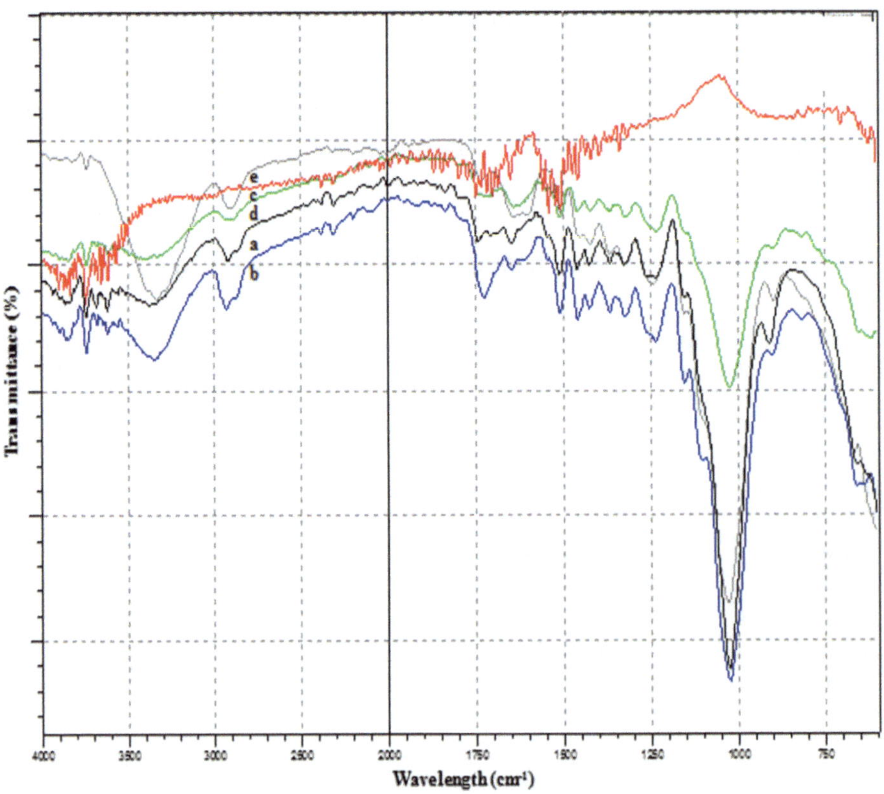

Fig. 1 FT-IR spectra of **a** RW, **b** AN-HNC WPNCs, **c** 50:50 AN-*co*-BMA-HNC WPNCs, **d** 70:30 AN-*co*-BMA-HNC WPNCs and **e** BMA-HNC WPNCs

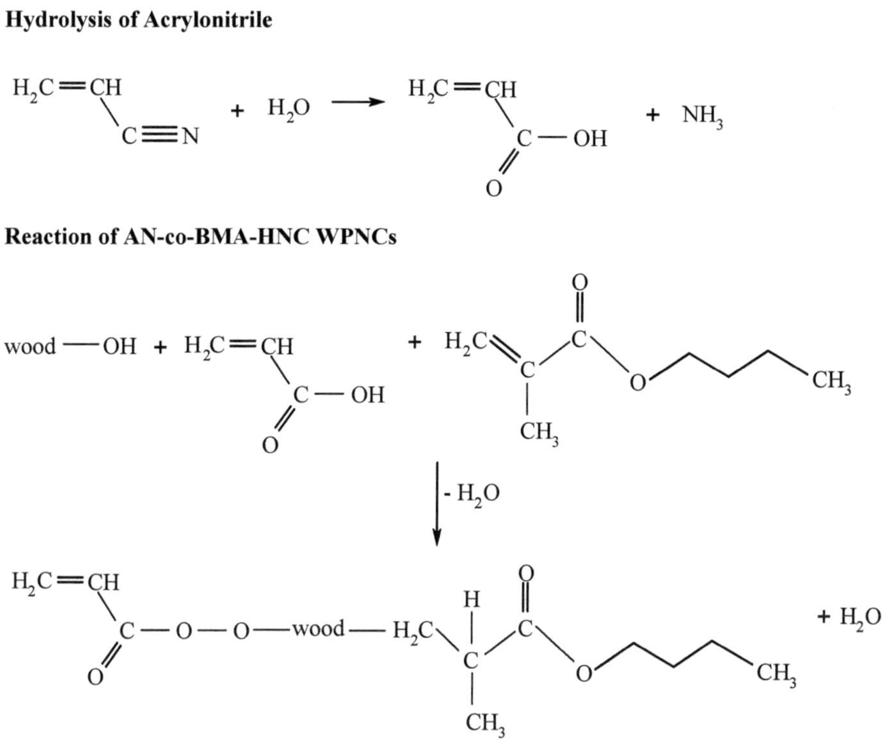

Fig. 2 Proposed schematic reaction diagram

3.3 *Scanning Electron Microscopy (SEM) Analysis*

The surface morphology of RW and AN-*co*-BMA-HNC WPNCs were investigated by Scanning Electron Microscopy (SEM) as shown in Fig. 3a–e. Figure 3a showed that the surface of RW contained a large number of void spaces which represented the unfilled cell wall and pore volume that absorbed moisture as reflected in moisture absorption test (Sultan et al. 2016). AN-HNC WPNCs showed rough and non-uniform wood surface as the AN-HNC partially filled the void spaces as presented in Fig. 3b. The void spaces of RW filled up on the impregnation of 50:50 AN-*co*-BMA-HNC, and 70:30 AN-*co*-BMA-HNC provided smoother wood surfaces as shown in Fig. 3c, d. The addition of halloysite nanoclay enhanced the impregnation of AN-*co*-BMA polymer matrix into the empty pits and capillaries of wood fiber (Hazarika and Maji 2014). Figure 3e confirmed the impregnation of BMA-HNC into RW. However, there were small void spaces detected due to the poor interaction between the polymer matrix and wood fiber (Li et al. 2011). Halloysite nanoclay crosslinked with all the polymer matrix to improve the interaction between polymer matrix and wood fiber (Sekharnath et al. 2015). Therefore, 50:50 AN-*co*-BMA-HNC and 70:30 AN-*co*-BMA-HNC showed better surface morphology which proved the stronger interaction between polymer matrix and wood fiber as well as the compatibility of polymer matrix impregnation to be applied in RW, as reflected in thermal and mechanical properties.

3.4 Three-Point Flexural Test

Figure 4 showed flexural properties of RW and AN-*co*-BMA-HNC WPNCs. A significant increment was observed when the AN-*co*-BMA-HNC was impregnated into the RW. This was due to the greater intercalation between the wood fiber and polymer matrix (Abdul Khalil et al. 2014). The value of flexural strength of RW, AN-HNC, 50:50 AN-*co*-BMA-HNC, 70:30 AN-*co*-BMA-HNC, and BMA-HNC WPNCs were 0.0053, 0.0653, 0.0950, 0.0668, and 0.0429 GPa, respectively. The well impregnation of AN-*co*-BMA-HNC reduced the formation of void on the wood surface that enhanced the stress transfer between the polymer matrix and fiber (Moodley 2007). With reduction of void formation, the effective cross-section area of the wood to interact with polymer matrix was greatly improved, which resulted in enhanced flexural strength of WPNCs compared to RW.

Modulus of elasticity of RW and AN-*co*-BMA-HNC WPNCs was presented in Fig. 5. It showed that optimum modulus of elasticity could be achieved at 50:50 AN-*co*-BMA-HNC. This was due to the well intercalation between AN-*co*-BMA-HNC and wood fiber that led to sufficient load transfer and enhanced the modulus of elasticity of WPNCs compared to RW (Kuan et al. 2004). Besides, small particle of AN-*co*-BMA-HNC had higher surface area, uniform surface structure, and good dispersion provided better adhesion strength to transfer stresses and reduced elastic deformation from AN-*co*-BMA-HNC to wood fiber (Zhang et al. 1993). Therefore, it was clearly showed that 50:50 AN-*co*-BMA-HNC WPNCs had higher flexural strength and modulus of elasticity compared to RW and other WPNCs.

3.5 Dynamic Mechanical Thermal Analysis (DMTA)

The dynamic mechanical thermal analysis (DMTA) properties such as storage modulus (E'), loss modulus (E''), and loss tangent ($\tan\delta$) were recorded over temperature from 30–200 °C as shown in Figs. 6, 7 and 8. The damping property is represented by the tan δ associated to the molecular motions and the bonding between the interface of the polymer matrix and wood fiber (Parshaei and Hosseinzadeh 2016).

The storage modulus of the RW and WPNCs were presented in Fig. 6. WPNCs especially 50:50 AN-*co*-BMA-HNC WPNCs were about 58% higher than RW. The storage modulus of WPNCs increased with the addition of polymer matrix which

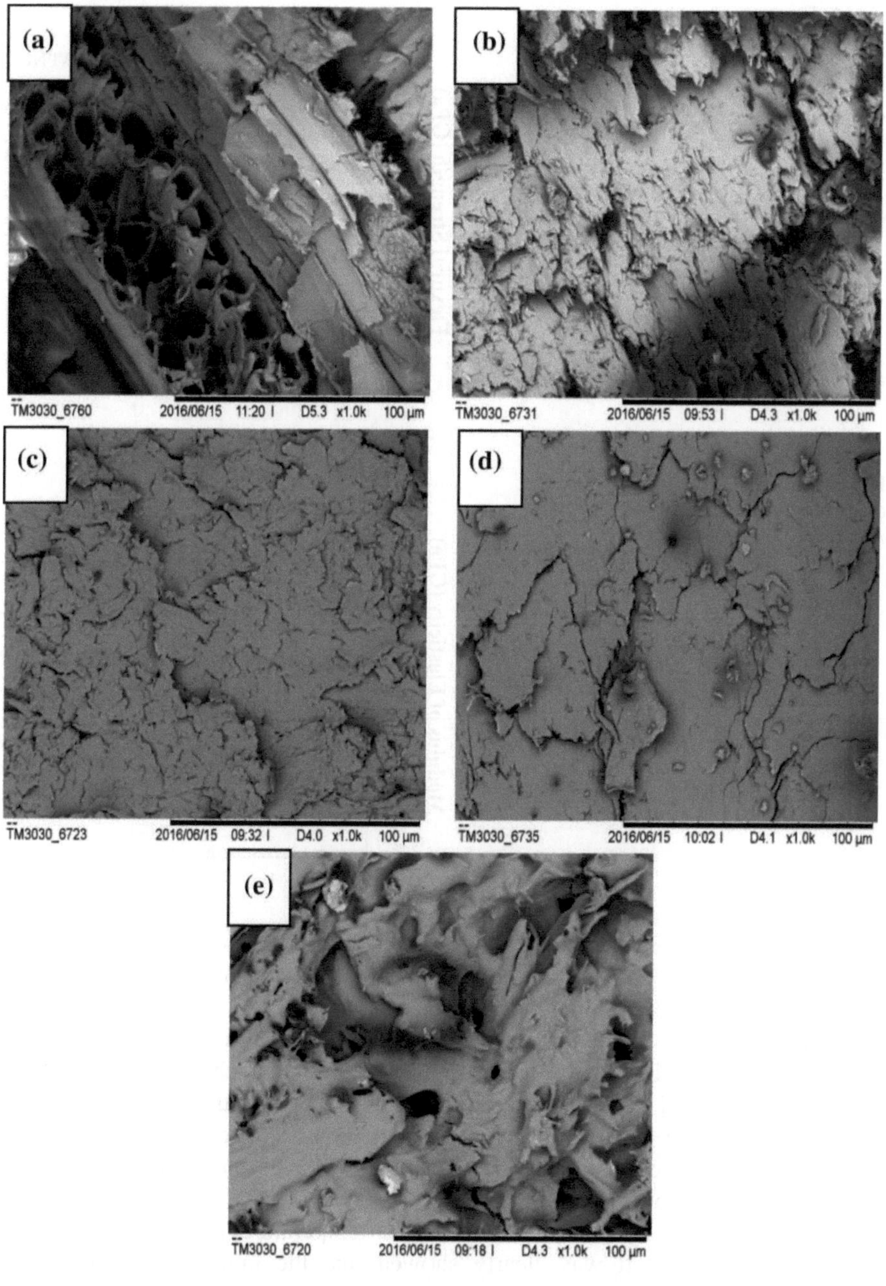

Fig. 3 SEM micrographs of **a** RW, **b** AN-HNC WPNCs, **c** 50:50 AN-*co*-BMA-HNC WPNCs, **d** 70:30 AN-*co*-BMA-HNC WPNCs and **e** BMA-HNC WPNCs

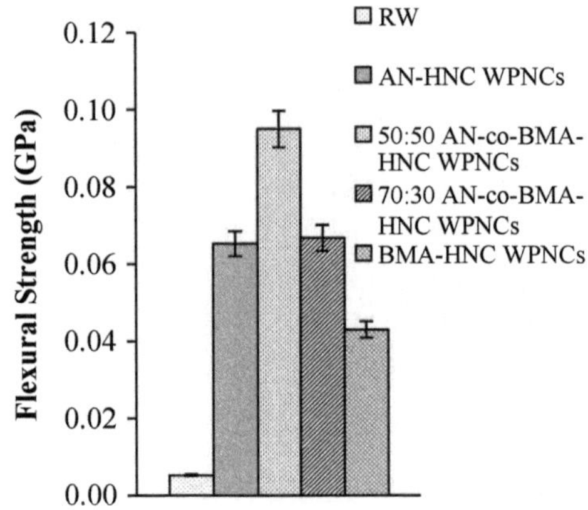

Fig. 4 Flexural strength of RW and different ratio of AN-*co*-BMA-HNC WPNCs

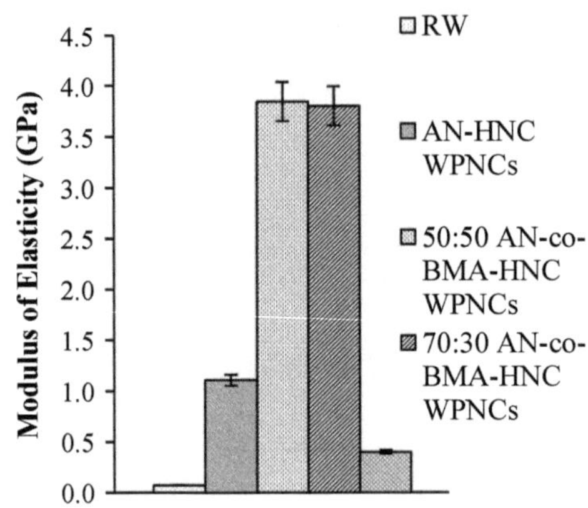

Fig. 5 Modulus of elasticity of RW and different ratio of AN-*co*-BMA-HNC WPNCs

improved the stiffness of the WPNCs (Parshaei and Hosseinzadeh 2016). The existence of strong interaction between the polymer matrix and wood fiber improved the storage modulus of WPNCs.

Figure 7 showed that 50:50 AN-*co*-BMA-HNC WPNCs had the highest loss modulus corresponding to the temperature. The polymer chains at the interface were tightly bound and their mobility was highly restricted (Parshaei and Hosseinzadeh 2016). Thus, larger viscous dissipation occurs easier which enhanced the impregnation into the wood fiber and thus improved the interface of AN-*co*-BMA-HNC and wood fiber.

Figure 8 showed tan δ as a function of temperature for RW and AN-*co*-BMA-HNC WPNCs. It was clearly showed that the impregnation of AN-*co*-BMA-HNC into wood fiber led to lower tan δ compared to RW. The addition of HNC with polymers shifted the glass transition temperature to a higher value. 50:50 AN-*co*-BMA-HNC WPNCs shifted the temperature from 50 to 80 °C, and the peak of the tan δ were broader compared to RW and other WPNCs.

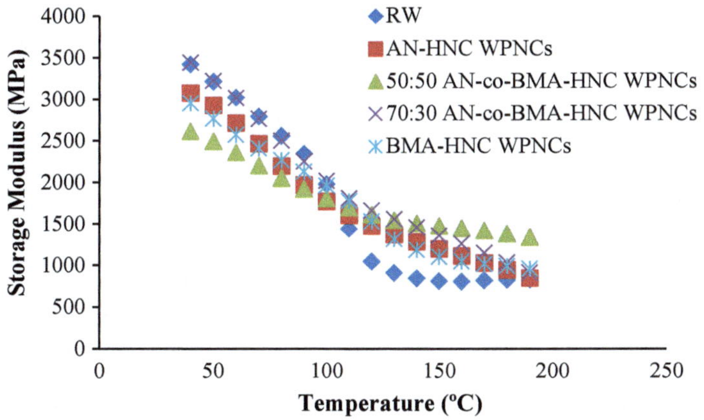

Fig. 6 Storage modulus versus temperature of RW and different ratio of AN-*co*-BMA-HNC WPNCs

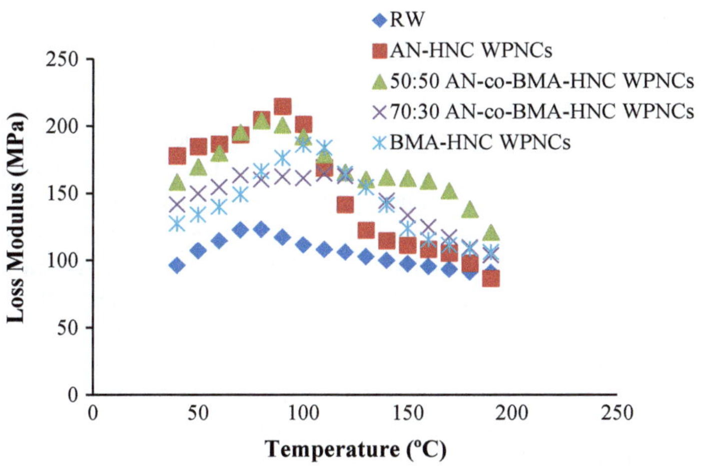

Fig. 7 Loss modulus versus temperature of RW and different ratio of AN-*co*-BMA-HNC WPNCs

Overall, storage modulus and loss modulus of WPNCs increased with the decreasing of loss tangent with the appropriate polymer matrix, as reflected in three-point flexural test.

3.6 Thermogravimetric Analysis (TGA)

The thermal stability of RW and AN-*co*-BMA-HNC WPNCs were carried out using TGA. There were three-step weight losses in AN-*co*-BMA-HNC WPNCs. The first

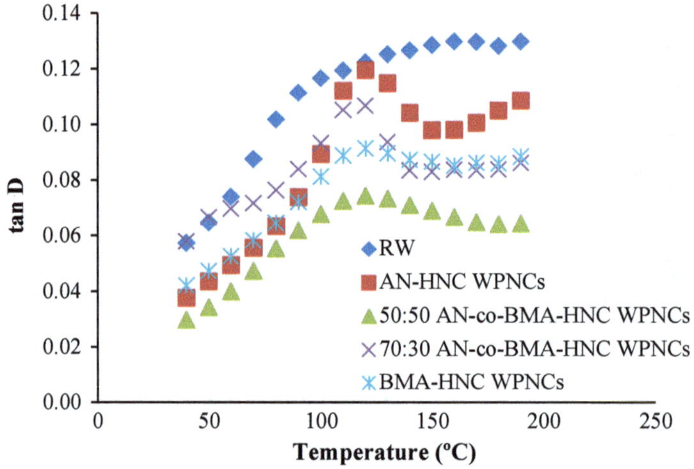

Fig. 8 Tangent δ versus temperature of RW and different ratio of AN-*co*-BMA-HNC WPNCs

step degradation, namely vaporization of moisture occurred up to 250 °C. This degradation occurred due to the changes of HCN and NH_3 (Bahrami et al. 2003). Second step degradation occurred between 250 and 400 °C due to the cyclization and oxidation reactions as well as the formation of ladder polymer structure in the polyacrylonitrile molecules (Gupta et al. 1995). The last step degradation started at 400–600 °C which was related to the residues of RW and WPNCs.

The weight loss in first step was 3.7, 5.5, 5.3, 5.7, and 5.7 wt% for RW, AN-HNC, 50:50 AN-*co*-BMA-HNC-clay, 70:30 AN-*co*-BMA-HNC-clay, and BMA-HNC WPNCs, respectively. For second step degradation, RW, AN-HNC, 50:50 AN-*co*-BMA-HNC, 70:30 AN-*co*-BMA-HNC, and BMA-HNC WPNCs showed weight loss about 45.2, 39.9, 39.6, 39.5, and 39.7 wt%. The last step degradation of RW, AN-HNC, 50:50 AN-*co*-BMA-HNC, 70:30 AN-*co*-BMA-HNC, and BMA-HNC WPNCs showed their weight loss was 13.1, 18.8, 26.9, 22.9, and 18.9 wt%.

Figure 9 showed that AN-*co*-BMA-HNC WPNCs had higher weight percent loss compared to RW. This was due to the effect of acrylate group that interrupted the nitrile sequence along the acrylonitrile which aided in forming a new chemical bond with butyl methacrylate and halloysite nanoclay (Md Jamil et al. 2014). This bond which was linked together impregnated into wood fiber to enhance the chain mobility of the WPNCs, and thus improved the thermal stability of WPNCs.

Table 3 showed the activation energy of AN-*co*-BMA-HNC WPNCs using Arrhenius equation (Chanmal and Jog 2008). It was found that the activation energy of AN-*co*-BMA-HNC WPNCs was significantly higher than RW and other WPNCs. The improved thermal stability of AN-*co*-BMA-HNC WPNCs was attributed to the clay acted as compatibilizer that retarded the motion of polymer chain (Benlikaya et al. 2009).

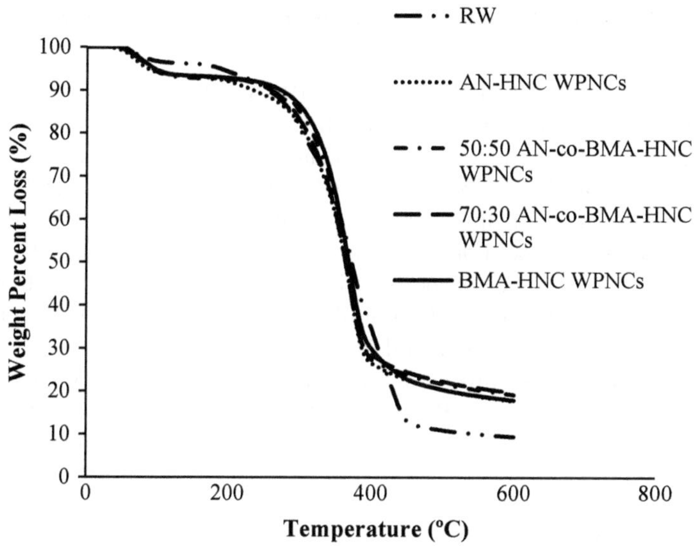

Fig. 9 TGA curves of RW and different ratio of AN-*co*-BMA-HNC WPNCs

Table 3 Activation energy of RW and different ratio of AN-*co*-BMA-HNC WPNCs determined by Arrhenius equation

Samples	RW	AN-HNC WPNCs	50:50 AN-*co*-BMA-HNC WPNCs	70:30 AN-*co*-BMA-HNC WPNCs	BMA-HNC WPNCs
T_i (°C)[a]	84.0	96.0	97.0	97.0	96.0
T_m (°C)[b]	360.0	362.0	366.0	366.0	364.0
T_f (°C)[c]	456.0	474.0	486.0	486.0	474.0
W_{Ti} (%)[d]	96.3	94.5	94.7	94.3	94.3
W_{Tm} (°C)[e]	51.1	54.6	55.1	54.8	54.6
W_{Tf} (°C)[f]	13.1	18.8	26.9	22.9	18.9
Activation energy, E_a (kJ/mol)	1120.9	1607.6	2021.1	1631.9	1613.2

[a]Temperature corresponding to the beginning of decomposition
[b]Temperature corresponding to the maximum rate of mass loss
[c]Temperature corresponding to the end of decomposition
[d]Mass loss at temperature corresponding to the beginning of decomposition
[e]Mass loss at temperature corresponding to the maximum rate of mass loss
[f]Mass loss at temperature corresponding to the end of decomposition

Overall, the thermal stability of AN-*co*-BMA-HNC WPNCs was higher due to the optimum ratio of AN-*co*-BMA-HNC. The impregnation of AN-*co*-BMA-HNC into wood fiber successfully improved thermal stability of AN-*co*-BMA-HNC WPNCs, as reflected in other results.

3.7 Differential Scanning Calorimetry (DSC) Analysis

The thermal behavior of the RW and AN-*co*-BMA-HNC WPNCs were determined using DSC analysis. DSC identifies the chemical activity occurring in the wood fiber as the temperature increased (Hamdan et al. 2010). Differential scanning analysis curves of RW and AN-*co*-BMA-HNC WPNCs are shown in Fig. 10. The exotherm peaks and enthalpy were shown in Table 4. Both RW and AN-*co*-BMA-HNC WPNCs showed broad endotherm between 120 and 160 °C which was associated with the existence of moisture in the wood. This confirmed that WPNCs were thermally stable compared to RW. Another peak at 210 °C proved the degradation of wood fiber (Akita and Kase 1967).

From Table 4, the crystallization enthalpy of all WPNCs was higher than RW, which proved that the AN-*co*-BMA-HNC-filled wood samples were more crystalline compared to RW, which led to better thermal stability of the WPNCs.

3.8 Moisture Absorption Analysis

Moisture absorption characteristics of AN-*co*-BMA-HNC WPNCs are presented in Fig. 11. The moisture absorption increased until a certain saturation point as the time taken increased. The hydrophilic characteristics of RW act as a limiting factor in the final application of the WPNCs. The amount of moisture absorption was

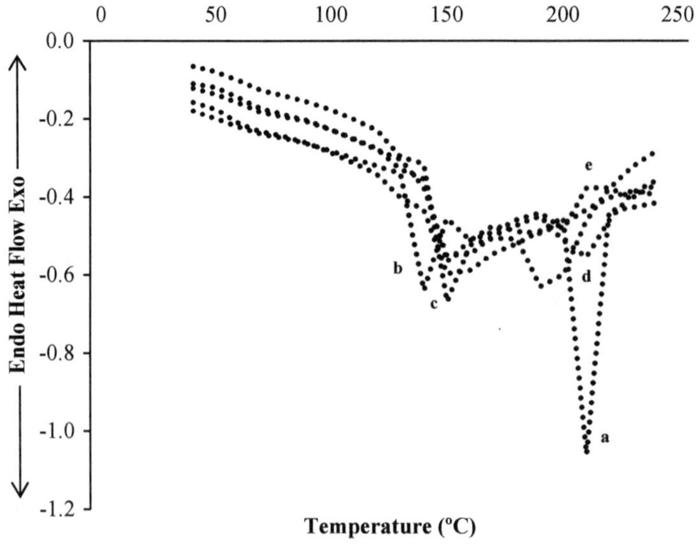

Fig. 10 DSC curves of **a** RW, **b** AN-HNC WPNCs, **c** 50:50 AN-*co*-BMA-HNC WPNCs, **d** 70:30 AN-*co*-BMA-HNC WPNCs and **e** BMA-HNC WPNCs

Table 4 DSC of RW and different ratio of AN-*co*-BMA-HNC WPNCs

Samples	Exotherm peaks (°C)	Crystallization enthalpy (J/g)
RW	139.33	7.64
AN-HNC WPNCs	145.81	21.40
50:50 AN-*co*-BMA-HNC WPNCs	158.98	59.59
70:30 AN-*co*-BMA-HNC WPNCs	204.74	42.67
BMA-HNC WPNCs	201.40	41.23

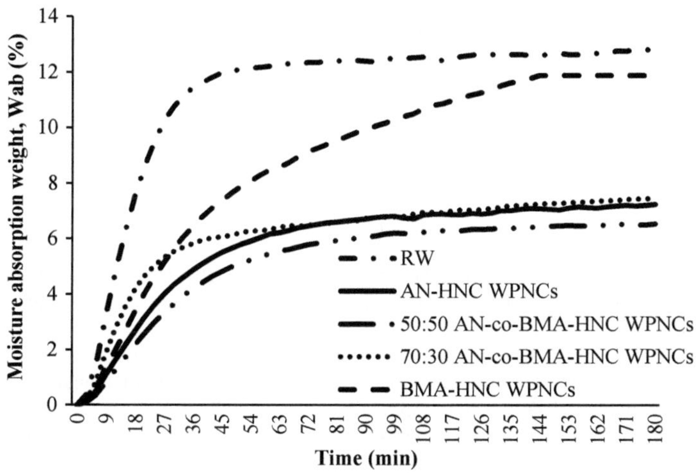

Fig. 11 Moisture absorption curves of RW and different ratio of AN-*co*-BMA-HNC WPNCs

strongly dependent on the molecular structure formed in WPNCs that were clearer over the time. The impregnation of AN-*co*-BMA-HNC into RW reduced the amount of moisture absorption in the WPNCs. This was due to new chemical bonding between the polar or hydrophilic nature of the lignocellulosic fibers and the hydrophobic nature of the polymer matrix that aided in reducing the moisture absorption in all the WPNCs (Malakani et al. 2015). Therefore, 50:50 AN-*co*-BMA-HNC WPNCs showed the lowest moisture absorption, followed by 70:30 AN-*co*-BMA-HNC WPNCs, BMA-HNC WPNCs, AN-HNC WPNCs, and RW, respectively.

4 Conclusion

From the present study, RW was impregnated with AN-*co*-BMA-HNC to produce WPNCs. It is concluded that the physical, mechanical, morphological, and thermal properties of AN-*co*-BMA-HNC WPNCs were greatly improved compared to the

RW. 50:50 AN-*co*-BMA-HNC WPNCs showed the highest weight percent gain with the least hydroxyl group in FT-IR. 50:50 AN-*co*-BMA-HNC WPNCs provided smoother morphology among all the samples. 50:50 AN-*co*-BMA-HNC WPNCs showed the highest flexural strength and modulus of elasticity. The storage modulus and loss modulus of AN-*co*-BMA-HNC WPNCs improved greatly with lower tan δ compared to RW. 50:50 AN-*co*-BMA-HNC WPNCs exhibited higher thermal stability with significantly lower moisture absorption. This has proved that the ratio 50:50 AN-*co*-BMA was the most suitable to be introduced in the RW to form WPNCs.

Acknowledgements The authors would like to acknowledge the financial support from Research and Innovation Management Centre, Universiti Malaysia Sarawak under Fund with Grant no. (F02/SpGS/1443/2016/25) during the research.

References

Abdul Khalil HPS, Jawaid M, Firoozian P, Alothman OY, Paridah MT, Zainudin ES (2014) Flexural properties of activated carbon filled epoxy nanocomposites. Malays J Anal Sci 18(2):391–397

Akita K, Kase M (1967) Determination of kinetic parameters for pyrolysis of cellulose and cellulose treated with ammonium phosphate by differential thermal analysis and thermal gravimetric analysis. J Polym Sci 5:833–848

Ashori A (2008) Wood–plastic composites as promising green-composites for automotive industries. Biores Technol 99:4661–4667

Bahrami SH, Bajaj P, Sen K (2003) Thermal behaviour of acrylonitrile carboxylic acid copolymers. J Appl Polym Sci 88:685–698

Bao YZ, Zhao WT, Huang ZM (2013) Preparation of mesoporous carbons from ac rylonitrile-methyl methacrylate copolymer/silica nanocomposites synthesized by in-situ emulsion polymerization. Chin J Chem Eng 21(6):691–697

Benlikaya R, Alkan M, Kaya I (2009) Preparation and characterization of sepiolite poly(ethyl methacrylate) and poly(2-hydroxyethyl methacrylate) nanocomposites. Polym Comp 30: 1585–1594

Bodirlau R, Teaca CA, Spiridon I (2009) Preparation and characterization of composites comprising modified hardwood and wood poly mers/poly(vinyl chloride). BioRes 4(4): 1285–1304

Chanmal CV, Jog JP (2008) Dielectric relaxations in PVDF/$BaTiO_3$ nanocomposites. eXPRESS Polym Lett 2(4):294–301

Gupta AK, Paliwal DK, Bajaj P (1995) Effect of an acidic comonomer on thermooxidative stabilization of polyacrylonitrile. J Appl Polym Sci 58:1161–1174

Hamdan S, Rahman MR, Ahmed AS, Talib ZA, Islam MS (2010) Influence of N, N-dimethylacetamid on the thermal and mechanical properties of polymer-filled wood. BioRes 5(4):2611–2624

Han N, Zhang XX, Wang XC (2010) Various conomoners in acrylonitrile based copolymers: effects on thermal behaviour. Iran Polym J 19(4):243–253

Hazarika A, Maji TK (2014) Modification of softwood by monomers and nanofillers. Defence Sci J 64(3):262–272

Herrera-Alonso JM, Sedlakova Z, Marand E (2010) Gas barrier properties of nanocomposites based on in situ polymerized poly(n-butyl methacrylate) in the presence of surface modified montmorillonite. J Memb Sci 349(1–2):251–257

Kallakas H, Shamim MA, Olutubo T, Poltimae T, Suld TM, Krumme A, Kers S (2015) Effect of chemical modification of wood flour on the mechanical properties of wood-plastic composites. Argo Res 13(3):639–653

Kobayashi M, Rharbi Y, Brauge L, Cao L, Winnik MA (2002) Effect of silica as fillers on polymer interdiffusion in poly(butyl methacrylate) latex films. Macromol 35(19):7387–7399

Kuan HC, Ma CCM, Chen KH, Chen SM (2004) Preparation, electrical, mechanical and thermal properties of composite bipolar plate for a fuel cell. J Power Source 134(1):7–17

Li YF, Liu YX, Wong XM, Wu QL, Yu HP, Li J (2011) Wood-polymer com posites prepared by in situ polymerization of monomers within wood. J Appl Polym Sci 119(6):3207–3216

Malakani M, Bazyar B, Talaiepour M, Hemmasi AH, Ghasemi I (2015) Effect of acetylation of wood flour and MAPP content during compounding on physical properties, decay resistance, contact angle, and morphology of polypropylene/wood flour composites. BioRes 10(2): 2113–2129

Md Jamil SNA, Daik R, Ahmad I (2014) Synthesis and thermal properties of acrylonitrile/butyl acrylate/fumaronitrile and acrylonitrile/ethyl hexyl acrylate/fumaronitrile terpolymers as a potential precursor for carbon fiber. Mater 7:6207–6223

Mekhemer WK, El-Ala AAA, El-Rafey E (2006) Clay as a filler in the thermoplastic compounding. Mol Crystals Liq Crystals Sci Technol 354(1):13–21

Mohy Eldin MS, Elaassar MR, Elzatahry AA, Al-Sabah MMB (2014) Poly(acrylonitrile-*co*-methyl methacrylate) nanoparticles: I. Arabian J Chem, Preparation and characterization. doi:10.1016/j.arabjc.2014.10.037

Moodley VK (2007) The synthesis, structure and properties of polypropylene nanocomposites. Durban University of Technology, South Africa

Pakeyangkoon P, Ploydee B (2013) Mechanical properties of acrylate-styrene-acrylonitrile/bagasse composites. Adv Mater Res 747:353–358

Parshaei S, Hosseinzadeh S (2016) Preparation of organo nanoclay incorporated polyamide/melamine cyanurate/nanoclay composites and study on thermal and mechanical behaviours. Iran Chem Comm 4:102–114

Pour RH, Soheilmoghaddam M, Hassan A, Bourbigot S (2015) Flammability and thermal properties of polycarbonate/acrylonitrile-butadiene-styrene nanocomposites reinforced with multilayer graphene. Polym Degrad Stab 120:88–97

Prochon M, Przepiorkowska A, Zaborki M (2007) Keratin as a filler for car boxylated acrylonitrile-butadiene rubber XNBR. J Appl Polym Sci 106(6):3674–3687

Rahman MR, Lai JCH, Hamdan S, Ahmed AS, Baini R, Saleh SF (2013) Combined styrene/MMA/nanoclay cross-linker effect on wood-polymer composites (WPCs). BioRes 8(3):4227–4237

Sekharnath KV, Rao SJ, Maruthi Y, Babu PK, Rao KC, Subha MCS (2015) Halloysite nanoclay-filled blend membranes of sodium carboxyl methyl cellulose/hydroxyl propyl cellulose for pervaporation separation of water- isopropanol mixtures. Ind J Adv Chem Sci 3(2):160–170

Siddiqui MN, Redhwi HH, Gkinis K, Achilias DS (2013) Synthesis and char acterization of novel nanocomposite materials based on poly(styrene-*co*- butyl methacrylate) copolymers and organo modified clay. Eur Polym J 49(2):353–365

Singh P, Ghosh AK (2014) Torsional, tensile and structural properties of acrylonitrile-butadiene-styrene clay nanocomposites. Mater Des 55:137–145

Smitinand T, Larsen K (1984) Flora of Thailand volume four part one. TISTR Press, Bangkok

Stark MN, Matuana ML (2007) Characterization of weathered wood-plastic composite surfaces using FTIR spectroscopy, contact angle and XPS. Polym Degrad Stab 92:1883–1890

Sultan MT, Rahman MR, Hamdan S, Lai JCH, Talib ZA (2016) Clay dispersed styrene-*co*-glycidyl methacrylate impregnated kumpang wood polymer nanocomposites: Impact on mechanical and morphological properties. BioRes 11(3):6649–6662

Tesinova P (2011) Advances in composite materials—analysis of natural and man-made material. InTech, European Union, p 584

Vorreiter L (1949) Holztechnologies Handbuch. Verlag Georg Fromme und Co, Wien, p 548

Zhang M, Zeng H, Zhang L, Lin G, Li RKY (1993) Fracture characteristics of discontinuous carbon fibre-reinforced PPS and PES-C composites. Polym Polym Comp 1:357–365

Studies on the Physical, Mechanical, Thermal and Morphological Properties of Impregnated Furfuryl Alcohol-*co*-Glycidyl Methacrylate/Nanoclay Wood Polymer Nanocomposites

M.R. Rahman, J.C.H. Lai and S. Hamdan

Abstract In this study, physical, morphological, mechanical, and thermal properties of furfuryl alcohol/glycidyl methacrylate/halloysite nanoclay wood polymer nanocomposites (FA-*co*-GMA-HNC WPNCs) were investigated. FA-*co*-GMA-HNC WPNCs were prepared via impregnation method, and the effect of different ratio between the polymers was subsequently investigated. The properties of nanocomposites were characterized using Fourier Transform Infrared Spectroscopy (FT-IR), Scanning Electron Microscopy (SEM), three-point flexural test, dynamic mechanical thermal analysis (DMTA), Thermogravimetric Analysis (TGA), differential Scanning calorimetry (DSC) analysis, and moisture absorption test. The weight percent gain for 50:50 FA-*co*-GMA-HNC WPNCs was the highest compared to raw wood (RW) and other WPNCs. FT-IR results confirmed the polymerization took place in the nanocomposites especially 50:50 FA-*co*-GMA-HNC WPNCs with reducing hydroxyl groups. SEM result revealed that the 50:50 FA-*co*-GMA-HNC WPNCs showed the best surface morphology among all the compositions. Besides, 50:50 FA-*co*-GMA-HNC WPNCs showed the highest flexural strength and modulus of elasticity. The DMA results revealed that the storage modulus and loss modulus of FA-*co*-GMA-HNC WPNCs were higher while the tan δ of FA-*co*-GMA-HNC WPNCs was lower compared to RW. FA-*co*-GMA-HNC WPNCs exhibited the higher thermal stability through TGA and DSC analysis. 50:50 FA-*co*-GMA-HNC WPNCs exhibited significantly lower moisture absorption compared to RW. From the analysis, 50:50 FA-*co*-GMA showed the best compatibility with RW among all the compositions.

Keywords Morphology · Strength · Thermal · Clay

M.R. Rahman (✉) · J.C.H. Lai · S. Hamdan
Faculty of Engineering, Universiti Malaysia Sarawak,
94300 Kota Samarahan, Sarawak, Malaysia
e-mail: rmrezaur@unimas.my

M.R. Rahman, *Wood Polymer Nanocomposites*, Engineering Materials,
DOI 10.1007/978-3-319-65735-6_14

1 Introduction

Wood is one of the natural resources that are important to be used in many applications such as fuels, furniture as well as raw materials for energy (Li et al. 2011). Wood is abundant in quantity, renewable, environmental-friendly and has outstanding mechanical properties (Li 2011). Wood consists of cell walls that are composed of biopolymer mainly cellulose, hemicellulose and lignin. The high strength-to-weight ratio of cellular wood provided it to be widely used as structural materials. However, wood components can be degraded easily due to microorganism attack (Fuller et al. 1997). Besides, the high amount of hydroxyl groups may lead to poor dimensional stability. Due to the disadvantages of wood, the applications on wood usage are limited. Therefore, wood modification is needed to improve the weakness of wood and strengthen the properties of wood.

One of the techniques used to enhance the wood properties is the fabrication of wood polymer nanocomposites (WPNCs) through impregnation. In general, the introduction of wood into polymer matrix improves the mechanical properties of the wood as well as the reducing the moisture attacks compared to raw wood (Baysal et al. 2007). The physical properties could be enhanced when the suitable polymer matrix is impregnated into the wood by strengthening the hydrophilic wood and hydrophobic polymer matrix (Wechsler et al. 2008).

Sindora wood is one of the species that are abundantly available in Borneo Island especially Sarawak. It is widely applied in the production of furniture, agriculture implements, and heave construction. However, there is limited usage of the wood due to the poor physical properties (Smitinand and Larsen 1984).

Furfuryl alcohol (FA) can be obtained by in situ polymerization that acted as a solvent of cellulose acetate butyrate and improved mechanical and thermal properties of the flax fiber nanocomposites (Toriz et al. 2003). Glycidyl methacrylate (GMA) is a monomer that retains its reactivity and the epoxy ring in the final products. GMA is widely used in biodegradable polymers such as polystyrene and polyethylene to improve the interactions between the polymer matrices (Kunita et al. 2006).

Clay is widely used as a crosslinking agent with polymer matrix in the fabrication of WPNCs. However, clay is not compatible with all types of polymers and it tends to agglomerate which leads to poor physical properties (Bouhelal et al. 2007).

From the previous research, it showed that the intercalation of molecules of salt in the clay layers as well as the well intercalation between the furfuryl alcohol and organically modified clay improved the thermal and mechanical properties of the nanocomposites by direct in situ polymerization (Kherroub et al. 2015). Montmorillonite nanoclays acted as matrix modifier in furfuryl alcohol nanocomposites through in situ polymerization had enhanced the thermal performance compared to the pure furfuryl alcohol (Pranger et al. 2012). Poly(furfuryl alcohol)/silica hybrid material had improved both mechanical and thermal stability due to the less molecular mobility compared to the unfilled matrix (Guigo et al. 2009).

The research in WPNCs has been increasing lately. However, there is no comprehensive work has been done on Sindora wood WPNCs. On the consideration of importance of wood modification and the preparation of the WPNCs, this article presents the detailed preparation and the characterizations of all the fabricated WPNCs as well as the raw wood (RW).

2 Experimental

2.1 *Materials*

Defect-free and straight-grained *Sindora* wood (RW) was obtained from Sematan Sarawak, Malaysia. Both control and modified wood samples were obtained from the same timber and machined at the same time with a dimension of 30 cm × 2 cm × 1 cm. The chemicals used to produce WPNCs were furfuryl alcohol (FA), glycidyl methacrylate (GMA), benzoyl peroxide and halloysite nanoclay (HN). Halloysite nanoclay was supplied by Sigma-Aldrich (USA). Furfuryl alcohol, glycidyl methacrylate, and benzoyl peroxide were supplied by Merck Millipore (USA). Halloysite nanoclay was nanopowder with the diameter of 30–70 nm and length between 1 and 3 μm.

2.2 *Preparation of Furfuryl Alcohol/Glycidyl Methacrylate/Halloysite Nanoclay Wood Nanocomposites (WPNCs) (FA-co-GMA-HNC WPNCs)*

The polymer system was prepared using FA-*co*-GMA-HNC in the presence of benzoyl peroxide. Benzoyl peroxide acted as an initiator to influence the reaction between furfuryl alcohol and glycidyl methacrylate. Furfuryl alcohol, glycidyl methacrylate, and halloysite nanoclay were mixed with different ratios as shown in Table 1. The mixture was covered with aluminum foil and placed in autoclave for 15 min to complete the reaction.

Table 1 Preparation of polymer system at different ratio

Volume of furfuryl alcohol (FA) (mL)	Volume of glycidyl methacrylate (GMA) (mL)	Amount of halloysite clay (HNC) (g)	Amount of benzoyl peroxide (g)
0	200	2	5
100	100	2	5
140	60	2	5
200	0	2	5

2.3 *Impregnation of FA-co-GMA-HNC WPNCs*

Five wood samples prepared according to ASTM D1037 were weighed using an electronic balance. The samples were then placed in the vacuum chamber. The mixtures were poured into the vacuum oven. The vacuum oven was operated in the pressure of 0.1 bar for 60 min. After impregnation, the excessive solution was removed from the wood samples with tissue paper. Wood samples were wrapped with aluminum foil and kept in an oven at 100 °C for 24 h for polymerization. After polymerization, the aluminum foil was removed and WPNCs were dried at 105 °C until constant weight was obtained. The weight percent gain (WPG) of all the WPNCs was then measured using Eq. (1),

$$\mathrm{WPG} = \frac{W_\mathrm{f} - W_\mathrm{o}}{W_\mathrm{o}} \times 100\% \qquad (1)$$

where W_f is oven-dried weight after modification of the WPNCs and W_o is oven-dried weight before modification of the WPNCs.

2.4 *Microstructural Characterizations*

2.4.1 Fourier Transform Infrared Spectroscopy (FT-IR)

The infrared spectra of the nanocomposite were recorded on a Shimadzu IRAffinity-1. The technique used was Attenuated Total Reflection (ATR). The spectral resolution of this equipment is 12,000 spectra. The transmittance range of the scan was 4000–600 cm^{-1}.

2.4.2 Scanning Electron Microscopy (SEM)

The interfacial bonding between the furfuryl alcohol, glycidyl methacrylate, and halloysite nanoclay was examined using a Scanning Electron Microscope (SEM) (JSM-6710F) supplied by JEOL Company Limited, Japan. The middle part of all the FA-*co*-GMA-HNC WPNCs was extracted in the dimension of 5 mm × 5 mm × 5 mm (length × width × height). The accelerating voltage of this instrument was 15 kV. The specimens were first fixed with Karnovsky's fixative and then take through a graded alcohol dehydration series. Once dehydrated, the specimen was coated with a thin layer of gold before being viewed microscopically. The micrographs were taken at magnifications ranging from 500 to 1000.

2.4.3 Three-Point Flexural Test

The static flexural tests of the FA-*co*-GMA-HNC WPNCs were carried out using AG-X Plus Series Precision Universal Testers (300 kN Floor Model) supplied by Shimadzu Corporation, Japan. Dimensions of flexural test specimen were 80 mm (length) × 30 mm (width) × 10 mm (thickness). Flexural tests were conducted following ASTM D 7900-00 at a crosshead speed of 10 mm/min. Five specimens of each sample were tested and the average values were reported. The flexural strength (Q_{fs}) and modulus of elasticity (E_m) were calculated using Eqs. (2) and (3).

$$Q_{fs} = \frac{3PL}{2bd^2} \tag{2}$$

where P is maximum load on the load-deflection curve; L is support span, 63.7 mm; b is width of beam tested, 10 mm; and d is thickness of beam, 4 mm.

$$E_m = \frac{L^3 m}{4bd^3} \tag{3}$$

where m is slope of the tangent to the initial straight-line portion of the load-deflection curve, N/mm of deflection.

2.4.4 Dynamic Mechanical Thermal Analysis (DMTA)

Dynamic mechanical thermal analysis measures the effect of temperature on the storage modulus ($\log E'$) and loss tangent ($\tan \delta$) of FA-*co*-GMA-HNC WPNCs. It is a mechanical test that allowed molecules in woods to interact with mechanical stress. This test was carried out using Perkin Elmer dynamic mechanical thermal analyzer (PE-DMTA) with frequency of 10 Hz, strain of ×4 and temperature rise of $2\,°\text{C}\,\text{min}^{-1}$. The rectangular specimens were tested using a dual-cantilever bending mode on a standard bending head. The chamber surrounding the specimen at 65% RH was cooled by liquid nitrogen. The system provided a simple thermal scan at $2\,°\text{C}\,\text{min}^{-1}$ with temperatures ranging from 0 to 200 °C. T_g was determined from the graph of $\tan \delta$ versus T.

2.4.5 Thermogravimetric Analysis (TGA)

Thermogravimetric Analysis (TGA) measurements were carried out on 5–10 mg of FA-*co*-GMA-HNC WPNCs at a heating rate of 10 °C/min in a nitrogen atmosphere using a Thermogravimetric Analyzer (TA Instrument SDT Q600). FA-*co*-GMA-HNC WPNCs were subjected to TGA in high purity nitrogen under a constant flow rate of 5 mL/min. Thermal decomposition of each sample occurred in a programmed temperature range of 0–700 °C. The continuous weight loss and temperature were recorded and analyzed.

2.4.6 Differential Scanning Calorimetry (DSC) Analysis

FA-*co*-GMA-HNC WPNCs were analyzed using a Perkin Elmer thermal analyzer. All the measurements were made under a nitrogen flow (30 mL/min), keeping a constant heating rate of 10 °C/min and using an aluminum crucible with a pinhole.

2.4.7 Moisture Absorption Test

Moisture absorption was carried out using electronic moisture balance (MOC-120H) supplied by Shimadzu Corporation, Kyoto, Japan at 110 °C for 3 h. Dry wood samples (dried at 25 °C) were immersed in distilled water. FA-*co*-GMA-HNC WPNCs were removed and placed on the pan and weighted. The heater lid had been closed firmly. The moisture absorbed percentage W_{ab} was calculated by using Eq. (4),

Moisture absorbed percentage, W_{ab} (%)

$$W_{ab} = \frac{W_w - W_d}{W_d} \times 100 \qquad (4)$$

where W_{ab} is absorbed weight, W_w is weight of wet WPNCs, and W_d is weight of dry WPNCs.

3 Results and Discussion

3.1 Weight Percent Gain (WPG %)

The weight percent gain (WPG) for RW and FA-*co*-GMA-HNC WPNCs is shown in Table 2. Values of WPG showed significant changes over the time. It was observed that WPG of WPNCs was higher than RW. Among all the nanocomposites, 50:50 FA-*co*-GMA-HNC WPNCs showed the highest weight percent gain that showed the impregnation of polymer matrix was higher than RW and other WPNCs (Dong et al. 2012).

Table 2 Average weight percent gain, WPG (%) of RW and different ratio of FA-*co*-GMA-HNC WPNCs

Samples	Average weight before impregnation (g)	Average weight after impregnation (g)	Weight percent gain, WPG (%)
RW	70.756	71.023	0.376
FA-HNC WPNCs	58.703	61.967	5.267
50:50 FA-*co*-GMA-HNC WPNCs	55.005	77.688	29.198
70:30 FA-*co*-GMA-HNC WPNCs	59.131	59.750	15.224
GMA-HNC WPNCs	54.600	55.861	2.257

3.2 Fourier Transform Infrared Spectroscopy (FT-IR)

Figure 1 showed the FT-IR spectra of RW and FA-*co*-GMA-HNC WPNCs over the 4000–600 cm^{-1} wave number range. The absorption peak at 3400 cm^{-1} was clearly detected for RW, FA-HNC, and GMA-HNC WPNCs as the peak was attributed to the stretching vibration of hydroxyl groups. The peaks at 1730 and 1600 cm^{-1} were attributed to C=O stretching vibration due to the ring opening of furan of furfuryl alcohol and C=O aromatic skeletal vibration of lignin, respectively (Ahmad et al. 2013). Another peak observed in Fig. 1 at 1250 cm^{-1} was associated with the C–O stretching vibration (Pandey 1999). The C=C stretching vibration of WPNCs at 1630 cm^{-1} was more significant compared to RW due to the chemical bond formed between FA and GMA (Qin et al. 2013). In addition, the smooth spectrum of 50:50 FA-*co*-GMA-HNC WPNCs showed the well reacted of GMA monomer with FA-*co*-HNC. The IR spectra of 50:50 FA-*co*-GMA-HNC and 70:30 FA-*co*-GMA-HNC WPNCs were similar due to the low amount of halloysite nanoclay particle impregnation. This confirmed that FA-*co*-BMA was successfully polymerized in the presence of halloysite nanoclay and well impregnated into RW (Dong et al. 2014). The reaction scheme was shown in Fig. 2.

3.3 Scanning Electron Microscopy (SEM) Analysis

The SEM micrographs of RW and FA-*co*-GMA-HNC WPNCs were shown in Fig. 3a–e. Figure 3a indicated that there was non-uniform size of void spaces in RW. The polymer matrix (FA-HNC) was successfully impregnated into wood cell wall which is shown in Fig. 3b. However, the clusters were formed and caused agglomeration at the wood surface of FA-HNC WPNCs (Dong et al. 2014). The impregnation of 50:50 FA-*co*-GMA-HNC and 70:30 FA-*co*-GMA-HNC showed

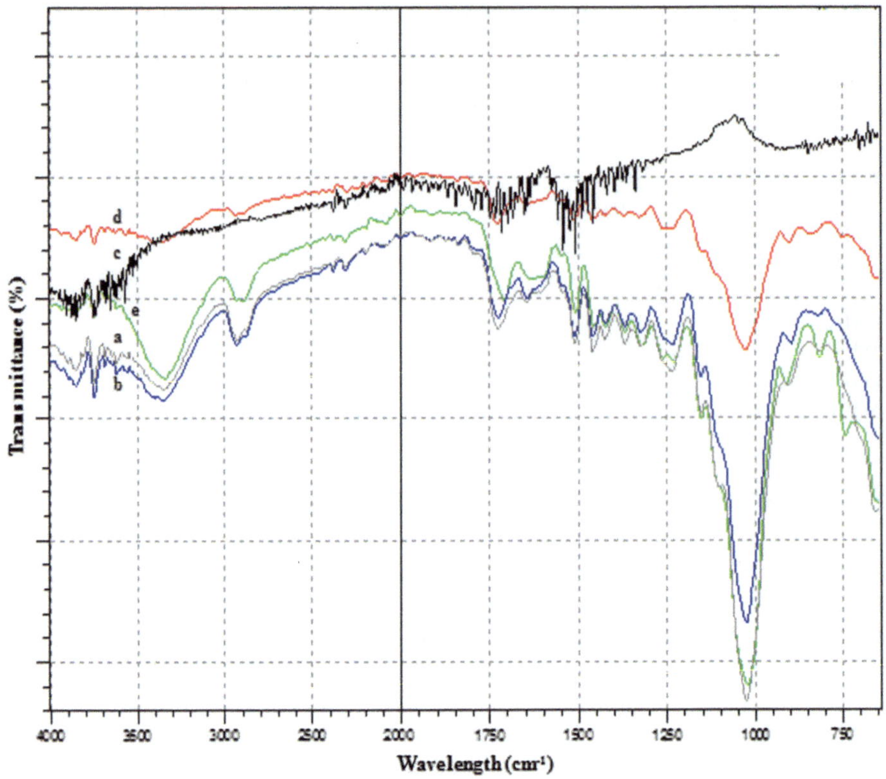

Fig. 1 FT-IR spectra of **a** RW, **b** FA-HNC WPNCs, **c** 50:50 FA-*co*-GMA-HNC WPNCs, **d** 70:30 FA-*co*-GMA-HNC WPNCs and **e** GMA-HNC WPNCs

uniform surfaces by improving the adhesion of polymer matrix to the wood cell wall. The strong wood fiber–polymer matrix adhesion reduced the wood fiber agglomeration as shown in Fig. 3c–d (Viet Cao et al. 2011). From Fig. 3e, the surface morphology of GMA-HNC WPNCs was disoriented with the presence of agglomeration due to poor adhesion between polymer matrix and wood fiber. In addition, halloysite nanoclay acted as compatibilizer to enhance the fiber–matrix interfacial bonding (Xue and Zhao 2008). The improved interfacial adhesion was clearly reflected in both three-point flexural test as well as dynamic mechanical thermal analysis that proved the compatibility of FA-*co*-GMA-HNC into RW.

3.4 Three-Point Flexural Test

Figure 4 summarized the flexural strength of RW and FA-*co*-GMA-HNC WPNCs. WPNCs showed higher flexural strength and modulus of elasticity compared to

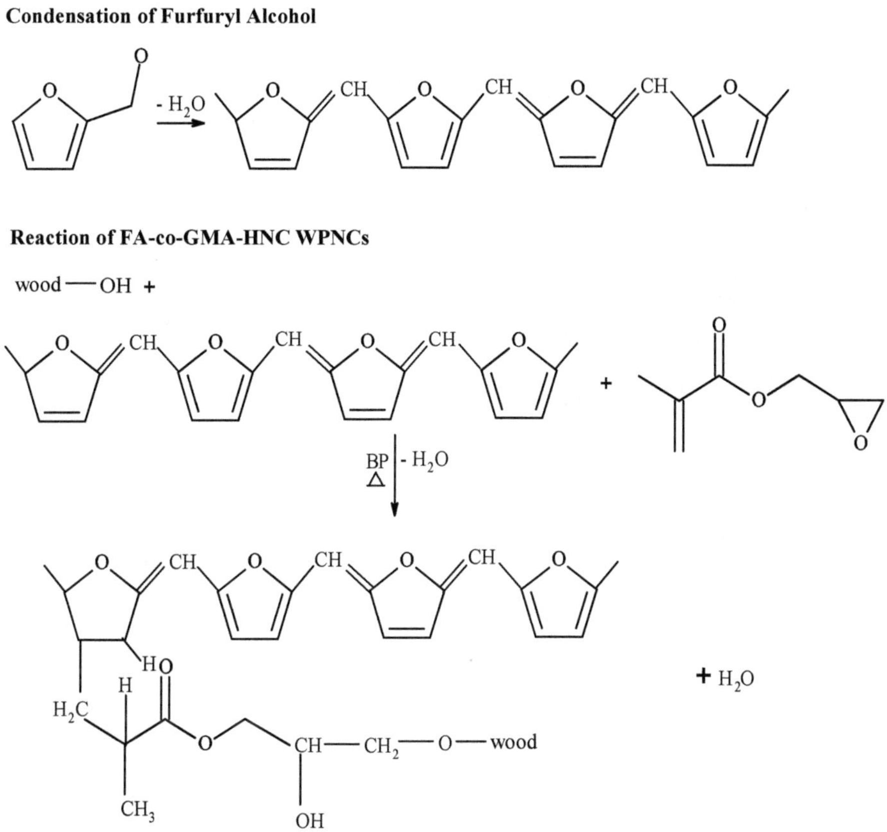

Fig. 2 Proposed schematic reaction diagram

RW. This was due to the formation of chemical bond. Strong interfacial bonding between the FA-*co*-GMA-HNC and wood fiber led to better adhesion and enhanced the bond strength, thus improved the flexural strength of WPNCs. In addition, the uniform distribution of stress from the FA-*co*-GMA-HNC to wood fiber was stronger enough for maximum orientation and hence optimum value of WPNCs was 50:50 FA-*co*-GMA-HNC WPNCs (Singha and Thakur 2008).

It was observed from Fig. 5 that the modulus of elasticity of FA-*co*-GMA-HNC WPNCs was approximately 21% higher than RW. This was due to the reinforcement of the FA-*co*-GMA-HNC with wood fiber that reduced the deformation in length of all WPNCs (Kabir et al. 2010). Besides, the spaces between RW were fully filled by the incorporation of FA-*co*-GMA-HNC molecules. Besides, the impregnation process improved the interfacial bonding between FA-*co*-GMA-HNC and wood fiber.

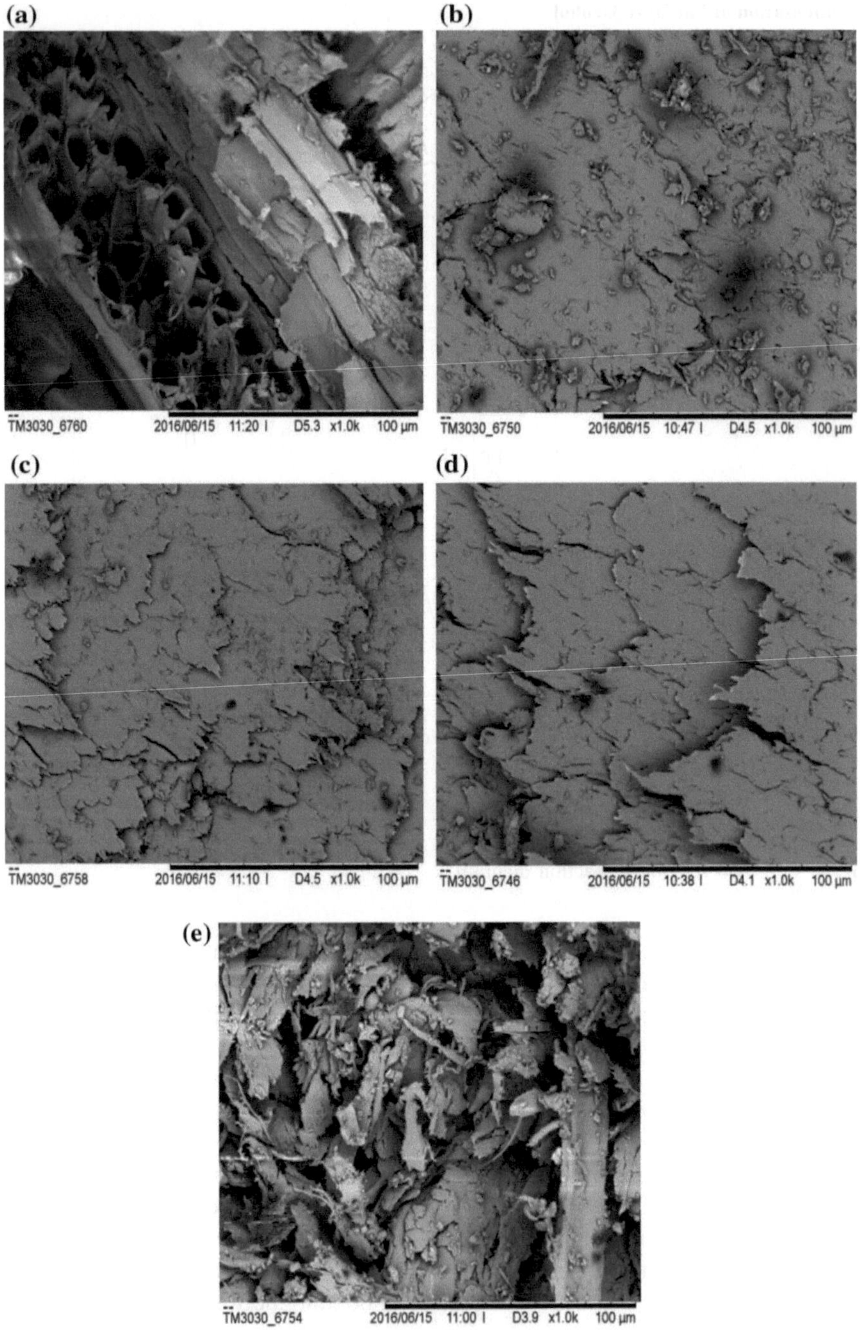

Fig. 3 SEM micrographs of **a** RW, **b** FA-HNC WPNCs, **c** 50:50 FA-*co*-GMA-HNC WPNCs, **d** 70:30 FA-*co*-GMA-HNC WPNCs and **e** GMA-HNC WPNCs

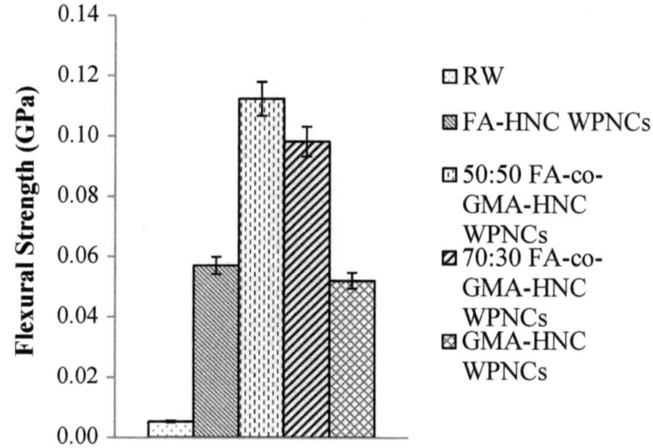

Fig. 4 Flexural strength of RW and different ratio of FA-*co*-GMA-HNC WPNCs

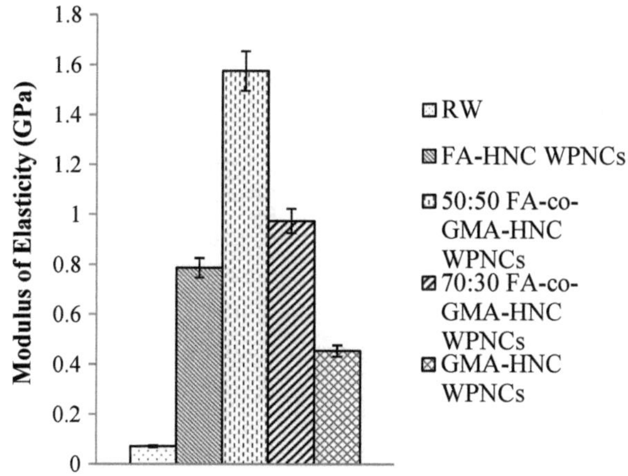

Fig. 5 Modulus of elasticity of RW and different ratio of FA-*co*-GMA-HNC WPNCs

3.5 Dynamic Mechanical Thermal Analysis (DMTA)

Dynamic mechanical thermal analysis (DMTA) was carried out to determine the viscoelastic behavior of RW and FA-*co*-GMA-HNC WPNCs as shown in Figs. 6, 7 and 8. The storage modulus of WPNCs was improved gradually compared to RW as shown in Fig. 6. This showed that strong interfacial bonding between the FA-co-GMA-HNC restricted the free molecular movement inside the WPNCs (Pothan et al. 2010).

Figure 7 showed that 50:50 FA-*co*-GMA-HNC WPNCs had the highest loss modulus compared with other WPNCs and RW. The polymer chains beyond the tightly bound chains were not restricted as the tightly bound chains at the interface. Thus, the loss modulus value of WPNCs over temperature was higher than RW. At

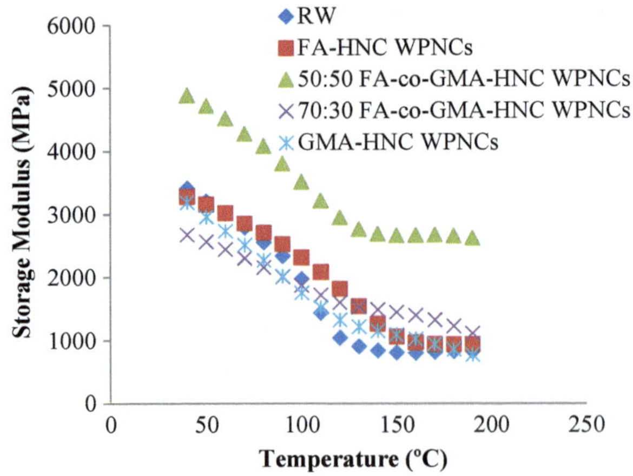

Fig. 6 Storage modulus versus temperature of RW and different ratio of FA-*co*-GMA-HNC WPNCs

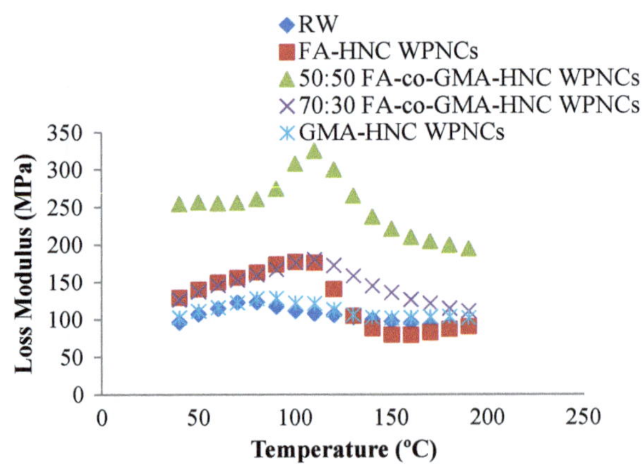

Fig. 7 Loss modulus versus temperature of RW and different ratio of FA-*co*-GMA-HNC WPNCs

the interface, the viscous dissipation occurred in WPNCs than RW, which resulted in a higher loss modulus value (Parshaei and Hosseinzadeh 2016).

The damping variation (tan δ) was shown in Fig. 8. Damping variation included the efficient energy loss for molecular rearrangement and internal friction in WPNCs (Liew et al. 2016). WPNCs showed lower tan δ that proved the high elastic recovery. This was due to the crosslinking of FA, GMA, and HNC that were effectively impregnated into the wood fiber, and the polymer matrix acted as plasticizers due to their nonpolar and high molecular weight (Hamdan et al. 2010).

Therefore, 50:50 FA-*co*-GMA-HNC exhibited strong interaction with wood fiber which reduced the mobility of the molecular chains at the interface, and thus, damping variation reduced significantly.

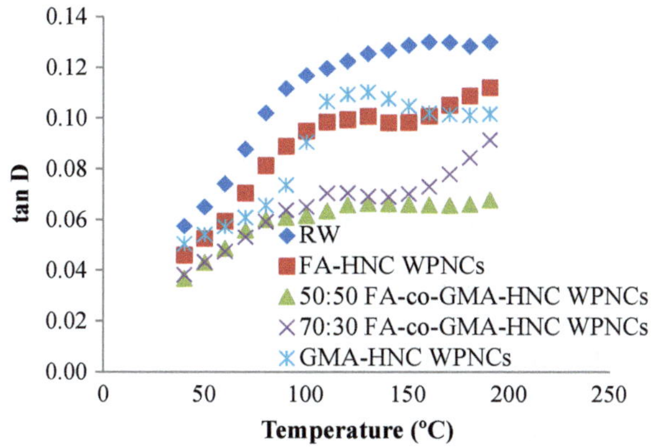

Fig. 8 Tangent δ versus temperature of RW and different ratio of FA-*co*-GMA-HNC WPNCs

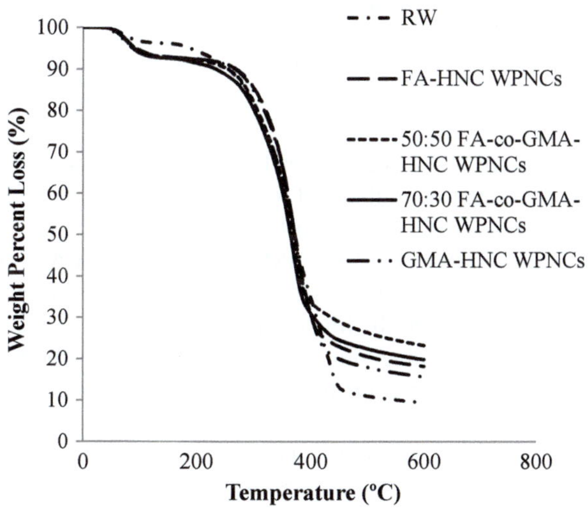

Fig. 9 TGA curves of RW and different ratio of FA-*co*-GMA-HNC WPNCs

3.6 Thermogravimetric Analysis (TGA)

Thermal stability and degradation temperature of RW and FA-*co*-GMA-HNC WPNCs were studied by TGA. The result was shown in Fig. 9. In Fig. 9, there were three distinct zones of thermal degradation observed. The first step was associated to the moisture evaporation, occurred at about 70 to 80 °C. The second step degradation occurred at about 250 to 400 °C. This step was associated to a significant weight loss in the thermal decomposition process due to the degradation of fiber (Saw and Datta 2009). The last step corresponded to the final thermal degradation, which started at 400 °C and ended at 600 °C. This step occurred due to the saturation of rings, rupture of lignin C–C bonds as well as the structural rearrangements (Rials and Glasser 1984).

Table 3 Activation energy of RW and different ratio of FA-*co*-GMA-HNC WPNCs determined by Arrhenius equation

Samples	T_i (°C)[a]	T_m (°C)[b]	T_f (°C)[c]	W_{Ti} (%)[d]	W_{Tm} (°C)[e]	W_{Tf} (°C)[f]	Activation energy, E_a (kJ/mol)
RW	84.0	360.0	456.0	96.3	51.1	13.1	1120.9
FA-HNC WPNCs	78.0	360.0	468.0	96.1	51.6	18.8	1399.3
50:50 FA-*co*-GMA-HNC WPNCs	84.0	366.0	474.0	95.4	55.1	27.5	2050.7
70:30 FA-*co*-GMA-HNC WPNCs	84.0	366.0	468.0	95.4	54.9	23.6	1883.0
GMA-HNC WPNCs	78.0	366.0	468.0	96.1	53.6	19.0	1333.4

[a]Temperature corresponding to the beginning of decomposition
[b]Temperature corresponding to the maximum rate of mass loss
[c]Temperature corresponding to the end of decomposition
[d]Mass loss at temperature corresponding to the beginning of decomposition
[e]Mass loss at temperature corresponding to the maximum rate of mass loss
[f]Mass loss at temperature corresponding to the end of decomposition

Figure 9 showed weight loss about 45.2, 44.5, 40.3, 40.5, and 42.5 wt% at the second step degradation for RW, FA-HNC, 50:50 FA-*co*-GMA-HNC, 70:30 FA-*co*-GMA-HNC, and GMA-HNC WPNCs. The last step degradation of RW, FA-HNC, 50:50 FA-*co*-GMA-HNC, 70:30 FA-*co*-GMA-HNC, and GMA-HNC WPNCs was 13.1, 18.8, 27.5, 23.6, and 19.0 wt%. 50:50 FA-*co*-GMA-HNC WPNCs showed the highest weight percent loss due to the crosslinking of glycidyl group and terminal double bond in GMA with the hydroxyl group of wood fiber as well as the high compatibility of furfuryl alcohol and halloysite nanoclay (Hazarika and Maji 2013; Kakiuchi 1964).

Table 3 showed the thermal characteristics such as initial temperature (T_i), maximum rate loss temperature (T_m), and final decomposition temperature (T_f) as well as the activation energy. The activation energy could be helpful in reaching conclusions about the thermal stability of RW and Fa-*co*-GMA-HNC WPNCs. Arrhenius equation was used to determine the activation energy. The higher activation energy implied higher thermal stability. 50:50 FA-*co*-GMA-HNC WPNCs showed the best thermal stability among all the composition. Therefore, it could be concluded that the thermal stability of RW was greatly improved by the impregnation of FA-*co*-GMA-HNC.

3.7 Differential Scanning Calorimetry (DSC) Analysis

The thermal stability of RW and FA-*co*-GMA-HNC WPNCs was characterized by DSC analysis as shown in Fig. 10. The bond dissociation enthalpy and exotherm peaks were summarized in Table 4.

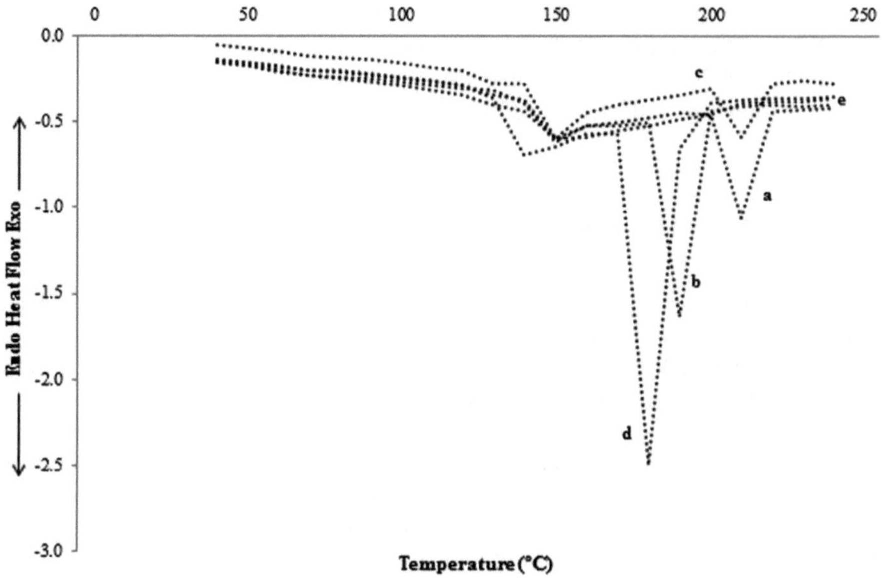

Fig. 10 DSC curves of **a** RW, **b** FA-HNC WPNCs, **c** 50:50 FA-*co*-GMA-HNC WPNCs, **d** 70:30 FA-*co*-GMA-HNC WPNCs and **e** GMA-HNC WPNCs

Table 4 DSC of RW and different ratio of FA-*co*-GMA-HNC WPNCs

Samples	Endotherm peaks (°C)	Bond dissociation enthalpy (J/g)
RW	139.33	7.64
FA-HNC WPNCs	196.00	19.51
50:50 FA-*co*-GMA-HNC WPNCs	178.54	85.13
70:30 FA-*co*-GMA-HNC WPNCs	185.82	57.87
GMA-HNC WPNCs	187.77	14.07

The significant decomposition of RW and WPNCs was divided into two stages. The first stage indicated the presence of moisture in the wood while the second stage confirmed the decomposition of wood fiber that started at about 200 °C (Aydemir et al. 2011). From Fig. 10 and Table 4, the first decomposition temperature of all WPNCs exhibited a higher value compared to RW which proved that the WPNCs were more thermally stable (Lee et al. 2006). WPNCs showed higher bond dissociation enthalpy than RW. This was due to the well interaction between the wood fiber and polymer matrix. As the impregnation of FA-*co*-GMA-HNC into the wood fiber interfered with bond dissociation, it showed that the increase of crystal size of fiber increased the bond dissociation enthalpy (Hamdan et al. 2010).

Overall, WPNCs especially 50:50 FA-*co*-GMA-HNC WPNCs had the highest thermal stability, as reflected in TGA results.

3.8 Moisture Absorption Analysis

Figure 11 showed the percentage of moisture absorbed in RW and FA-*co*-GMA-HNC WPNCs. It was observed that WPNCs exhibited lower water absorption compared to RW. This confirmed that the impregnation of FA-*co*-GMA-HNC into wood fiber decreased the moisture absorption in WPNCs. The hydrophilic character of the fibers was significantly reduced by the introduction of polymer matrix on the fiber surface that broke down the hydroxyl bonds in the moisture (Mulinari et al. 2010). Therefore, 50:50 FA-*co*-GMA-HNC WPNCs showed the lowest absorption compared to other WPNCs as well as RW, as confirmed in three-point bending as surface morphology.

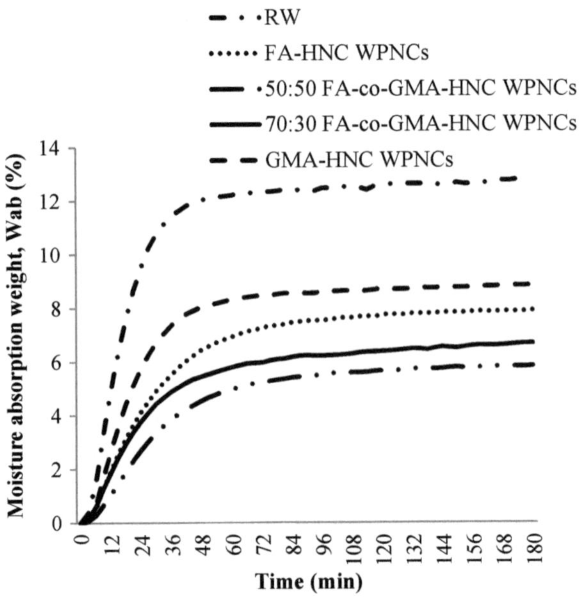

Fig. 11 Moisture absorption curves of RW and different ratio of FA-*co*-GMA-HNC WPNCs

4 Conclusion

In this study, RW was impregnated with FA-*co*-GMA-HNC to produce WPNCs. It is concluded that the physical, mechanical, morphological, and thermal properties of FA-*co*-GMA-HNC WPNCs were greatly improved compared to the RW. 50:50 FA-*co*-GMA-HNC WPNCs showed the highest weight percent gain with the least hydroxyl group in FT-IR analysis. 50:50 FA-*co*-GMA-HNC WPNCs provided smoother morphology among all the compositions. 50:50 FA-*co*-GMA-HNC WPNCs showed the highest flexural strength and modulus of elasticity. The storage modulus and loss modulus of FA-*co*-GMA-HNC WPNCs were improved greatly with lower tan δ compared to RW. 50:50 FA-*co*-GMA-HNC WPNCs exhibited higher thermal stability with significantly lower moisture absorption. This has proved that the ratio 50:50 FA-*co*-GMA was the most suitable to be introduced in the RW to form WPNCs.

Acknowledgements The authors would like to acknowledge the financial support from Research and Innovation Management Centre, Universiti Malaysia Sarawak under fund with Grant No. (F02/SpGS/1443/2016/25).

References

Ahmad EEM, Luyt AS, Djokovic V (2013) Thermal and dynamic mechanical properties of bio-based poly(furfuryl alcohol)/sisal whiskers nanocomposites. Polym Bull 70(4):1265–1276

Aydemir D, Gunduz G, Altuntas E, Ertas M, Sahin HT, Alma AH (2011) Investigating changes in the chemical constituents and dimensional stability of heat-treated hornbeam and uludag fir wood. BioRes 6(2):1308–1321

Baysal E, Yalinkilic M, Altinok M, Sonmezc A, Pekerd H, Colaka M (2007) Some physical, biological, mechanical, and fire properties of wood polymer composite (WPC) pretreated with boric acid and borax mixture. Constr Build Mater 21(9):1879–1885

Bouhelal S, Cagiao ME, Benachour D, Calleja FJC (2007) Structure modification of isotactic polypropylene through chemical crosslinking: toughening mechanism. J Appl Polym Sci 103(5):2968–2976

Dong Y, Qin Y, Wang K, Yan Y, Zhang S, Li J, Zhang S (2012) Assessment of the performance of furfurylated wood and acetylated wood: comparison among four fast-growing wood species. BioRes 11(2):3679–3690

Dong Y, Yan Y, Zhang S, Li J (2014) Wood/polymer nanocomposites prepared by impregnation with furfuryl alcohol and nano-SiO_2. BioRes 9(4):6028–6040

Fuller B, Ellis W, Rowell R, US 5605767, 25 Feb 1997

Guigo N, Mija A, Zavaglia R, Vincent L, Sbirrazzuoli N (2009) New insights on the thermal degradation pathways of neat poly(furfuryl alcohol) and poly(furfuryl alcohol)/SiO_2 hybrid materials. Polym Degrad Stab 94(6):908–913

Hamdan S, Talib ZA, Rahman MR, Ahmed AS, Islam MS (2010) Dynamic Young's modulus measurement of treated and post-treated tropical wood polymer composites (WPC). BioRes 5(1):324–342

Hazarika A, Maji TK (2013) Effect of different crosslinkers on properties of melamine formaldehyde-furfuryl alcohol copolymer/montmorillonite impregnated softwood (Ficus hispida). Polym Eng Sci 53(7):1394–1407

Kabir MA, Huque MM, Islam MR, Bledzki AK (2010) Mechanical properties of jute fiber reinforced polypropylene composites: Effect of chemical treatment by benzenediazonium salt in alkaline medium. BioRes 5(3):1618–1625

Kakiuchi H (1964) Manufacture and application of epoxy resins. Macromolecule Chemistry Publication Society, Kyoto

Kherroub DE, Belbachir M, Lamouri S (2015) Synthesis of poly(furfuryl alcohol)/montmorillonite nanocomposites by direct in-situ polymerization. Arabian J Sci Eng 40(1):143–150

Kunita MH, Girotto EM, Muniz EC, Rubira AF (2006) Polypropylene grafted with glycidyl methacrylate using supercritical CO_2 medium. Braz J Chem Eng 23(2):267–271

Lee SY, Doh GH, Kang IA (2006) Thermal behaviour of hwangto and wood flour reinforced high density polyethylene (HDPE) composites. Mokchae Konghak 34:59–66

Li Y (2011) Wood-polymer composites. Trans Tech Publications Inc., Switzerland

Li Y, Liu Y, Wang X, Wu Q, Yu H, Li J (2011) Wood-polymer composites prepared by in-situ polymerization of monomers within wood. J Appl Polym Sci 119(6):3207–3216

Liew FK, Hamdan S, Rahman MR, Mahmood MR, Rahman MM, Lai JCH, Sultan MT (2016) 4-methylcatechol-treated jute-bamboo hybrid composites: effects of pH on thermo-mechanical and morphological properties. BioRes 11(3):6880–6895

Mulinari DR, Voorwald HJC, Cioffi MOH, Rocha GJ, Silva MLCPD (2010) Surface modification of sugarcane bagasse cellulose and its effect on mechanical and water absorption properties of sugarcane bagasse cellulose/HDPE composites. BioRes 5(2):661–671

Pandey K (1999) A study of chemical structure of soft and hardwood and wood polymers by FTIR spectroscopy. J Appl Polym Sci 71(12):1969–1975

Parshaei S, Hosseinzadeh S (2016) Preparation of organo nanoclay incorporated polyamide/melamine cyanurate/nanoclay composites and study on thermal and mechanical behaviours. Iranian Chem Comm 4:102–114

Pothan LA, George CN, John MJ, Thomas S (2010) Dynamic mechanical and dielectric behaviour of banana-glass hybrid fiber reinforced polyester composites. J Reinf Plastics Compos 29(8):1131–1145

Pranger LA, Nunnery GA, Tannenbaum R (2012) Mechanism of the nanoparticle-catalyzed polymerization of furfuryl alcohol and the thermal and mechanical properties of the resulting nanocomposites. Compos Part B 43:1139–1146

Qin Z, Gao Q, Zhang S, Li J (2013) Glycidyl methacrylate grafted onto enzyme-treated soybean meal adhesive with improved wet shear strength. BioRes 8(4):5369–5379

Rials TG, Glasser WG (1984) Engineering plastics from lignin and enthalpy relaxation of prepolymers. J Wood Chem Technol 4(3):331–345

Saw SS, Datta C (2009) Thermomechanical properties of jute/bagasse hybrid fibre reinforced epoxy thermoset composites. BioRes 4(4):1455–1476

Singha AS, Thakur VK (2008) Fabrication and study of lignocellulosic hibiscus sabdariffa fiber reinforced polymer composites. BioRes 3(4):1173–1186

Smitinand T, Larsen K (eds) (1984) Flora of Thailand. Royal Forest Department, Bangkok

Toriz G, Arvidsson R, Westin M, Gatenholm P (2003) Novel cellulose ester-poly(furfuryl alcohol)-flax fiber biocomposites. J Appl Polym Sci 88(2):337–345

Viet Cao X, Ismail H, Rashid AA, Takeichi T, Vo-Huu T (2011) Mechanical properties and water absorption of kenaf powder filled recycled high density polyethylene/natural rubber biocomposites using mape as a compatibilizer. BioRes 6(3):3260–3271

Wechsler A, Hiziroglu S, Ballerini AA (2008) Some of the properties of wood plastic composites. Paper presented at the proceedings of the 51st International Convention of Society of Wood Science and Technology, Concepcion, Chile, 10–12 Nov 2008

Xue F, Zhao G (2008) Optimum preparation technology for Chinese fir wood/Ca-montmorillonite (Ca-MMT) composite board. Forest Stud Chin 10(3):2581–2590

Nanoclay Dispersed Furfuryl Alcohol-*co*-Ethyl Methacrylate Wood Polymer Nanocomposites: The Enhancement on Physico-mechanical and Thermal Properties

M.R. Rahman, J.C.H. Lai and S. Hamdan

Abstract In this study, physical, morphological, mechanical, and thermal properties of furfuryl alcohol/glycidyl methacrylate/halloysite nanoclay wood polymer nanocomposites (FA-*co*-EHMA-HNC WPNCs) were investigated. FA-*co*-EHMA-HNC WPNCs were prepared via impregnation method, and the effect of different ratio between the polymers was subsequently investigated. The properties of nanocomposites were characterized by weight percent gain, Fourier Transform Infrared Spectroscopy (FT-IR), Scanning Electron Microscopy (SEM), three-point flexural test, dynamic mechanical thermal analysis (DMTA), Thermogravimetric Analysis (TGA), differential scanning calorimetry (DSC) analysis, and moisture absorption test. The weight percent gain in 50:50 FA-*co*-EHMA-HNC WPNCs was the highest compared to raw wood (RW) and other WPNCs. FT-IR results confirmed the polymerization took place in the nanocomposites especially 50:50 FA-*co*-EHMA-HNC WPNCs with reducing hydroxyl groups. SEM result revealed that the 50:50 FA-*co*-EHMA-HNC WPNCs among all the nanocomposites. Through the three-point flexural test, 50:50 FA-*co*-EHMA-HNC WPNCs showed the highest flexural strength and modulus of elasticity. The results revealed that the storage modulus and loss modulus of FA-*co*-EHMA-HNC WPNCs were higher while the tan δ of FA-*co*-EHMA-HNC WNPCs was lower compared to RW. FA-*co*-EHMA-HNC WPNCs exhibited the higher thermal stability through TGA and DSC analysis. 50:50 FA-*co*-EHMA-HNC WPNCs exhibited significantly lower moisture absorption compared to RW. Overall, this study proved that the ratio 50:50 FA-*co*-EHMA was the most suitable to be introduced in the RW.

Keywords Morphology · Strength · Thermal · Clay

M.R. Rahman (✉) · J.C.H. Lai · S. Hamdan
Faculty of Engineering, Universiti Malaysia Sarawak,
94300 Kota Samarahan, Sarawak, Malaysia
e-mail: rmrezaur@unimas.my

M.R. Rahman, *Wood Polymer Nanocomposites*, Engineering Materials,
DOI 10.1007/978-3-319-65735-6_15

1 Introduction

The rapid growing on the demand for resources leads to the importance of making sustainable and efficient resource utilization is proportionally growing (Brown et al. 2011). The constituents of wood include lignin with hemicelluloses that acted as soft polymer matrix and lignocellulosic fibers that acted as rigid cellulosic microfibrils as reinforcement (Rong et al. 2001). Wood fiber is known to be the most broadly used as composite materials reinforcement. However, wood is a hygroscopic materials that will adsorb surrounding moisture (Rose 2002). Besides, wood is an organic materials that will be easily deteriorated as well as fungi attack (Rose 2002). Therefore, the introduction of wood fibers in polymer matrix to form wood polymer nanocomposites (WPNCs) has been widely used in many applications.

WPNCs are thermoplastic polymers which have attracted both research and industrialists due to their environmental friendliness (Markarian 2005). WPNCs product shows good durability in wet environments due to the hydrophobic polymer matrix that allowed them to substitute wood in outdoor applications (Cheung et al. 2009). WPNCs product can be manufactured easily and rapidly using forming techniques that are typical of thermoplastic polymers (Zini and Scandola 2011). The growing usage of renewable resources helps to encounter various environmental problems due to the climate change and biodiversity threats (Lenzen et al. 2012).

Furfuryl alcohol (FA) is a derived renewable monomer which was added into sisal whiskers had improved the mechanical and thermal properties of the nanocomposites compared to pure FA (Ahmad et al. 2013). Besides, the incorporation of modified clay-furfuryl alcohol into wood exhibited better mechanical properties and lower water uptake of the nanocomposites compared to RW (Hazarika and Maji 2013).

2-ethylhexyl methacrylate (EHMA) is a monomer of gelling nature that usually applied in biomedical applications (Karlsson and Gatenholm 1996; Ende and Peppas 1997). The past research showed that the combination of poly(EHMA) and methyl methacrylate was introduced into wood cellulose to improve the functional properties for biomedical applications (Sharma and Chauhan 2009).

Clay acts as crosslinking agent had enhanced the morphological, tensile, and flexural properties of the WPNCs (Hazarika and Maji 2014). Freeze-dried FA-clay nanocomposites enhanced the mechanical properties and thermal stability (Wang et al. 2016).

There is very little research work that has been reported on FA-based WPNCs and EHMA-based WPNCs especially with the combination of FA and EHMA in WPNCs. Therefore, the aim of this present study is to fabricate FA-*co*-EHMA-HNC WPNCs via impregnation technique and to investigate the effect of different ratios of the polymer matrix on physical, mechanical, morphological, and thermal properties of wood polymer nanocomposites.

2 Experimental

2.1 *Materials*

Defect-free and straight-grained *Sindora glabra* wood was obtained from Forest Farm, Sarawak. Both control and modified wood samples were obtained from the same timber and machined at the same time with a dimension of 30 cm × 2 cm × 1 cm. The chemicals used to produce WPNCs were furfuryl alcohol (FA), 2-ethylhexyl methacrylate (EHMA), benzoyl peroxide, and halloysite nanoclay (HN). Halloysite nanoclay was supplied by Sigma Aldrich (USA) while furfuryl alcohol, 2-ethylhexyl methacrylate, and benzoyl peroxide were supplied by Merck Millipore (USA). Halloysite nanoclay was nanopowder with the diameter of 30–70 nm and length was in between 1 and 3 μm.

2.2 *Methods*

2.2.1 Preparation of Furfuryl Alcohol/2-Ethylhexyl Methacrylate/Halloysite Nanoclay Wood Polymer Nanocomposites (FA-*co*-EHMA-HNC WPNCs)

The polymer system was prepared using FA-*co*-EHMA-HNC in the presence of benzoyl peroxide. Benzoyl peroxide acted as an initiator to influence the reaction between furfuryl alcohol and 2-ethylhexyl methacrylate. Furfuryl alcohol, 2-ethylhexyl methacrylate, and halloysite nanoclay were mixed with different ratios as shown in Table 1. The mixtures were covered with aluminum foil and undergo autoclave for 15 min to complete the reaction.

2.2.2 Impregnation of Wood Specimens with FA-*co*-EHMA-HNC

Four wood samples prepared according to ASTM D1037 were weighed using an electronic balance. The samples were then placed in the vacuum drying chamber.

Table 1 Preparation of polymer system at different ratio

Volume of furfuryl alcohol (FA) (mL)	Volume of 2-ethylhexyl methacrylate (EHMA) (mL)	Amount of halloysite clay (HNC) (g)	Amount of benzoyl peroxide (g)
0	200	2	5
100	100	2	5
140	60	2	5
200	0	2	5

The mixtures were poured into the vacuum drying oven with pressure of 0.1 bar for 60 min. After impregnation, the excessive solution was removed from the wood samples with tissue paper. Wood samples removed were wrapped with aluminum foil, and it was kept in an oven at 100 °C for 24 h for polymerization and formed wood polymer nanocomposites (WPNCs). After polymerization, the aluminum foil was removed, and WPNCs were dried at 105 °C until constant weight was obtained. The weight percent gain (WPG) of all the WPNCs was then measure using Eq. (1),

$$\mathrm{WPG} = \frac{W_{\mathrm{f}} - W_{\mathrm{o}}}{W_{\mathrm{o}}} \times 100\% \qquad (1)$$

where W_{f} is oven-dried weight after modification of the WPNCs and W_{o} is oven-dried weight before modification of the WPNCs.

2.3 Microstructural Characterizations

2.3.1 Fourier Transform Infrared Spectroscopy (FT-IR)

The infrared spectra of the wood polymer nanocomposite were recorded on a Shimadzu IRAffinity-1. The technique used was Attenuated Total Reflection (ATR). The spectral resolution of this equipment is 12,000 spectra. The transmittance range of the scan was 4000–600 cm^{-1}.

2.3.2 Scanning Electron Microscopy (SEM)

The furfuryl alcohol, 2-ethylhexyl methacrylate, and halloysite nanoclay were examined using a Scanning Electron Microscope (SEM) (JSM-6710F) supplied by JEOL Company Limited, Japan. The middle part of all the FA/EHMA/HN WPNCs was extracted in the dimension of 5 mm × 5 mm × 5 mm (length × width × height). The accelerating voltage of this instrument was 15 kV. The specimens were first fixed with Karnovsky's fixative and then take through a graded alcohol dehydration series. Once dehydrated, the specimen was coated with a thin layer of gold before being viewed microscopically. The micrographs were taken at magnifications ranging from 500 to 1000.

2.3.3 Three-Point Flexural Test

The static flexural tests of the FA-*co*-EHMA-HNC WPNCs were carried out using AG-X Plus Series Precision Universal Testers (300 kN Floor Model) supplied by Shimadzu Corporation, Japan by only changing the attachment. Dimensions of flexural test specimen were 80 mm (length) × 30 mm (width) × 10 mm (thickness). Flexural tests were conducted following ASTM D 7900-00 at a crosshead speed of 10 mm/min. Four specimens of each composition were tested, and the average values were reported. The-load displacement curves were obtained from the instrument, which could be recalled after the completion of test. The flexural strength (Q_{fs}) and modulus of elasticity (E_{m}) were calculated using Eqs. (2) and (3).

$$Q_{\text{fs}} = \frac{3PL}{2bd^2} \tag{2}$$

where P is maximum load on the load-deflection curve, L is support span, 63.7 mm, b is width of beam tested, 10 mm, and d is thickness of beam, 4 mm.

$$E_{\text{m}} = \frac{L^3 m}{4bd^3} \tag{3}$$

where m is slope of the tangent to the initial straight-line portion of the load-deflection curve, N/mm of deflection.

2.3.4 Dynamic Mechanical Thermal Analysis (DMTA)

Dynamic mechanical thermal analysis was applied to study the effect of the temperature on the storage modulus ($\log E'$) and loss tangent ($\tan \delta$) of FA-*co*-EHMA-HNC WPNCs. It is a mechanical test that allowed molecules in woods to interact with mechanical stress. This test was carried out using Perkin Elmer dynamic mechanical thermal analyser (PE-DMTA) supplied by Perkin Elmer, the USAwith frequency of 10 Hz, strain of ×4 and temperature rise of 2 °C min^{-1}. The rectangular specimens with moisture content around 15% were tested using a dual-cantilever bending mode on a standard bending head. The chamber surrounding the specimen at 65% RH was cooled by liquid nitrogen, and the system provided a simple thermal scan at 2 °C min^{-1} with various temperatures ranging from 0 to 200 °C. T_g was determined from the graph of $\tan \delta$ versus T.

2.3.5 Thermogravimetric Analysis (TGA)

Thermogravimetric Analysis (TGA) measurements were carried out on 5–10 mg of FA-*co*-EHMA-HNC WPNCs at a heating rate of 10 °C/min in a nitrogen atmosphere using a Thermogravimetric Analyzer (TA Instrument SDT Q600) supplied by TA Instruments, the USA. FA-*co*-EHMA-HNC WPNCs were subjected to TGA in high purity nitrogen under a constant flow rate of 5 mL/min. Thermal decomposition of each sample occurred in a programmed temperature range of 0–700 °C. The continuous weight loss and temperature were recorded and analyzed.

2.3.6 Differential Scanning Calorimetry (DSC) Analysis

FA-*co*-EHMA-HNC WPNCs were analyzed using a Perkin Elmer thermal analyser. All the measurements were made under an N_2 flow (30 mL/min), keeping a constant heating rate of 10 °C/min and using an alumina crucible with a pinhole.

2.3.7 Moisture Absorption Test

Moisture absorption was carried out by using electronic moisture balance (MOC-120H) supplied by Shimadzu Corporation, Kyoto, Japan at 110 °C for 3 h. Dry wood samples (dried at 25 °C) were immersed in distilled water. FA-*co*-EHMA-HNC WPNCs were removed and were placed on the pan and weighted. The heater lid had been closed firmly, and the display was switched from a display of the weight to a percentage display and the measuring time was also displayed. The moisture absorbed, W_{ab} was calculated by using Eq. (4),

Moisture absorbed percentage, W_{ab} (%)

$$W_{ab} = \frac{W_w - W_d}{W_d} \times 100 \tag{4}$$

where W_{ab} is absorbed weight, W_w is weight of wet WPNCs and W_d is weight of dry WPNCs.

3 Results and Discussion

3.1 Weight Percent Gain (WPG %)

The weight percent gain of RW and FA-*co*-EHMA-HNC WPNCs is shown in Table 2. The WPG values of RW, FA-HNC, 50:50 FA-*co*-EHMA-HNC, 70:30 FA-*co*-EHMA-HNC, and EHMA-HNC WPNCs were 0.376, 5.267, 28.655, 25.132, and 12.959%, respectively. It was due to the types of polymer matrix characteristics which impregnated into the wood (Islam et al. 2012). Therefore, it confirmed that the impregnation of FA-*co*-EHMA-HNC WPNCs improved the mechanical and thermal properties.

3.2 Fourier Transform Infrared Spectroscopy (FT-IR)

The formation of new chemical bonds between furfuryl alcohol, 2-ethylhexyl methacrylate, clay, and wood fiber of FA-*co*-EHMA-HNC WPNCs and RW is shown in Fig. 1. RW had specific IR spectra such as 3400, 2900, and 1735 cm^{-1}

Table 2 Average weight percent gain, WPG (%) of RW and different ratio of FA-*co*-EHMA-HNC WPNCs

Samples	Average weight before impregnation (g)	Average weight after impregnation (g)	Weight percent gain, WPG (%)
RW	70.756	71.023	0.376
FA-HNC WPNCs	58.703	61.967	5.267
50:50 FA-*co*-EHMA-HNC WPNCs	43.079	60.381	28.655
70:30 FA-*co*-EHMA-HNC WPNCs	48.479	64.753	25.132
EHMA-HNC WPNCs	46.810	53.779	12.959

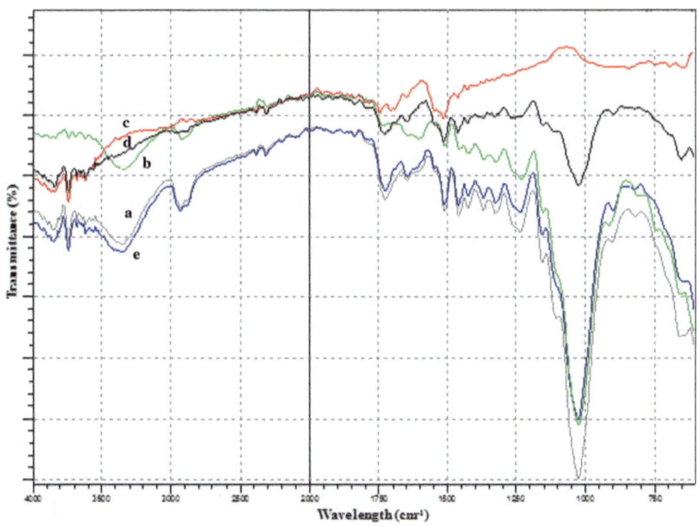

Fig. 1 FT-IR spectra of **a** RW **b** FA-HNC WPNCs **c** 50:50 FA-*co*-EHMA-HNC WPNCs **d** 70:30 FA-*co*-EHMA-HNC WPNCs **e** EHMA-HNC WPNCs

which were attributed to the –OH, C–H, and C=O stretching vibrations, respectively. All these absorption bands were significant due to the hydroxyl group in fiber and carbonyl group of acetyl ester in fiber (Ismail et al. 2002). However, the IR spectrum at 3400 cm^{-1} significantly reduced when the polymer matrix was well impregnated into the raw wood especially at 50:50 FA-*co*-EHMA-HNC and 70:30 FA-*co*-EHMA-HNC WPNCs. IR peak at 1735 cm^{-1} was shifted toward 1730 cm^{-1} in WPNCs due to the fracture of ester carbonyl bonds in the fiber in the impregnation system (Mohd Idrus et al. 2011). In addition, the peak intensity at around 1350 cm^{-1} that attributed to C–O–H stretching vibration was reduced in WPNCs

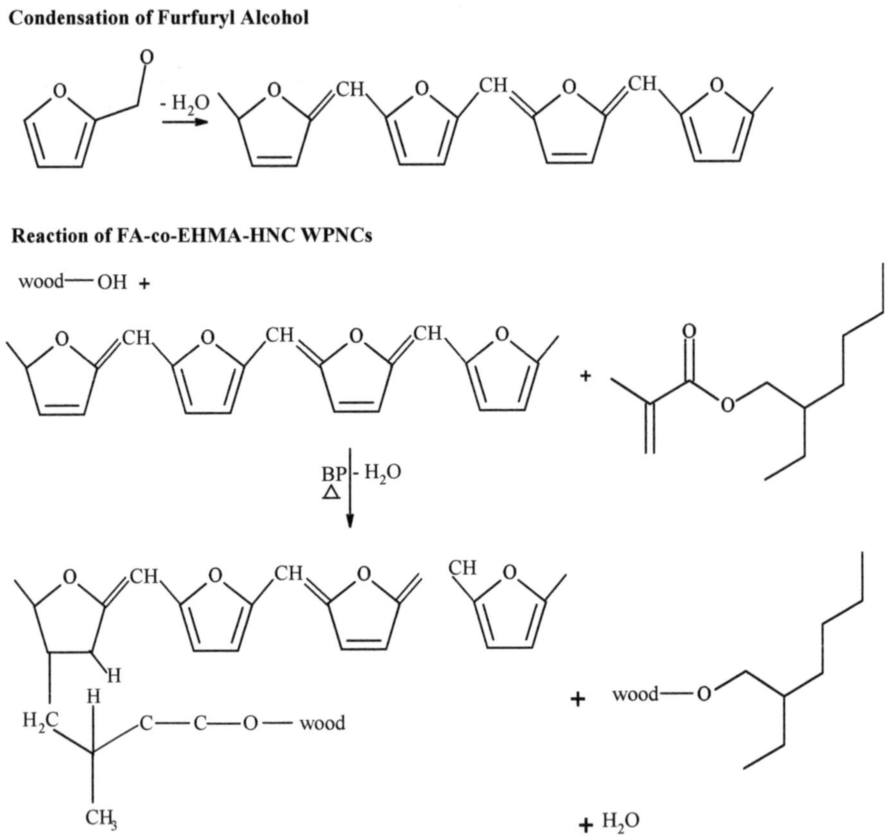

Fig. 2 Proposed schematic reaction diagram

due to good chemical bond formation between the polymer matrix and hydroxyl group in fiber. Therefore, it could be concluded that RW was well impregnated with FA-*co*-EHMA-HNC that reduced the hydroxyl groups in the wood cell. The reaction scheme is shown in Fig. 2.

3.3 Scanning Electron Microscopy (SEM) Analysis

Scanning Electron Microscopy (SEM) was carried out to examine the compatibility between the wood cell wall and the polymer matrix. Figure 3a–e showed the surface morphology of RW and FA-*co*-EHMA-HNC WPNCs. Significant difference was observed in all the samples as the interfacial bonding of wood cell-polymer matrix was different. Figure 3a clearly showed the void spaces inside RW. The void spaces detected due to the weak fiber-fiber interaction in the wood that would

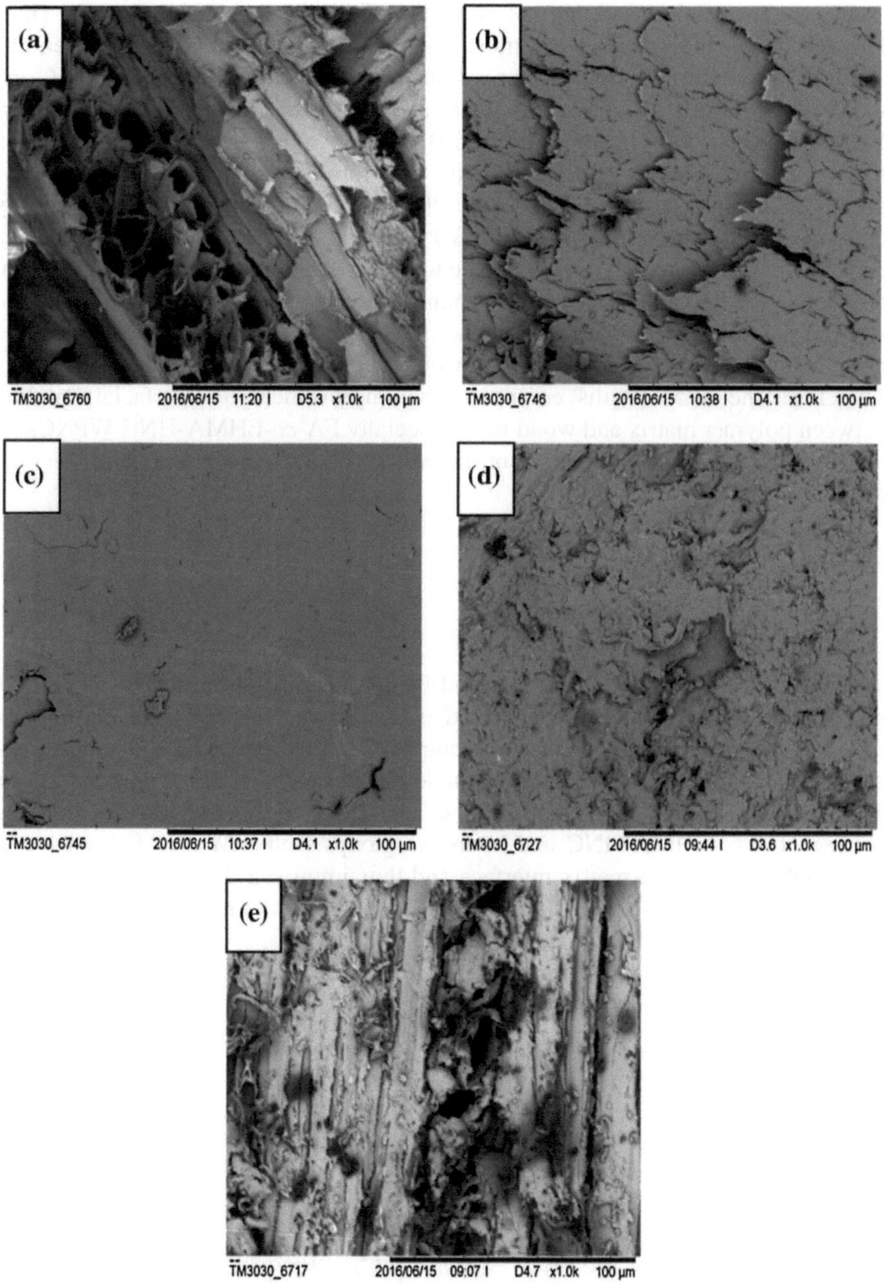

Fig. 3 SEM micrographs of **a** RW **b** FA-HNC WPNCs **c** 50:50 FA-*co*-EHMA-HNC WPNCs **d** 70:30 FA-*co*-EHMA-HNC WPNCs **e** EHMA-HNC WPNCs

reduce the mechanical and thermal properties. With the impregnation of FA-HNC into RW, the void spaces were filled up by the polymer matrix. However, the surface was not uniform due to the weak interfacial bonding between FA-HNC and wood fiber. Figure 3c showed smooth and uniform surface of 50:50 FA-*co*-EHMA-HNC WPNCs (Rahman et al. 2015). The strong interfacial bond between the wood fiber and FA-*co*-EHMA-HNC was formed that reduced the hydroxyl groups in the wood fiber (Liew et al. 2016). The surface in Fig. 3d was fully covered due to the impregnation of 70:30 FA-*co*-EHMA-HNC. However, the surface was not as smooth as Fig. 3c due to the different of ratio that led to different reorganization of the polymer matrix chains (Ndiaye and Tidjani 2014). Figure 3e showed the non-uniform surface with some void spaces. This could be due to the poor adhesion of EHMA-HNC into wood cell wall (Sheshmani et al. 2010). Therefore, the SEM results confirmed the improvement of interfacial bonding between polymer matrix and wood fiber especially FA-*co*-EHMA-HNC WPNCs by impregnation method, which contributed to better mechanical and thermal properties.

3.4 *Three-Point Flexural Test*

The mechanical properties of RW and WPNCs are shown in Figs. 4 and 5, respectively. The flexural strength increased with the incorporation of different ratios of polymer matrix. Flexural strength of WPNCs increased about 8–20% compared to the RW. This was due to the less agglomeration of the polymer matrix at the surface of the WPNCs (Ismail et al. 2001). Besides, hydrogen bonding between FA-*co*-EHMA-HNC and hydroxyl groups from RW was formed led to better adhesion at fiber-matrix interface and thus improved the flexural strength of WPNCs. Low flexural strength was detected in RW due to the abundant hydroxyl groups on the surface (Lu et al. 2005).

Fig. 4 Flexural strength of RW and different ratio of FA-*co*-EHMA-HNC WPNCs

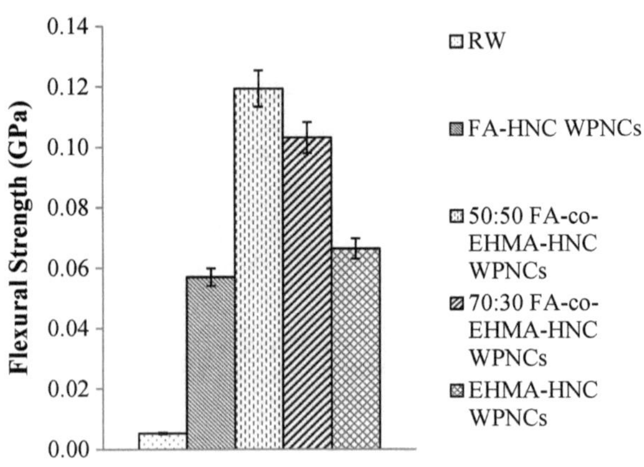

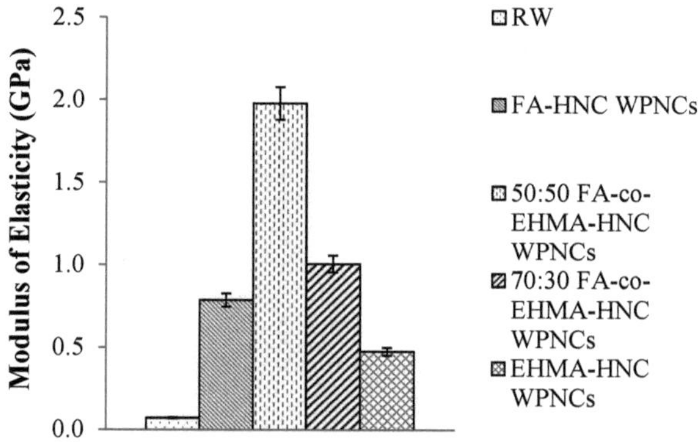

Fig. 5 Modulus of elasticity of RW and different ratio of FA-*co*-EHMA-HNC WPNCs

The modulus of elasticity of RW and FA-*co*-EHMA-HNC WPNCs is presented in Fig. 5. WPNCs showed higher modulus of elasticity compared to RW due to the better interactions between the polymer matrix and wood cell wall. Besides, the impregnation of FA-*co*-EHMA-HNC into RW increased the modulus of elasticity due to the inclusion of rigid matrix particles in the RW. The better interfacial bonding of WPNCs reduced the interfacial stress concentration that led to better modulus of elasticity (Cao et al. 2011). Therefore, 50:50 FA-*co*-EHMA-HNC WPNCs showed the highest flexural strength and modulus of elasticity compared to other WPNCs and RW.

3.5 Dynamic Mechanical Thermal Analysis (DMTA)

Dynamic mechanical thermal analysis (DMTA) was carried out to study the storage modulus (E'), loss modulus (E''), and damping behavior (tan δ).

There were three observed stages, namely glassy region, transition region, and rubbery region (Komadan et al. 2007). Figure 6 showed the storage modulus of RW and FA-*co*-EHMA-HNC WPNCs decreased over the temperature. The value of storage modulus represented the stiffness of the WPNCs. It was significant that changes in the storage modulus were less around glass transition region due to the semi-crystalline nature of FA-*co*-EHMA-HNC. The crystallinity of both RW and WPNCs reduced as storage modulus decreased with increase in temperature and finally leveled off at high temperature. The polymer matrix component was greatly affected the storage modulus over temperature (Quintens et al. 1990).

The loss modulus peak associated with the maximum heat dissipation per unit deformation (Komadan et al. 2007). The peak in Fig. 7 represented the glass transition temperature. Figure 7 showed that the loss modulus of RW and WPNCs increased over temperature. RW had the lowest loss modulus in the rubbery stage. There was only one curve shown for all the WPNCs which indicated the miscibility of the polymer matrix (Varughese et al. 1988).

Fig. 6 Storage modulus versus temperature of RW and different ratio of FA-*co*-EHMA-HNC WPNCs

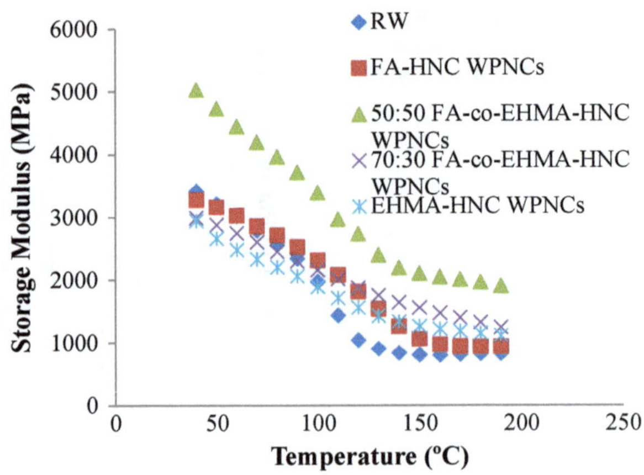

Fig. 7 Loss modulus versus temperature of RW and different ratio of FA-*co*-EHMA-HNC WPNCs

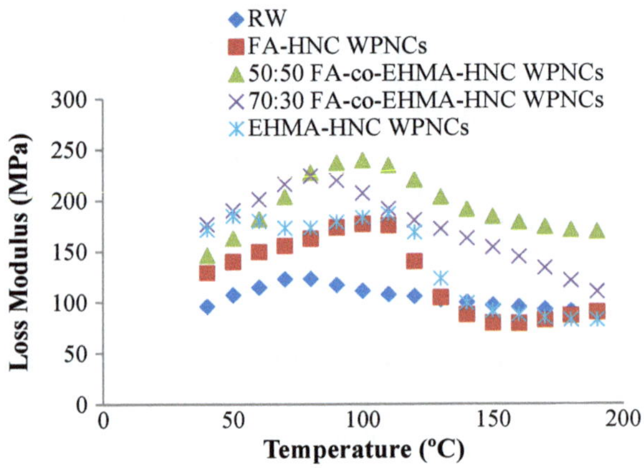

Figure 8 showed the variation of damping behavior (tan δ) over temperature for RW and WPNCs. The damping was the highest when the samples were in transition region. Both glassy and rubbery region showed low tan δ as the molecular movement was either in frozen stage or free to move around (Komadan et al. 2007). WPNCs had lower damping variation compared to RW due to the rubbery nature of WPNCs. Besides, the removal of stress on RW caused more permanent deformation and thus produced higher tan δ values.

Overall, storage modulus and loss modulus increased with reduction in damping behavior of WPNCs due to the significant changes in the molecular motion in the transition region.

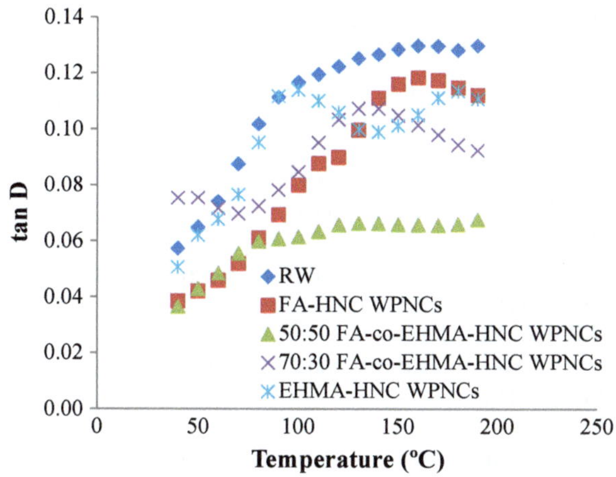

Fig. 8 Tangent δ versus temperature of RW and different ratio of FA-*co*-EHMA-HNC WPNCs

3.6 *Thermogravimetric Analysis (TGA)*

The TGA curves of RW, FA-HNC, 50:50 FA-*co*-EHMA-HNC, 70:30 FA-*co*-EHMA-HNC, EHMA-HNC WPNCs were shown in Fig. 9. TGA curves showed that the decomposition of RW and FA-*co*-EHMA-HNC WPNCs occurred in three stages. First stage occurred within the temperature range of 0–200 °C. The second stage of thermal degradation started at 200 °C and ended at 350 °C. Fiber was decomposed at this stage. The third stage was occurred at the temperature range of 350–450 °C that associated to the decomposition wood cell wall (Hasnan et al. 2016). From Fig. 9, the weight loss percentage of thermal degradation was lower for all the WPNCs, compared to RW below 400 °C. However, the weight loss of WPNCs was much higher than RW at the temperature range of 400–600 °C. This indicated that the WPNCs had higher thermal stability than RW due to the better

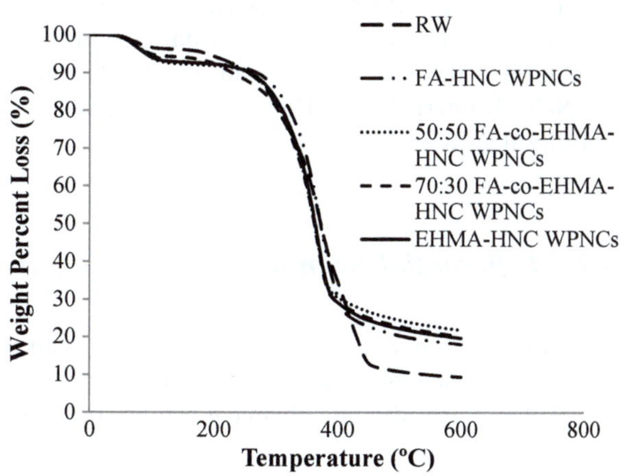

Fig. 9 TGA curves of RW and different ratio of FA-*co*-EHMA-HNC WPNCs

Table 3 Activation energy of RW and different ratio of FA-*co*-EHMA-HNC WPNCs determined by Arrhenius equation

Samples	T_i (°C)[a]	T_m (°C)[b]	T_f (°C)[c]	W_{Ti} (%)[d]	W_{Tm} (°C)[e]	W_{Tf} (°C)[f]	Activation energy, E_a (kJ/mol)
RW	84.0	360.0	456.0	96.3	51.1	13.1	1120.9
FA-HNC WPNCs	78.0	360.0	468.0	96.1	51.6	18.8	1399.9
50:50 FA-*co*-EHMA-HNC WPNCs	78.0	360.0	486.0	96.4	57.3	22.9	2211.9
70:30 FA-*co*-EHMA-HNC WPNCs	78.0	360.0	480.0	96.2	53.0	23.1	2021.6
EHMA-HNC WPNCs	78.0	360.0	468.0	96.1	51.7	21.9	1701.0

[a]Temperature corresponding to the beginning of decomposition
[b]Temperature corresponding to the maximum rate of mass loss
[c]Temperature corresponding to the end of decomposition
[d]Mass loss at temperature corresponding to the beginning of decomposition
[e]Mass loss at temperature corresponding to the maximum rate of mass loss
[f]Mass loss at temperature corresponding to the end of decomposition

interfacial adhesion of FA-*co*-EHMA-HNC in the wood cell wall. Besides, the polymer matrix-cell wall interaction was significantly stronger compared to the fiber-fiber interaction (Hasnan et al. 2016).

Table 3 showed the thermal characteristics such as initial temperature (T_i), maximum rate loss temperature (T_m), and final decomposition temperature (T_f) as well as the activation energy. Arrhenius equation was used to determine the activation energy (Chanmal and Jog 2008). The higher activation energy implied higher thermal stability. It was found that the activation energy of 50:50 FA-*co*-EHMA-HNC WPNCs was significantly higher followed by 70:30 FA-*co*-EHMA-HNC, EHMA-HNC, FA-HNC WPNCs, and RW. The impregnation of FA-*co*-EHMA-HNC increased the thermal stability because the particles of polymer matrix filled the void appeared in wood cell wall and provided better surface of WPNCs (Kumari 2008). Therefore, impregnated wood performed better thermal stability than RW.

3.7 *Differential Scanning Calorimetry (DSC) Analysis*

The DSC thermograms of RW and WPNCs with different ratios were shown in Fig. 10. DSC analysis was carried out to determine the thermal behavior of the RW and WPNCs. Besides, the chemical activity occurring within the RW and WPNCs

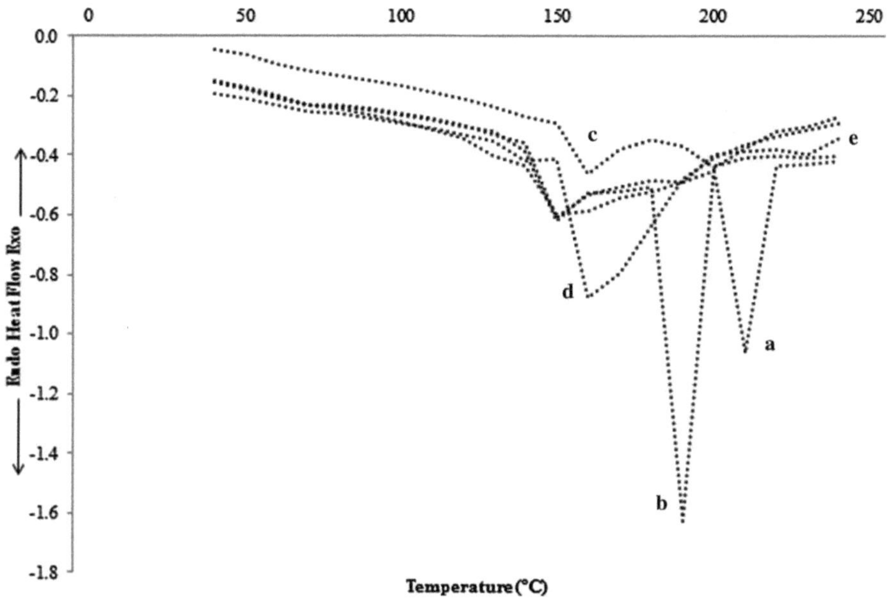

Fig. 10 DSC curves of **a** RW **b** FA-HNC WPNCs **c** 50:50 FA-*co*-EHMA-HNC WPNCs **d** 70:30 FA-*co*-EHMA-HNC WPNCs **e** EHMA-HNC WPNCs

could be identified. The result showed that chemical reaction changes over temperature. However, the properties could be affected due to the moisture content in the wood (Tay et al. 2016). There were two significant stages of thermal decomposition occurred between 50 and 200 °C.

From Fig. 10, it was clearly detected that broad endotherm was observed in the temperature range of 50–200 °C due to the presence of water molecules in the wood cell wall. At the temperature at above 200 °C, it was clearly indicated the decomposition of cellulose in the wood cell wall (Akita and Kase 1967).

All the first endotherm peaks were detected at about 60 °C. The second endotherm peak of RW, FA-HNC, 50:50 FA-*co*-EHMA-HNC, 0:30 FA-*co*-EHMA-HNC, and EHMA-HNC WPNCs occurred at 210, 190, 160, 160, and 150 °C. From the second decomposition temperature, it was significantly indicated that the WPNCs were more thermally stable compared to RW. Both RW and WPNCs differed widely as the wood cell wall at the second stage as wood cell wall contained large amount of hemicellulose that could be decomposed at about 170 °C. From Table 4, WPNCs showed higher bond dissociation enthalpy that proved the higher crystallinity of the WPNCs compared with the RW due to the well impregnation of polymer matrix into wood cell wall (Sulaiman et al. 2013). Overall, WPNCs especially 50:50 FA-*co*-EHMA-HNC WPNCs had the highest thermal stability, as reflected in TGA results.

Table 4 DSC of RW and different ratio of FA-*co*-EHMA-HNC WPNCs

Samples	Endotherm peaks (°C)	Bond dissociation enthalpy (J/g)
RW	139.33	7.64
FA-HNC WPNCs	196.00	19.51
50:50 FA-*co*-EHMA-HNC WPNCs	183.18	83.17
70:30 FA-*co*-EHMA-HNC WPNCs	192.16	61.98
EHMA-HNC WPNCs	191.86	55.30

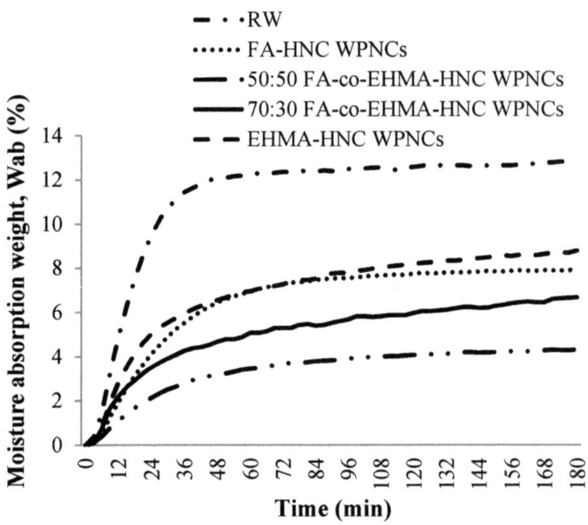

Fig. 11 Moisture absorption curves of RW and different ratio of FA-*co*-EHMA-HNC WPNCs

3.8 *Moisture Absorption Analysis*

The moisture absorption of RW and FA-*co*-EHMA-HNC WPNCs as a function of heating time is shown in Fig. 11. This was very clear that the moisture absorption decreased as different types of polymer matrix was introduced into RW as RW was highly hydrophilic in nature. Incorporation of FA-*co*-EHMA-HNC polymer matrix into wood fiber decreased the rate of moisture sorption ability by breaking hydrogen bond of moisture to form new chemical bonds between the free hydrogen molecules and the polymer matrix (Cao et al. 2011). As the polymer matrix introduced, the number of hydrogen bonds between the moisture greatly reduced. Impregnation of polymer matrix content in RW filled up most of the voids leading lower moisture accumulating at the interface between the wood fiber and the polymer matrix (Jacob et al. 2005). Therefore, WPNCs especially 50:50 FA-*co*-EHMA-HNC WPNCs showed the lowest moisture absorption percentage among the WPNCs and RW.

4 Conclusion

From the present study, RW was impregnated with FA-*co*-EHMA-HNC to produce WPNCs. It is concluded that the physical, mechanical, morphological, and thermal properties of FA-*co*-EHMA-HNC WPNCs were greatly improved compared to the RW. 50:50 FA-*co*-EHMA-HNC WPNCs showed the highest weight percent gain with the least hydroxyl group in FT-IR. Besides, 50:50 FA-*co*-EHMA-HNC WPNCs provided smoother morphology among all the samples. 50:50 FA-*co*-EHMA-HNC WPNCs showed the highest flexural strength and modulus of elasticity. The storage modulus and loss modulus of FA-*co*-EHMA-HNC WPNCs were improved greatly with lower tan δ compared to RW. 50:50 FA-*co*-EHMA-HNC WPNCs exhibited the higher thermal stability with significantly lower moisture absorption. This has proved that the ratio 50:50 FA-*co*-EHMA was the most suitable to be introduced in the RW to form WPNCs.

Acknowledgements The authors would like to acknowledge the financial support from Research and Innovation Management Centre, Universiti Malaysia Sarawak under Fund with Grant no. (F02/SpGS/1443/2016/25) during the research.

References

Ahmad EEM, Luyt AS, Djokovic V (2013) Thermal and dynamic mechanical properties of bio-based poly(furfuryl alcohol)/sisal whiskers nanocomposites. Polym Bull 70(4):1265–1276

Akita K, Kase M (1967) Determination of kinetic parameters for pyrolysis of cellulose and cellulose treated with ammonium phosphate by differential thermal analysis and thermal gravimetric analysis. J Polym Sci Part A 5:833–848

Brown JH, Burnside WR, Davidson AD, DeLong JP, Dunn WC, Hamilton MJ (2011) Energetic limits to economic growth. BioSci 61:19–26

Cao XV, Ismail H, Rashid AA, Takeichi T, Vo-Huu T (2011) Mechanical properties and water absorption of kenaf powder filled recycled high density polyethylene/natural rubber biocomposites using mape as a compatibilizer. BioRes 6(3):3260–3271

Chanmal CV, Jog JP (2008) Dielectric relaxations in PVDF/$BaTiO_3$ nanocomposites. eXPRESS Polym Lett 2(4):294–301

Cheung H, Ho M, Lau K, Cardona F, Hui D (2009) Natural fibre-reinforced composites for bioengineering and environmental engineering applications. Compos Part B Eng 40:655–663

Ende MT, Peppas NA (1997) Transport of ionisable drugs and proteins in crosslinked poly(acrylic acid) and poly(acrylic acid-*co*-2-hydroxyethyl methacrylate) hydrogels. II. Diffusion and release studies. J Control Release 48(1):47–56

Hasnan MAKD, Husseinsyah S, Lim BY, Abdul Rahman MF (2016) Chemical modification of palm kernel shell filled polylactic acid biocomposite films. BioRes 11(3):6639–6648

Hazarika A, Maji TK (2013) Study on the properties of wood polymer nanocomposites based on melamine formaldehyde-furfuryl alcohol copolymer and modified clay. J Wood Chem Technol 33(2):103–124

Hazarika A, Maji TK (2014) Properties of softwood polymer composites impregnated with nanoparticles and melamine formaldehyde furfuryl alcohol copolymer. Polym Eng Sci 54(5):1019–1029

Islam MS, Hamdan S, Rusop M, Rahman MR, Ahmed AS, Mohd Idrus MAM (2012) Dimensional stability and water repellent efficiency measurement of chemically modified tropical light hardwood. BioRes 7(1):1221–1231

Ismail H, Nizam JM, Abdul Khalil HPS (2001) The effect of a compatibilizer on the mechanical properties and mass swell of white rice husk ash filled natural rubber/linear low density polyethylene blends. Polym Test 20(2):125–133

Ismail H, Edyhan M, Wirjosentono B (2002) Bamboo fiber filled natural rubber composites: the effect of filler loading and bonding agent. J Polym Test 21:139–144

Jacob M, Varughese KT, Thomas S (2005) Water sorption studies of hybrid biofiber-reinforced natural rubber biocomposites. Biomacromol 6(6):2969–2979

Karlsson JO, Gatenholm P (1996) Solid-supported wettable hydrogels prepared by ozone induced grafting. Polymer 37(19):4251–4256

Komadan C, George KE, Kumar PAS, Varughese KT, Thomas S (2007) Dynamic mechanical analysis of binary and ternary polymer blends based on nylon copolymer/EPDM rubber and EPM grafted maleic anhydride compatibilizer. eXPRESS Polym Lett 1(10):641–653

Kumari R (2008) Dissertation, Kinki University

Lenzen M, Moran D, Kanemoto K, Foran B, Lobefaro L, Geschke A (2012) International trade drives biodiversity threats in developing nations. Nature 486:107–111

Liew FK, Hamdan S, Rahman MR, Mahmood MR, Rahman MM, Lai JCH, Sultan MT (2016) 4-methylcatechol-treated jute-bamboo hybrid composites: Effects of pH on thermo-mechanical and morphological properties. BioRes 11(3):6880–6895

Lu JZ, Negulescu JI, Wu Q (2005) Maleated wood-fiber/high density polyethylene composites: coupling mechanisms and interfacial characterization. Compos Interf 12(1–2):125–140

Markarian J (2005) Wood-plastic composites: current trends in materials and processing. Plastic Addit Compound 4:20–25

Mohd Idrus MAM, Hamdan S, Rahman MR, Islam MS (2011) Treated tropical wood sawdust-polypropylene polymer composite: mechanical and morphological study. J Biomater Nanobiotechnol 2:435–444

Ndiaye D, Tidjani A (2014) Physical changes associated with gamma doses on wood/polypropylene composites. IOP Conf Series Mater Sci Eng 62(1):012025

Quintens D, Groeninckx G, Guest M, Aerts L (1990) Mechanical behaviour related to the phase morphology of PC/SAN polymer blends. Polym Eng Sci 30:1474–1483

Rahman MR, Hamdan S, Hashim DMA, Islam MS, Takagi H (2015) Bamboo fiber polypropylene composites: effect of fiber treatment and nanoclay on mechanical and thermal properties. J Vinyl Addit Technol 21(4):253–258

Rong MZ, Zhang MQ, Liu Y, Yang GC, Zeng HM (2001) The effect of fiber treatment on the mechanical properties of unidirectional sisal-reinforced epoxy composites. Compos Sci Technol 61:1437–1447

Rose WB (2002) Structure and its enclosure. Paper presented at 2002 ACSA technology conference, Portland Oregon, 10–13 Oct 2002

Sharma RK, Chauhan G (2009) Synthesis and characterization of graft copolymers of 2-hydroxyethyl methacrylate and some comonomers onto extracted cellulose for use in separation technologies. BioRes 4(3):986–1005

Sheshmami S, Ashori A, Hamzeh Y (2010) Physical properties of polyethylene-wood fiber-clay nanocomposites. J Appl Polym Sci 118(6):3255–3259

Sulaiman NS, Hashim R, Amini MHM, Sulaiman O, Hiziroglu S (2013) Evaluation of the properties of particleboard made using oil palm starch modified with epichlorohydrin. BioRes 8(1):283–301

Tay CC, Hamdan S, Osman MS (2016) Urea formaldehyde composites reinforced with sago fibres analysis by FTIR, TGA and DSC. Adv Mater Sci Eng 2016:1–10

Varughese KT, Nando GB, De PP, De SK (1988) Miscible blends from rigid poly (vinyl chloride) and epoxidized NR. J Mater Sci 23:3894–3900

Wang T, Sun H, Long J, Wang YZ, Schiraldi D (2016) Biobased poly(furfuryl alcohol)/clay aerogel composite prepared by a free-drying process. ACS Sustainable Chem Eng 4(5): 2601–2605

Zini E, Scandola M (2011) Green composites: an overview. Polym Compos 32:1905–1915

Sustainable Application of Various Monomer/Clay Dispersed Wood Polymer Nanocomposites

M.R. Rahman, S. Hamdan and J.C.H. Lai

Abstract The goal of this study was to evaluate the decay resistance of different ratio of ST/MMA/clay monomer system impregnated batai wood polymer nanocomposites (WPNCs) against *Trametes versicolor* (white-rot) and *Chaetomium globosum* (soft rot) fungi. Besides, *Kumpang* wood was impregnated by styrene, 3-(trimethoxysilyl) propyl methacrylate (MSPMA), ethylene glycol dimethacrylate (EGDMA), maleic acid (MA), glycidyl methacrylate (GMA), and nanoclay. Overall, both fungi very lightly attacked to the 50:50:5 ST/MMA/clay monomers impregnated WPNCs. 50:50:5 ST/MMA/clay monomer impregnated WPNCs greatly enhance the decay resistance against the both fungi. For Kumpang wood, the raw wood was well impregnated and improved the decay resistance toward *Trametes versicolor* and *Coniphora puteana.* Therefore, it is recommended that 50:50:5 ST/MMA/clay monomers impregnated WPNCs as well as ST-*co*-MSPA-WPC, ST-clay-WPNC, ST-*co*-MA-WPC, and ST-*co*-GMA-WPCs are technically suitable for exterior use where both moisture and favorable conditions for fungi development are present. Besides, WPCs at pH8 and pH9 showed higher decay resistance toward *Coniphora puteana* and *Trametes versicolor*, respectively.

Keywords Styrene · Clay · *Trametes versicolor* · *Chaetomium globosum* · *Coniphora puteana*

1 Introduction

Wood is a three-dimensional natural material, polymeric composite that is made of cellulose, hemicellulose, and lignin. It is a remarkable material of great value and importance in the world's economy (Zabel and Morrell 1992). According to Boryniec and Przygocki (1999), nanocomposites can be defined as a group of

M.R. Rahman (✉) · S. Hamdan · J.C.H. Lai
Faculty of Engineering, Universiti Malaysia Sarawak,
94300 Kota Samarahan, Sarawak, Malaysia
e-mail: rmrezaur@feng.unimas.my

M.R. Rahman, *Wood Polymer Nanocomposites*, Engineering Materials,
DOI 10.1007/978-3-319-65735-6_16

materials with a combination of various raw materials that generate a great possibility to obtain the selected desired properties. Nowadays, wood polymer nanocomposites (WPNCs) are frequently used especially in construction industry. Those are composite materials based on polymer as a warp and various wood fractions could be the fillers.

Wood polymer nanocomposites (WPNCs) bring more advantages than solid wood. According to Wu et al. (2012), the new generation wood-based nanocomposites usually offer enhanced long-term durability for structural purposes that have been specially constructed with natural wood products. Most of the WPNCs offer some inherent technical advantages over conventional nanocomposites, such as lower costs, lower weight, and less energy used in their production (Clemons 2002; Lu et al. 2000; Rahman et al. 2013). All these technical advantages which can be grouped together would qualify them to be called sustainable and "environmentally friendly" or "green". Besides, WPNCs can be manufactured in a variety of colors, shapes, and sizes, and also different surface textures. Usually WPNCs do not require painting or other finishes, and they normally will not wrap or rot like wood does.

Although there are a lot of advantages, WPNCs are not maintenance-free, and they can be degraded in outdoor environments. The most important point is the wood in the WPNCs can still be attacked by termites, rot, and mold fungi (Laks et al. 2000; Verhey et al. 2001). Fungi act as decomposers in the forest and play an important role in the ecosystem. Wood cell wall components are usually attacked by different kinds of fungi with differences in behavior. There are three major groups of decay fungi, the white-rot, brown-rot and soft rot fungi. White-rot fungi have the ability to degrade lignin (Pointing et al. 2003). Therefore, white-rot fungi have been used as biological agents in pretreatment to degrade the lignin and disrupt the crystalline structure of cellulose. The term soft rot was first coined by Savoy (1954) to describe a type of wood decay initiated by fungi imperfect and ascomycetes in moist and aquatic environment. Soft decay is divided into two groups, type 1 and type 2, where type 1 results in the development of cavities in the secondary walls of wood fibers (primarily the S2 layer), and type 2 results in the erosion of the wood cell wall outward from the cell lumen. Wood extractives, and to lesser extent lignin, naturally retard fungal attack in wood (Eaton and Hale 1993). Decay fungi and termites cause wood biodeterioration and biodegradation (Zabel and Morrell 1992).

For this reason, chemical modification and hinders penetration of low molecular weight diffusible polymers are required for against the fungal degradation into the wood cell wall (Hill et al. 2005; Papadopoulos and Hill 2002). In order to reduce wood deterioration, chemical treatment is one of the most effective ways to induce dimensional stability, UV resistance, and biological resistance in wood (Chang and Chang 2001a, b; Evans et al. 2000; Hill 2006; Plakette et al. 1996; Rowell 1983, 2005, 2006; Williams 2005). Chemical agents are commonly employed to control wood degradation caused by stain and decay fungi. Phenol-formaldehyde (PF) resin provided more resistance to fungal degradation than urea-formaldehyde (UF) resin due to its high pH and the presence of non-condensed phenols (Schmidt et al. 1978). According to Cao et al. (2011), wood fungal and insect resistance had improved by thermal treatment. Boric acid, ammonium hydroxide, alkoxysilanes, and immobilize copper impregnated wood increased the resistance against decay fungi (Feci et al. 2009; Hien and Li 2012; Reinsch et al. 2002; Solar et al. 2005; Terziev et al. 2009).

The styrene (ST), methyl methacrylate (MMA), and butyl acrylate (BA) were used to synthesize a polyacrylic emulsion by core-shell emulsion polymerization that played an important role in the water-resistant ability of the fiber-based materials (Chen and Wu 2014). Styrene-butadiene lattices are widely used as binders in pigmented coatings for the paper and board industry (Piltonen et al. 2013). Poly (methyl methacrylate)-functionalized grapheme/polyurethane (MG-PU) dielectric elastomer nanocomposites were synthesized and used as conducting filler (Chen et al. 2014). According to our previous research, we found that the ST and MMA ratio need to be concerned as different ratio of ST and MMA will produce different types of polymers which might affect the wood impregnation and the quality of wood polymer nanocomposites (Rahman et al. 2013).

Besides, 3-(trimethoxysilyl)propyl methacrylate (MSPA) acted as coupling agent was bonded onto colloidal silica to be introduced into the polyester-based waterborne polyurethane hybrid films led to a significant improvement in thermal decomposition temperature as well as mechanical properties (Lin et al. 2012). Multiresponsive polymer composite microspheres with core-shell structure which synthesized in water by precipitation polymerization were modified with MSPA which improved the glass transition temperature (Cai et al. 2007). MSPA was used to modified the wood species which had improved the mechanical properties of the wood composites (Rangel-Vazquez and Leal-Garcia 2010). In addition, graft copolymerization of MSPA onto styrene-butadiene-styrene (SBS) triblock copolymer improved both adhesive property as well as the tensile strengths due to strong Si-O-Si bonding between SBS-g-MSPA (Lee and Chen 2015).

Ethylene glycol dimethacrylate (EGDMA) improved the interpenetrating polymer network (IPN) in the presence of exfoliated clay as well as the mechanical properties of IPN (Esmaeil et al. 2014). EGDMA crosslinker was introduced into a wood cell wall and block the water uptake as well as enhanced the ultimate strength (Duanmu et al. 2007). A polymer-impregnated wood composite with the introduction of GMA/EGDMA in the ratio of 2:1 had significantly improved mechanical properties and durability relative to raw wood (Dong et al. 2015).

Maleic anhydride (MA)-grafted polypropylene was introduced into bamboo powder/polypropylene foamed composites showed significant improvement in mechanical properties and water resistance (Zhou et al. 2013). Moreover, MA was combined with glycidyl methacrylate/methyl methacrylate (GMA/MMA) through graft copolymerization and was introduced into the wood cell wall to improve the compatibility between polymer and wood cell wall which improve the decay resistance of the resultant composite (Li et al. 2011a, b). MA was employed as compatibilizer to improve the adhesion between polyethylene matrix and bamboo fibers (Eze et al. 2013).

In addition, glycidyl methacrylate (GMA) functioned as the crosslinking monomer to be combined with styrene to be impregnated into the wood has greatly improved the water uptake percentage as well as the modulus of elasticity and modulus of rupture. The thermal properties of ST-GMA/wood composites were enhanced significantly compared to raw wood or styrene-treated composites (Devi and Maji 2007). ST-*co*-GMA impregnated wood polymer nanocomposites (WPNC) carried out by Sultan et al. (2016) showed that the WPNC was improved physically, mechanically, and morphologically. In situ polymerization of ST-GMA into poplar wood showed strong crosslinking reaction between the wood and polymers due to the presence of both terminal double bond and the epoxy group of GMA (He et al. 2014).

Nanoclay plays an important role in the monomer mixture in the composite system. It is used as a reinforcing material in polymer–clay nanocomposites, and it exhibited substantial improvement in many physical properties including mechanical properties, thermal resistance, ionic conductivity, and flame resistance (Pinnavaia and Beall 2000; Rahman et al. 2010). The ratio of ST to MMA is important to differentiate the monomer systems prepared and to investigate the effect of the impregnation of higher loading of styrene with lower loading of methyl methacrylate on the batai wood. Clay was kept at certain level as the amount of clay will effect on the mechanical properties.

However, the application of chemical fungicides has become limited in order to meet the requirement for low pesticide residues in products and to increasing problems associated with resistance to the active principles (Mommaerts et al. 2008). Based on our previous work, we have set up the condition that is not excited the pesticide limit (Rahman et al. 2013).

Thus, the aim of present study is to prepare the combined monomer system such as 20:80:5, 30:70:5, 40:60:5, and 50:50:5 ST/MMA/clay, respectively, to impregnate batai wood specimen and investigate the decay resistance. Besides, this study also works on tropical wood impregnation and chemical modification concurrently with styrene-*co*-3-(trimethoxysilyl) propyl methacrylate, styrene-co-ethylene glycol dimethacrylate, styrene-*co*-maleic acid impregnated, styrene-co-glycidyl methacrylate as well as 4-methyl catechol at different pH levels of NaOH to evaluate the decay resistance toward fungi.

2 Experimental

2.1 Materials

The batai wood (local name) species were collected for this study. The dimensions of the wood specimen were 9 mm (L-longitudinal) × 25 mm (T-tangential) × 25 mm (R-radial). The chemicals used to produce WPNCs were styrene (ST), methyl methacrylate (MMA), benzoyl peroxide, and Nanoclay Nanomer 1.31PS were supplied by Merck, Germany, and Sigma-Aldrich, respectively. Catalog number of the clay was 682632-500G. It was montmorillonite clay surface modified with 15–35 wt% octadecylamine and 0.5–5 wt% aminopropyl-triethoxysilane. The bulk density of the clay was 200–500 kg/m^{3,} and average particle size was around 20 μm.

Defect-free and straight-grained Kumpang raw wood (RW) was obtained from Forestry Department, Sarawak, Malaysia. The chemicals used to produce WPCs and WPNCs were styrene (ST), 3-(trimethoxysilyl) propyl methacrylate (MSPMA), ethylene glycol dimethacrylate (EGDMA), maleic acid (MA), glycidyl methacrylate (GMA), 4-methyl catechol (C), nanoclay (Nanomer®1.28E), ethanol, benzoyl peroxide and sodium hydroxide. Styrene was supplied by A. S. Joshi and Company, Goregaon West, Mumbai, India. Ethylene glycol dimethacrylate (EGDMA), sodium hydroxide, and benzoyl peroxide were supplied by Merck Schuchardt OHG, Germany. 3-(trimethoxysilyl) propyl methacrylate, 4-methyl catechol, glycidyl methacrylate, ethanol, and Nanoclay (NanomerR1.28E) that had been modified with 25–30 wt% trimethyl stearyl ammonium were supplied by Sigma-Aldrich (USA). Benzoyl peroxide (BP) was used as a free radical catalyst to induce polymerization of monomers. Ethanol was used as a solvent for 4-methyl catechol. Sodium hydroxide acted as a catalyst for polymerization of 4-methyl catechol. In this study, all chemicals were of analytical grade.

2.2 *Preparation of ST*-co-*MMM-Nanoclay*

The monomer system was prepared using ST/MMA/nanoclay in the presence of benzoyl peroxide. Benzoyl peroxide acts as an initiator to influence the reaction between styrene and methyl methacrylate. Two grams of benzoyl peroxide and 5 g of nanoclay were added into each monomer system. The mixtures were covered with aluminum foil and put in an autoclave for 15 min to complete the reaction. Styrene-co-Methyl Methacrylate with nanoclay monomer system was mixed with different ratios as shown in Table 1. The reaction scheme of the Styrene-co-Methyl Methacrylate with nanoclay cross-linker had shown Fig. 1.

Table 1 Preparation of monomer system at different ratio

Volume of styrene (mL)	Volume of methyl methacrylate (mL)	Amount of nanoclay (g)
20	80	5
30	70	5
40	60	5
50	50	5

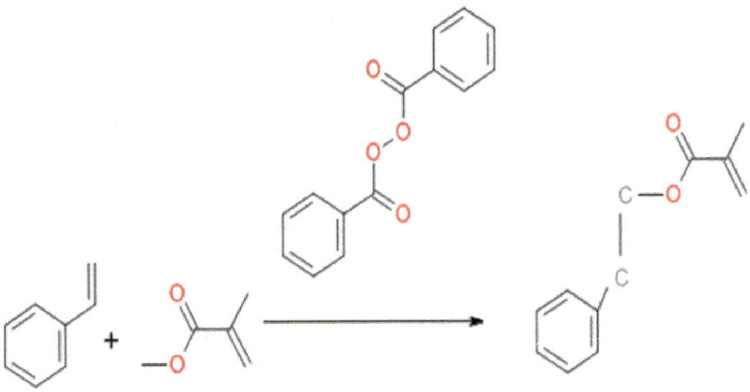

Fig. 1 Reaction scheme for synthesis of ST/MMA/nanoclay crosslinker

2.3 Impregnation of Wood Specimens with ST-co-MMM-Nanoclay

The specimens were immersed into different monomer systems and were subjected to a vacuum process for 1 h. After 1 h, the samples were taken out and covered with aluminum foil for 48 h at 70 °C to carry out the polymerization with wood cell wall and produce wood polymer nanocomposites (WPNCs). After 48 h, the WPNCs were taken out of the oven and the unreacted chemicals were removed. The coupling reaction between wood cell wall and Styrene-co-Methyl Methacrylate is shown in Fig. 2.

2.4 Specimen Preparation

A Kumpang tree was cut into three bolts that were 1.2 m in length. Each bolt was sawn into quarters in order to produce planks that were 4 cm thick. These planks were subsequently conditioned by air drying in a room with a relative humidity of 60% and ambient temperature of approximately 25 °C for one month prior to

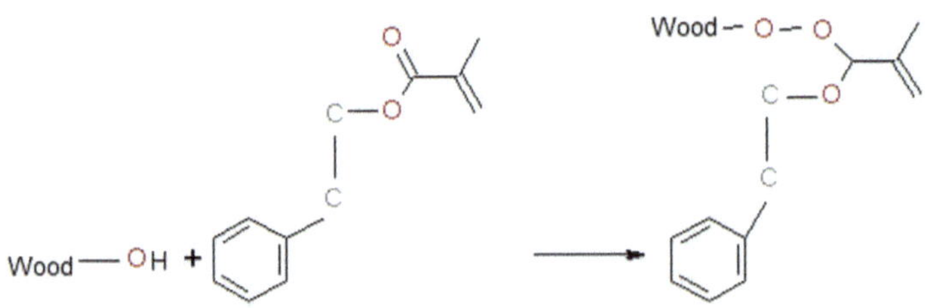

Fig. 2 Coupling reaction of wood with ST/MMA/nanoclay

testing. The planks were ripped and machined to 340 mm (L) × 20 mm (T) × 10 mm (R) for three-point bending and free-free vibration tests.

2.5 Preparation of Different WPNCs and WPCs

In this study, for preparation of different types of five WPNCs and nine WPCs, oven-dried raw wood samples have been impregnated with either one monomer, or a mixture of styrene with one monomer, or a mixture of styrene with clay or a mixture of styrene with clay and a monomer. The compositions of the mixtures for impregnation of wood are described in Tables 2 and 3. The impregnation of the raw wood by this mixture or a single monomer was completed in a vacuum chamber (75 mm Hg) for 30 min. These impregnated wood samples were then removed from the chamber, and the excess chemicals were removed. Specimens were wrapped with aluminum foil and placed into a 105 °C oven for 24 h for polymerization to take place. In each impregnation, ten replicate samples were produced.

2.6 Decay Tests for Wood Specimens with ST-co-MMM-Nanoclay

Samples were packed in an argon atmosphere and sterilized by irradiation (2.5 Mrad.) prior to decay testes, using the methods described in DD ENV 12038:1996. Laboratory pure strains of the brown-rot fungi *Chaetomium globosum* and white-rot fungi *Trametes versicolor* were used, grown on malt agar. Blocks were planted on sterile specimen supports placed on the cultures of the test fungus actively growing on 5% malt agar in 500 ml capacity jars. An additional set of sterile control samples were used to assess operational control losses. The closed jars were incubated for 16 weeks at 22 ± 1 °C and 75 ± 5% relative humidity to evaluate the efficacy of the treatments. After incubation, the samples were removed from the jars, cleaned, weighted, conditioned to weight as above and re-weighed. A weight loss (WL) was expressed as a percentage of the initial weight of the sample. Weight losses from sterile controls were subtracted from the decay results to give corrected data.

2.7 Laboratory Fungal Decay Resistance Test for WPCs and WPNC

The decay resistance test was carried out using the Standard Method of Accelerated Laboratory test of natural decay resistance of wood ASTM D2017 (2001). Decay

Table 2 The composition of materials for the manufacturing of WPCs and WPNCs

Name of the prepared WPCs/WPNCs	Amount of the corresponding materials					Matrix
	ST (mL)	MSPM (mL)	EGDMA (mL)	MA (g)	Clay (g)	
ST-WPC (ST impregnated wood)	1000	–	–	–	–	Raw wood
ST-clay-WPNC (ST with clay impregnated wood)	1000	–	–	–	10	
ST-*co*-MSPMA-WPC (ST with MSPMA impregnated by)	1000	200	–	–	–	
ST-*co*-MSPMA-clay-WPNC (ST with MSPMA and clay impregnated wood)	1000	200	–	–	10	
ST-*co*-EGDMA-WPC (ST with EGDMA impregnated wood)	1000	–	500	–	–	
ST-*co*-EGDMA-clay-WPNC (ST with EGDMA and clay impregnated wood)	1000	–	500	–	10	
ST-*co*-MA-WPC (ST with MA impregnated wood)	1000	–	–	35	–	
ST-*co*-MA-clay-WPNC (ST with MA and clay impregnated wood)	1000	–	–	35	10	
ST-*co*-GMA-WPC (ST with GMA impregnated wood)	1000	250	–	–	–	
ST-*co*-GMA-clay-WPNC (ST with GMA and clay impregnated wood)	1000	250	–	–	10	

10 g BP was used as an initiator in each impregnation

Table 3 Wood impregnation by 4-methyl catechol at different pH levels

pH levels of 5% 4-methyl catechol in 1000 mL methanol	Name of prepared composites
8	CWPC at pH 8
9	CWPC at pH 9
10	CWPC at pH 10
11	CWPC at pH 11

CWPC (4-methyl catechol impregnated wood polymer composite)

resistance was classified using the scale described in ASTM Standard D2017 where highly resistant heartwood experiences (0–10)% weight loss, resistant wood (11–24)% weight loss, moderately resistant wood (25–44)% weight loss, and nonresistant wood experiences weight loss greater than 45%.

The specimens were air dried and after conditioning to constant weight, they were weighed accurately in the laboratory and transferred into a large totally dark container maintained at 20 ± 1 °C and a relative humidity of 65 ± 4%. Two types of fungus, white-rot (*Trametes versicolor*) and brown-rot (*coniphora puteana*) were used to study the decay resistance of WPCs and WPNCs against fungi. There were six replications for each specimen. The decay test was terminated after 14 weeks. Mycelium was brushed off and test specimens were air dried and again conditioned to constant weight. The weight was recorded for each specimen. The difference in weights of specimens before and after the decay test gave the rate of decay in test specimen.

3 Results and Discussion

*3.1 Decay Test for Wood Specimens with ST-*co*-MMM-Nanoclay*

Weight loss due to fungal attack for control sample and WPNCs are shown in Fig. 3. The result revealed that both control and WPNCs were affected by the decay fungi *Trametes versicolor* (white-rot) and *Chaetomium globosum* (brown-rot), respectively. The control wood samples were highly affected by both fungi compare with 20:80:5, 30:70:5, 40:60:5, and 50:50:5 (ST-*co*-MMA/clay) monomer impregnated WPNCs, respectively. These results indicate that after 12 week incubation period for both fungi very lightly attack to the 50:50:5 ST-*co*-MMA/clay impregnated WPNCs. From Fig. 3, it showed that most of the control and modified samples were less decay resistance toward *Chaetomium globosum* fungus compared to *Trametes versicolor* fungus. However, 50:50:5 ST/MMA/clay monomer system impregnated WPNCs showed high resistance toward both the fungi which proved that the system itself can inhibit fungal growth on agar medium and it was established by our previous work that showed the higher thermal stability and mechanical properties (Rahman et al. 2013; Akbarpour et al. 2013; Sapkota 2011). The ratio 50:50:5 ST-*co*-MMA/clay was highly protect the batai softwood species against the decay fungi due to their complete copolymerization and there was no excess ST or MMA. Visual examination also proved the decay resistance of the different monomer systems as shown in Figs. 4 and 5, respectively. A slight presence of mycelium in the surface of the 40:60:5 and 50:50:5 (ST-*co*-MMA/clay) impregnated WPNCs tested samples. Similar observation has also been found by Rahman et al (2010) in cement-bonded particleboards. From Fig. 4, it showed that both of the control and 50:50:5 (ST-*co*-MMA/clay) samples for week 1 to week 12. It was clearly seen that combined monomer system (ST-*co*-MMA/clay) impregnated WPNCs showed high resistance against decay fungi compared with control sample. From Fig. 5, there was an enormous visual difference between Fig. 5b, d. It showed that control samples were fully attacked by the fungus, whereas the 50:50:5

ST/MMA/clay monomer systems were slightly attacked by the fungus. Due to the clay addition, it acts as reinforcement filler for all ratios and enhances not only the decay resistance but also improves the mechanical and thermal properties (Rahman et al. 2013).

The result also showed that generally control wood samples were nonresistant to decay exposure. How, the monomer systems 40:60:5 and 50:50:5 (ST-*co*-MMA/clay) enhance the decay resistance and dramatically reduce the weight losses against the both fungi. From our previous research, it has been proven that 50:50:5 (ST-*co*-MMA/clay) monomer impregnated WPNCs showed better mechanical, thermal, and morphological properties compared with 40:60:5, 30:70:5, 20:80:5 ((ST-*co*-MMA/clay) monomer impregnated WPNCs and control wood sample, respectively (Rahman et al. 2013). The result also proved that different ratio of the monomer systems may affect the decay resistance toward the both fungal attack.

Furthermore, it concludes that the bonding interaction between wood cell wall and the monomer systems 50:50:5(ST-*co*-MMA/clay) and 40:60:5(ST-*co*-MMA/clay) ratios were strong enough resistant to the attack by both decay fungi.

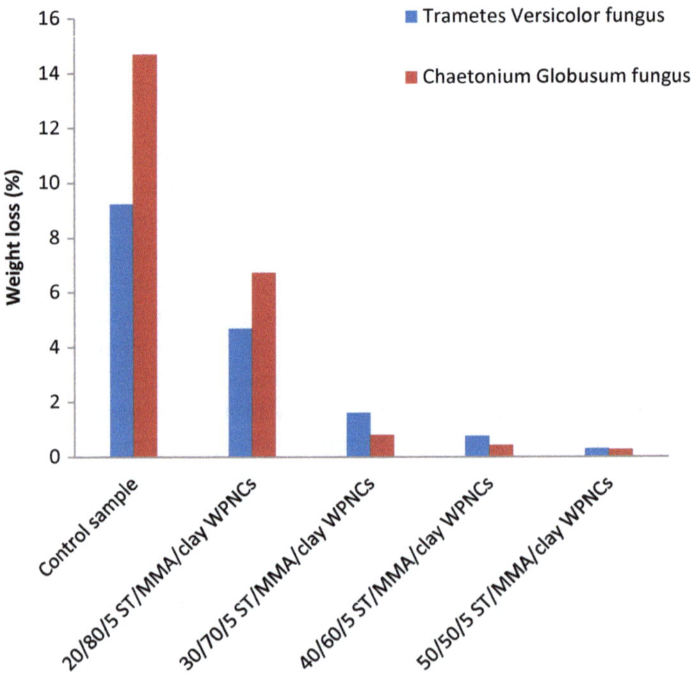

Fig. 3 Weight losses of control wood and WPNCs after exposure to the decay fungus *Trametes versicolor* (white-rot) fungus and *Chaetomium globosum* (brown-rot) for 12 weeks

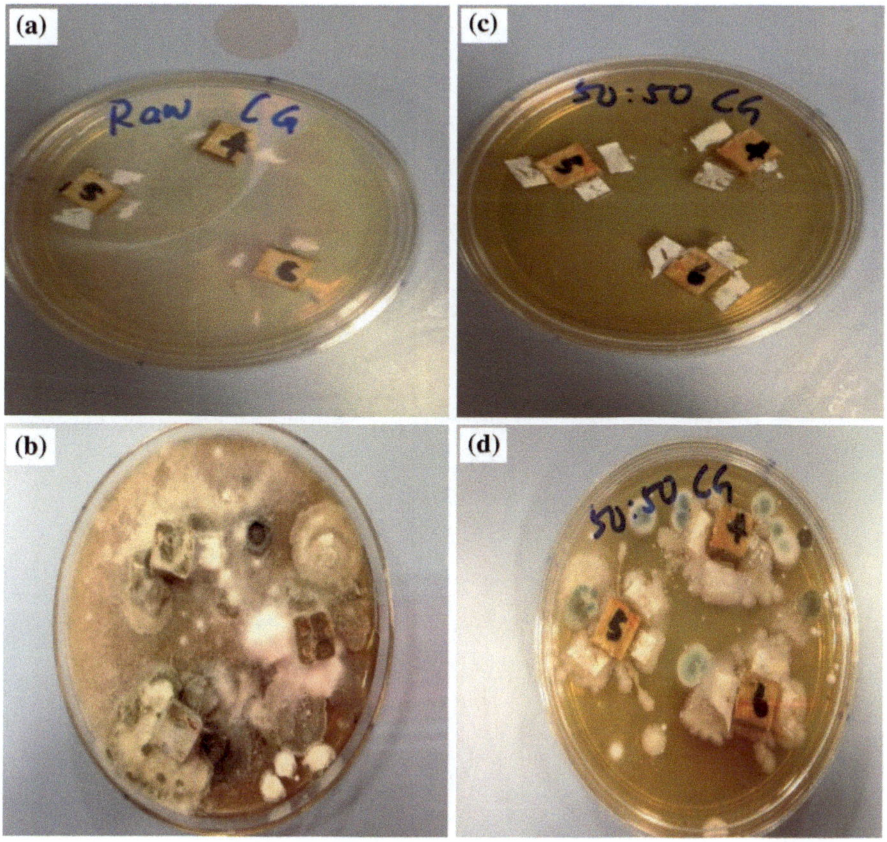

Fig. 4 *Chaetomium globosum* fungus exposure petri dish for **a** control sample (week 1), **b** control sample (week 12) **c** 50:50:5 WPNCs (week 1) and **d** 50:50:5 WPNCs (week 12)

3.2 Decay Resistance of Styrene-co-3-(Trimethoxysilyl) Propyl Methacrylate with Clay Impregnated Wood Polymer Nanocomposites

Weight loss due to fungal attack for raw wood, WPCs, and WPNC was given in Table 4. The results showed that raw wood, WPCs, and WPNCs were affected by the exposure to the decay fungi of *Trametes versicolor* and *Coniphora puteana*, respectively. ST-*co*-MSPA-WPC was less affected by *Trametes versicolor*, compared to *Coniphora puteana*. The results also showed that generally all raw wood species were low resistant to decay exposure (Rahman et al. 2010). However, styrene enhanced the decay resistance and decreased the weight losses to the fungi attack. According to the results, ST-*co*-MSPA-WPC was the highest resistant toward *Trametes versicolor* fungi decay exposure, followed by ST-clay-WPNC,

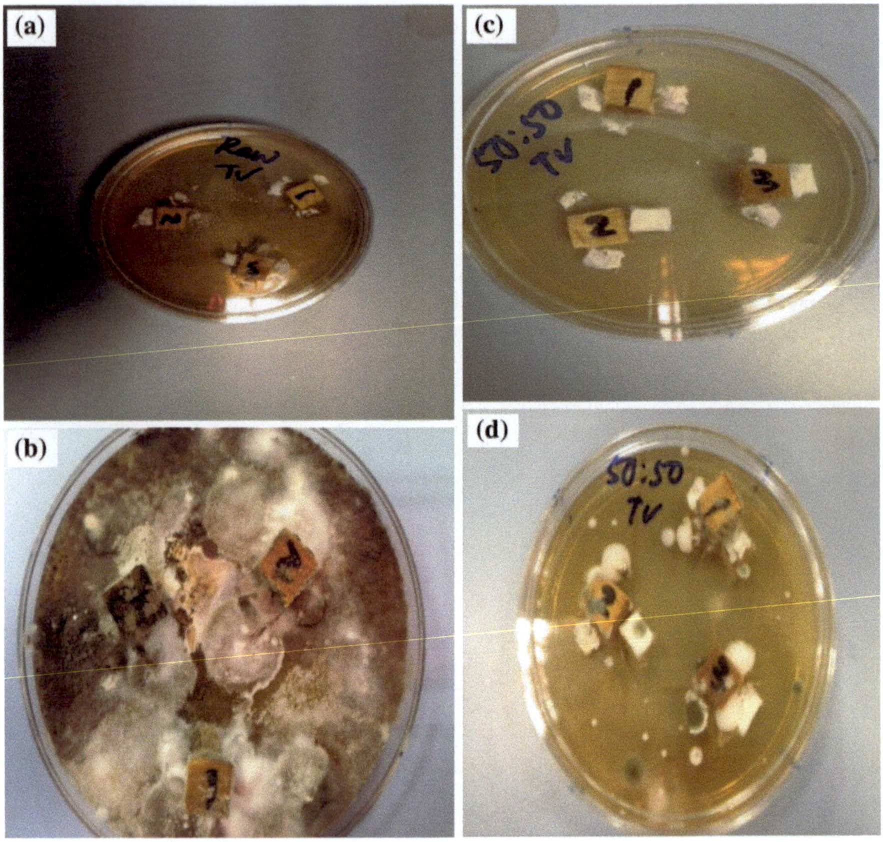

Fig. 5 *Trametes versicolor* fungus exposure petri dish for **a** control sample (week 1), **b** control sample (week 12) **c** 50:50:5 WPNCs (week 1) and **d** 50:50:5 WPNCs (week 12)

ST-*co*-MSPA-clay-WPNC, ST-WPC, and raw wood. For *Coniphora puteana* fungi decay exposure, all the specimens showed moderately resistant with the range of 7–17%.

Wood treated with different silanes showed improved decay resistance and more improvement were noticed when the silanes contained larger alkyl group (Donath et al. 2004). In this study, 1% weight loss observed in ST-co-MSPA modified wood whereas silane treated wood loss 5% weight at the same fungi attack (Donath et al. 2004). It may attribute that ST-*co*-MSPA has antifungal efficiency and this efficiency decreases in presence of clay. ST impregnated wood showed lower resistance against fungi due to a barrier of water up taking compared to RW. It could be concluded that chemical modification was highly effective on the decay resistance test to improve the raw wood (Yalinkilic et al. 1998).

Table 4 *t*-test analysis for decayed samples for exposing to brown or white fungi

Samples	Fungus	Weight loss (%)	t-test grouping[a]	Decay resistance according to ASTM D2017
RW	*C. putena*	13.79 ± 0.01	A	Resistant
	T. versicolor	58.00 ± 0.00	A1	Nonresistant
ST-WPC	*C. putena*	17.40 ± 0.02	B	Resistant
	T. versicolor	52.00 ± 0.02	B1	Nonresistant
ST-*co*-MSPA-WPC	*C. putena*	10.26 ± 0.04	C	Highly resistant
	T. versicolor	1.80 ± 0.03	C1	Highly resistant
ST-clay-WPNC	*C. putena*	9.20 ± 0.00	D	Highly resistant
	T. versicolor	15.74 ± 0.06	D1	Resistant
ST-*co*-MSPA-clay-WPNC	*C. putena*	7.30 ± 0.04	E	Highly resistant
	T. versicolor	25.00 ± 0.03	E1	Moderately resistant

[a]The same letters are not significantly different at $\alpha = 5\%$

3.3 *Investigation of Decay Resistance Properties of Clay Dispersed Styrene-*co*-Ethylene Glycol Dimethacrylate Impregnated Wood Polymer Nanocomposites*

Table 5 showed the weight loss due to fungal attack for raw wood, WPCs, and WPNC toward the decay fungi of *Trametes versicolor* and *Coniphora puteana*, respectively. All the samples showed higher fungi decay resistance toward the *Coniphora puteana* fungi compared to *Trametes versicolor* fungi. From Table 5, ST-clay-WPNC showed better fungi resistance toward both the fungi attacks followed by ST-*co*-EGDMA-WPC, ST-*co*-EGDMA-clay-WPNC, ST-WPC, and RW. For *Coniphora puteana* fungi decay exposure, all the specimens showed moderately resistant with the range of 7–17% while the all the specimens showed weight loss range toward *Trametes versicolor* fungi from 15–58%.

Among the WPCs and WPNCs, ST-WPC showed the higher deterioration by fungi because the impregnation of raw wood by ST was lower and the hydrophilicity of ST-WPC was the highest (Emeydan et al. 2014). The *Coniphora puteana* fungi resistance of WPCs and WPNCs were higher due to the effects of esterification of wood by EGDMA and encapsulation by clay (Doczekalska et al. 2014; Berix 2012). Therefore, it can be concluded that wood esterified by EGDMA was less effective to resist *Trametes versicolor* fungi compared to clay loaded WPNC.

Table 5 *t*-test analysis for decayed samples for exposing to brown or white fungi

Samples	Fungus	Weight loss (%)	t-test grouping[a]	Decay resistance according to ASTM D2017
RW	*C. putena*	13.79 ± 0.01	A	Resistant
	T. versicolor	58.00 ± 0.00	A1	Nonresistant
ST-WPC	*C. putena*	17.40 ± 0.02	B	Resistant
	T. versicolor	52.00 ± 0.02	B1	Nonresistant
ST-*co*-EGDMA-WPC	*C. putena*	12.11 ± 0.06	C	Resistant
	T. versicolor	18.00 ± 0.10	C1	Resistant
ST-clay-WPNC	*C. putena*	9.20 ± 0.00	D	Highly resistant
	T. versicolor	15.74 ± 0.06	D1	Resistant
ST-*co*-EGDMA-clay-WPNC	*C. putena*	7.80 ± 0.03	E	Highly resistant
	T. versicolor	20.33 ± 0.03	E1	Resistant

[a]The same letters are not significantly different at $\alpha = 5\%$

3.4 *Clay Dispersed Styrene-*co*-Maleic Acid Impregnated Wood Polymer Nanocomposites: Impact on Decay Resistance Properties*

The weight loss due to fungal attack for raw wood, WPCs, and WPNC were given in Table 6. All the specimens were significantly affected by the exposure to the decay fungi of *Trametes versicolor* and *Coniphora puteana*, respectively. ST-*co*-MA-WPC and ST-*co*-MA-clay-WPNC were less affected by *Trametes versicolor*, compared to *Coniphora puteana* among all the specimens. From Table 6, it was clearly showed that the raw wood was significantly improved by the introduction of ST, MA, and clay. However, all the specimens showed better decay resistance toward *Coniphora puteana* fungi compared to *Trametes versicolor* fungi. This was due to some unreacted –COOH groups of MA which have lower resistance against *Trametes versicolor* fungi than *Coniphora puteana* fungi (Oldertroen et al. 2016). Rubber wood treated with maleic anhydride showed higher weight losses which were 36.04 and 7.09% against *Coriolus versicolor* (white-rot fungi) and *Polyporus meliae* (brown-rot fungi) compared to ST-*co*-MA-WPC and ST-*co*-MA-clay-WPNC (Nagaveni et al. 2005). Therefore, it could be concluded that the ST-*co*-MA-WPC and ST-*co*-MA-clay-WPNC showed better resistance against white-rot fungi and brown-rot fungi among all the specimens.

3.5 Clay Dispersed Styrene-co-Glycidyl Ethacrylate Impregnated Wood Polymer Nanocomposites: Impact on Decay Resistance Properties

The results that indicated that the type of fungus and fungal exposure on the specimens had significant effect on weight loss was given in Table 7. *Trametes versicolor* caused a higher weight loss of 58% while *Coniphora puteana* caused a higher weight loss of 52% on the raw wood. The great weight loss of raw wood specimen exposed to *Trametes versicolor* could indicate a difference in decay mechanism (Zabihzadeh et al. 2009). This showed that raw wood has lower fungi decay resistance compared to WPCs and WPNCs (Gupta et al. 2004). Wood etherified by isopropyl glycidyl ether showed drastic reduction in weight loss due to fungal decay by white-rot and brown-rot fungi (Chang and Chang 2001a, b). ST-*co*-GMA-clay-WPNC showed the highest decay resistance against *Coniphora puteana* fungi compared to ST-*co*-GMA-WPC and ST-clay-WPNC. This was due to the synergic effect of ST-*co*-GMA and clay in wood (Li et al. 2011a, b). Besides, the presence of clay decrease the antifungal efficiency of WPC and WPNC as clay had low catalytic effect to resist both fungi attacks (Liu et al. 2016). Therefore, it could be concluded that chemical modification was significantly effective to improve the raw wood toward fungi attack.

Table 6 *t*-test analysis for decayed samples for exposing to brown or white fungi

Samples	Fungus	Weight loss (%)	t-test grouping[a]	Decay resistance according to ASTM D2017
RW	*C. putena*	13.79 ± 0.01	A	Resistant
	T. versicolor	58.00 ± 0.00	A1	Nonresistant
ST-WPC	*C. putena*	17.40 ± 0.02	B	Resistant
	T. versicolor	52.00 ± 0.02	B1	Nonresistant
ST-*co*-MA-WPC	*C. putena*	8.70 ± 0.05	C	Highly resistant
	T. versicolor	1.00 ± 0.10	C1	Highly resistant
ST-clay-WPNC	*C. putena*	9.20 ± 0.00	D	Highly resistant
	T. versicolor	15.74 ± 0.06	D1	Resistant
ST-*co*-MA-clay-WPNC	*C. putena*	9.10 ± 0.09	D	Highly resistant
	T. versicolor	5.6 ± 0.04	E1	Highly resistant

[a]The same letters are not significantly different at $\alpha = 5\%$

Table 7 *t*-test analysis for decayed samples for exposing to brown or white fungi

Samples	Fungus	Weight loss (%)	t-test grouping[a]	Decay resistance according to ASTM D2017
RW	*C. putena*	13.79 ± 0.01	A	Resistant
	T. versicolor	58.00 ± 0.00	A1	Nonresistant
ST-WPC	*C. putena*	17.40 ± 0.02	B	Resistant
	T. versicolor	52.00 ± 0.02	B1	Nonresistant
ST-*co*-GMA-WPC	*C. putena*	9.30 ± 0.02	C	Highly resistant
	T. versicolor	7.17 ± 0.12	C1	Highly resistant
ST-clay-WPNC	*C. putena*	9.20 ± 0.00	D	Highly resistant
	T. versicolor	15.74 ± 0.06	D1	Resistant
ST-*co*-GMA-clay-WPNC	*C. putena*	7.30 ± 0.04	E	Highly resistant
	T. versicolor	21.00 ± 0.06	E1	Resistant

[a]The same letters are not significantly different at $\alpha = 5\%$

3.6 *Decay Resistance Characterization of Wood Polymer Composites Impregnated by 4-Methyl Catechol at Various pH Levels*

The weight losses on raw wood and WPCs with different pH were presented in Table 8. The degree of fungi resistance of WPCs depended on the weight percentage gain by the impregnation raw wood with 4-methyl catechol. The results showed that WPCs at lower pH level (pH8) was less affected by *Trametes versicolor* and *Coniphora puteana* fungi. For *Coniphora puteana* fungi decay exposure, WPCs at pH9 showed the highest resistance with the lowest weight loss while

Table 8 *t*-test analysis for decayed samples for exposing to brown or white fungi

Samples	Fungus	Weight loss (%)	t-test grouping[a]	Decay resistance according to ASTM D2017
Raw wood	*C. putena*	13.79 ± 0.01	A	Resistant
	T. versicolor	58.00 ± 0.00	A1	Nonresistant
WPCs at pH8	*C. putena*	9.32 ± 0.02	B	Highly resistant
	T. versicolor	7.80 ± 0.12	B1	Highly resistant
WPCs at pH9	*C. putena*	6.50 ± 0.04	C	Highly resistant
	T. versicolor	12.40 ± 0.05	C1	Resistant
WPCs at pH10	*C. putena*	9.29 ± 0.04	D	Highly resistant
	T. versicolor	14.00 ± 0.06	D1	Resistant
WPCs at pH11	*C. putena*	10.00 ± 0.06	E	Highly resistant
	T. versicolor	17.40± 0.01	E1	Resistant

[a]The same letters are not significantly different at $\alpha = 5\%$

WPCs at pH8 showed the lowest weight loss which indicated the highest fungi decay resistance toward *Trametes versicolor* fungi.

All the WPC showed lower weight losses due to the polymerized 4-methylcatechol which created strong covalent bond with the matrix and significantly able to resist *Trametes versicolor* and *Coniphora puteana* fungi (Billes 2015). It could be concluded that chemical modification was highly effective on the decay resistance test, as found by previous researchers (Yalinkilic et al. 1998).

4 Conclusion

Trametes versicolor and *Chaetomium globosum* fungi used in this work were very lightly attacked to the 50:50:5 (ST-*co*-MMA/clay) monomer impregnated WPNCs except others ratio and control wood. 50:50:5 (ST-*co*-MMA/clay) monomer impregnated WPNCs greatly enhance the decay resistance against the both fungi. It is recommended that 50:50:5 (ST-*co*-MMA/clay) monomers impregnated WPNCs are technically suitable for exterior use where both moisture and favorable condition for fungi development are present. Besides, *Trametes versicolor* and *Coniphora puteana* slightly attack WPC and WPNC compared to raw wood due to the better combination between the matrices with the wood cell. WPCs at pH8 showed better decay resistance toward *Coniphora puteana* fungi attack while WPCs at pH9 showed higher decay resistance toward *Trametes versicolor* fungi exposure. For this research, it showed that nanoclay was compatible with ST combined monomer system to improve the quality of the monomer system and also the crosslinking between the monomers.

Acknowledgements The authors would like to acknowledge the financial support from Ministry of Higher Education, Malaysia with Grant no. (ERGS/02(08)/860/2912(12)) during the research.

References

Akbarpour I, Ghaffari M, Ghasemian A (2013) Deinking of different furnishes of recycled MOW, ONP, and OMG pulps in silicate-free conditions using organic complex of PHASS. BioRes 8(1):31–44

Berix MJA (2012) Encapsulation of clay through non-aqueous dispersion polymerizations. Technische Universitelt Eindhoven, Eindhoven

Billes E (2015) Cellulose oligomers preparation by depolymerization for the synthesis of new bio-based amphiphilic compounds. Universite de Bordeaux, France

Boryniec S, Przygocki W (1999) Procesyspalaniapolimerow.Cz III. Opoznianiespalaniamaterial owpolimerowych. Polim 10:656–665

Cai J, Guo J, Ji M, Yang W (2007) Preparation and characterization of multiresponsive polymer composite microspheres with core-shell structure. Colloid Polym Sci 285:1607–1615

Cao Y, Lu J, Huang R, Zhao Y, Wu Y (2011) Evaluation of decay resistance for steam-heat-treated wood. BioRes 6(4):4696–4704

Chang HT, Chang ST (2001a) Inhibition of the photodiscoloration of wood by butyrylation. Holz 55:255–259
Chang ST, Chang HT (2001b) Comparisons of the photostability of esterified wood. Polym Degrad Stabil 71:261–266
Chen F, Wu H (2014) Water-resistant material from recovered fibers and acrylic emulsion terpolymer. BioRes 9(1):1148–1158
Chen T, Qiu J, Zhu K, He X, Kang X, Dong E (2014) Poly(methyl methacrylate)-functionalized graphene/polyurethane dielectric elastomer composites with superior electric field induced strain. Mater Lett 128:19–22
Clemons CM (2002) Wood-plastic composites in the United States: the interfacing of two industries. Forest Prod J 52(6):10–18
Devi RR, Maji TK (2007) Effect of glycidyl methacrylate on the physical properties of wood-polymer composites. Polym Compos 28(1):1–5
Doczekalska B, Batkowiak M, Zakrzewski R (2014) Esterification of willow wood with cyclic acid anhydrides. Wood Res 59(1):85–96
Donath S, Militz H, Mai C (2004) Wood modification with alkoxysilanes. Wood Sci Technol 38:555–566
Dong X, Sun T, Liu Y, Li C, Li Y (2015) Structure and properties of polymer-impregnated wood prepared by in situ polymerization of relative monomers. BioRes 10(4):7854–7864
Duanmu J, Gamstedt EK, Rosling A (2007) Hydromechanical properties of composites of crosslinked allylglycidyl-ether modified starch reinforced by wood fibres. Compos Sci Technol 67:3090–3097
Eaton R, Hale MDC (1993) Wood, decay, pests and protection. Chapman and Hall, London
Emeydan MA, Cabane E, Gierlinger N, Koetz J, Burgert I (2014) Improvement of wood material properties via in situ polymerization of styrene into tosylated cell walls. RSC Adv 4: 12981–12988
Esmaeil SP, Qazvini NT, Mahdavi H (2014) Interpenetrating polymer networks (IPN) based on gelatin poly(ethylene glycol) dimethacrylate/clay nanocomposites: Structure-properties relationship. Mater Chem Phys 143(3):1396–1403
Evans PD, Wallis AFA, Owen NL (2000) Weathering of chemically modified wood surfaces: natural weathering of scots pine acetylated to different weight gains. Wood Sci Technol 34: 151–155
Eze IO, Obidiegwu MU, Eyarefe SO (2013) The effects of maleic acid on some mechanical and end-use properties of bamboo powder filled low density polyethylene composites. Acad Res Int 4(2):208–217
Feci E, Nunes L, Palanti S, Duarte S, Predieri G, Vignali F (2009) Effectiveness of sol-gel treatments coupled with copper and boron against subterranean termites. Stockholm, Sweden
Gupta SK, Gupta NK, Singhal R, Nagpal AK (2004) Dimensional stability and fungal decay of rubber wood-polymer composites processed by gamma radiation. Ind J Chem Technol 11: 243–247
He W, Zhang QS, Jiang SX (2014) Modification of fast growing poplar with styrene and glycidyl methacrylate. Int Wood Prod J 5(2):98–102
Hien NTT, Li S (2012) Effects of water-borne rosin on the fixation and decay resistance of copper-based preservative treated wood. BioRes 7(3):3573–3584
Hill CAS (2006) In wood modification chemical, thermal and other processes. Wiley, England
Hill CAS, Forster SC, Farahani MRM, Hale MDC, Ormondroyd GA, Williams GR (2005) An investigation of cell wall micropore blocking as a possible mechanism for the decay resistance of anhydride modified wood. Int Biodeterior Biodegrad 55:69–76
Laks PE, Ritcher DL, Larkin GL (2000) Biological deterioration of wood-based composite panels. Forest Prod Lab 11(4):7–14
Lee WF, Chen CY (2015) Graft copolymerization of 3-(trimethylsilyl) propyl methacrylate onto styrene-butadiene-styrene triblock copolymer. J Elastomers Plast 47(2):103–116
Li Y, Dong X, Liu Y, Li J, Wang F (2011a) Improvement of decay resistance of wood via combination treatment on wood cell wall: swell-bonding with maleic anhydride and graft

copolymerization with glycidyl methacrylate and methyl methacrylate. Int Biodeterior Biodegrad 65(7):1087–1094
Li YF, Liu YX, Wang XM, Wu QL, Yu HP, Li J (2011b) Wood-polymer composites prepared by in situ polymerization of monomers within wood. J Appl Polym Sci 119(6):3207–3216
Lin WC, Yang CH, Wang TL, Shieh YT, Chen WJ (2012) Hybrid thin films derived UV-curable acrylate-modified waterborne polyurethane and monodispersed colloidal silica. eXPRESS Polym Lett 6(1):2–13
Liu PG, Pu WF, Ni JH (2016) Catalytic effect analysis of clay minerals on low-temperature oxidation of crude oil through combined thermal analysis method. Petrol Sci Technol 34(4): 343–349
Lu JZ, Wu Q, McNabb HS Jr (2000) Chemical coupling in wood fiber and polymer composites: a review of coupling agents and treatments. Wood Fiber Sci 32(1):88–104
Mommaerts V, Platteau G, Boulet J, Sterk G, Smagghe G (2008) Trichoderma-based biological control agents are compatible with the pollinator Bombusterrestris: a laboratory study. Biol Control 46:463–466
Nagaveni HC, Vijayalakshmi G, Chauhan SS (2005) Decay resistance of esterified and oligoesterified rubberwood (*hevea brasiliensis*). J Trop For Sci 17:588–595
Oldertroen K, Kittikun AH, Phongpaichit S, Riyajan S, Teanpaisal R (2016) Treatment of rubberwood (Hevea brasiliensis) (Willd. Ex. A. Juss) Mull. Arg. with maleic anhydride to prevent moulds. J For Sci 62(7):314–321
Papadopoulos AN, Hill CAS (2002) The biological effectiveness of wood modified with linear chain carboxylic acid anhydrides against Coniophoraputeana. Holz als Roh-und Werkstoff 60(5):329–332
Piltonen P, Stoor T, Niinimaki J (2013) The effect of rewetting on the adhesion tendency of styrene-butadiene latices on steel surfaces. Int J Adhes Adhes 40:108–111
Pinnavaia TJ, Beall GW (2000) Polymer-clay nanocomposites. Wiley, United Kingdom
Plakette DV, Dunningham EA, Singh AP (eds) (1996) Weathering of chemically modified wood. In: Hons DNS (ed) Chemical modification of lignocellulosic materials. Marcel Dekker, New York, pp 277–294
Pointing SB, Parungao MM, Hyde KD (2003) Production of wood decay enzymes, mass loss and lignin solubility in wood tropical Xylariaceae. Mycol Res 107(2):231–235
Rahman MR, Hamdan S, Saleh AA, Islam MS (2010) Mechanical and biological performance of sodium mataperiodate impregnated plasticized wood (PW). BioRes 5(2):1022–1035
Rahman MR, Lai JCH, Hamdan S, Ahmed AS, Baini R, Saleh SF (2013) Combined styrene/mma/nanoclay cross-linker effect on wood-polymer composites (WPCs). BioRes 8(3):4227–4237
Rangel-Vazquez NA, Leal-Garcia T (2010) Spectroscopy analysis of chemical modification of cellulose fibers. J Mex Chem Soc 54(4):192–197
Reinsch S, Bocker W, Bucker M, Seeger S, Unger B (2002) Development of wood-inorganic composites with enhanced properties and environment stability. Paper presented at the proceeding of 4th international wood and fiber symposium, Kassel, Germany, 10–11 April 2002
Rowell RM (1983) Chemical modification of wood. Forest Prod Abs 6:363–382
Rowell RM (2005) Chemical modification of wood. In: Rowell RM (ed) Handbook of wood chemistry and wood composite. Taylor and Francis, CRC Press, pp 381–420
Rowell RM (2006) Chemical modification of wood: a short review. Wood Mater Sci Eng 1:29–33
Sapkota J (2011) Influence of clay modification on curing kinetics of natural rubber nanocomposites. Tampere University of Technology, Finland
Savoy JG (1954) Breakdown of timber by ascomycetes and fungi imperfecti. Ann Appl Biol 41(2): 336–347
Schmidt EL, French DW, Gerthejensen RO, Hermann J, Hall H (1978) Strength reductions in particleboard caused by fungi. Forest Prod J 28(2):26–31
Solar R, Reinperchet L, Lang R (2005) Decay resistance of untraditionally compressed beech wood. Faculty of Wood Sciences and Technology Slovakia, Zvolen

Sultan MT, Rahman MR, Hamdan S, Lai JCH, Talib ZA (2016) Clay dispersed styrene-*co*-glycidyl methacrylate impregnated kumpang wood polymer nanocomposites: impact on mechanical and morphological properties. BioRes 11(3):6649–6662

Terziev N, Panov D, Temiz A, Palanti S, Feci E, Daniel G (2009) Laboratory and above ground exposure efficacy of silicon-boron treatments. Stockholm, Sweden

Verhey S, Laks PE, Ritcher DL (2001) Laboratory decay resistance of wood fiber/thermoplastic composites. Forest Prod J 51(9):44–49

Williams RS (2005) Weathering of wood. In: Rowell RM (ed) Handbook of wood chemistry and wood composites. CRC Press, United States of America, pp 142–185

Wu Q, Lei Y, Lian K, Qi Y (2012) Copper/Carbon core shell nanoparticles as additive for natural fiber/wood plastic blends. BioRes 7(3):3213–3222

Yalinkilic MK, Tsunoda K, Takahashi M, Gezer ED, Dwianto W, Nemoto H (1998) Enhancement of biological and physical properties of woof by boric acid-vinyl monomer combination treatment. Holz 52(6):667–672

Zabel RA, Morrell JJ (1992) Wood microbiology—decay and its prevention. Academic Press Inc, United States of America

Zabihzadeh SM, Hosseini Hashemi SK, Mehregan Nikoo H, Sepidehdam SMJ (2009) Influence of fungal decay on physico-mechanical properties of a commercial extruded bagasse/PP composite. J Reinf Plast Compos 29(11):1750–1756

Zhou X, Yu Y, Lin Q, Chen L (2013) Effects of maleic anhydride-grafted polypropylene (MAPP) on the physico-mechanical properties and rheological behavior of bamboo powder-polypropylene foamed composites. BioRes 8(4):6263–6279

MIX
Papier aus verantwortungsvollen Quellen
Paper from responsible sources
FSC® C105338

If you have any concerns about our products,
you can contact us on
ProductSafety@springernature.com

In case Publisher is established outside the EU,
the EU authorized representative is:
Springer Nature Customer Service Center GmbH
Europaplatz 3, 69115 Heidelberg, Germany

Printed by Libri Plureos GmbH
in Hamburg, Germany